Lecture Notes in Mathematics

Volume 2291

This series reports on new developments in all areas of mathematics and their applications - quickly, informally and at a high level. Mathematical texts analysing new developments in modelling and numerical simulation are welcome. The type of material considered for publication includes:

1. Research monographs
2. Lectures on a new field or presentations of a new angle in a classical field
3. Summer schools and intensive courses on topics of current research.

Texts which are out of print but still in demand may also be considered if they fall within these categories. The timeliness of a manuscript is sometimes more important than its form, which may be preliminary or tentative.

Titles from this series are indexed by Scopus, Web of Science, Mathematical Reviews, and zbMATH.

More information about this series at http://www.springer.com/series/304

Cornelia Schneider

Beyond Sobolev and Besov

Regularity of Solutions of PDEs and Their
Traces in Function Spaces

 Springer

Cornelia Schneider (iD)
Department of Mathematics
University Erlangen-Nuremberg
Erlangen, Germany

ISSN 0075-8434 ISSN 1617-9692 (electronic)
Lecture Notes in Mathematics
ISBN 978-3-030-75138-8 ISBN 978-3-030-75139-5 (eBook)
https://doi.org/10.1007/978-3-030-75139-5

Mathematics Subject Classification: M12155, M12066, M14050, M12015, M12023

This Springer imprint is published by the registered company Springer Nature Switzerland AG.
The registered company address is: Gewerbestrasse 11, 6330 Cham, Switzerland

To my son Pascal

Preface

The monograph *Beyond Sobolev and Besov* grew out of the habilitation thesis of the author written in 2019. In view of the substantial material, decision was taken to collect the obtained results in a book rather than a number of papers.

The aim pursued in this book is twofold: On the one hand, we present studies of the regularity of solutions of partial differential equations (PDEs). These investigations are motivated by fundamental questions arising in the context of the numerical treatment of such PDEs. Special attention is being paid to the regularity of solutions in function spaces build upon the scales of Sobolev and Besov spaces

$$H_p^s(\Omega) \quad \text{and} \quad B_{p,q}^s(\Omega),$$

since they are closely related with the convergence order of uniform and adaptive numerical algorithms, respectively. In particular, we study regularity estimates of PDEs of elliptic, parabolic, and hyperbolic type on non-smooth domains. Linear as well as nonlinear equations are considered and special attention is paid to PDEs of parabolic type. For the classes of PDEs investigated, we aim at justifying the use of adaptive methods for solving PDEs. Therefore, we show that the smoothness of the solution—displayed by the parameter s above—in the specific adaptivity scale of Besov spaces is higher than its smoothness in the respective Sobolev scale.

On the other hand—related to these regularity studies but of interest on its own—we investigate traces in various function spaces and different settings. Here, we go (as the title suggests) far beyond the scales of Sobolev and Besov and study also traces in quite sophisticated generalized Smoothness Morrey spaces, denoted by $\mathcal{A}_{\mathcal{M}_{p,q}^\varphi}^s(\Omega)$ and $A_{p,q}^{s,\tau}(\Omega)$. These scales generalize the Sobolev and Besov spaces (and many others such as Triebel-Lizorkin, Hölder-Zygmund, Slobodeckij, and Bessel-potential spaces) and allow a unified treatment of all of these well-known spaces. For an overview regarding the relations and connections between the different spaces considered in this monograph, we refer to the mindmap in Fig. 1 on page viii.

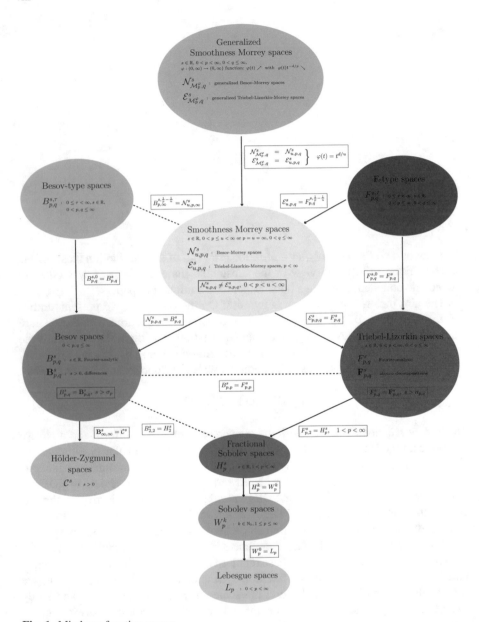

Fig. 1 Mindmap function spaces

Having a look beyond the spaces at the real persons, whose inspirations flow through this book and led to its title, it turns out that S.L. Sobolev (*1908–†1989) and O.V. Besov (*1933) in real life knew each other quite well. When asked about their relation, Professor Besov kindly revealed the following information:

> I knew S.L. Sobolev well. I graduated from Moscow University in 1955 under his supervision. Afterwards (under the leadership of the academician S.M. Nikolsky) I was engaged in spaces of differentiable functions of many variables, S.L. Sobolev was the founder in this field of research. He knew my works, and my publications in the reports of the Academy of Sciences were presented to him. For 25 years, S.L. Sobolev was the director of the Institute of Mathematics in the city of Novosibirsk (now this institute bears his name). I participated in many conferences organized by this institute, and often and fruitfully communicated with him. We both also participated in one project, jointly with the Czechoslovak Mathematical Institute. In the last years of his life S.L. lived in Moscow, and together with S.M. Nikolsky he led a seminar on the theory of differentiable functions at the Steklov Mathematical Institute, of which I was a participant.

—O.V. Besov, March 2021

This book is destined as a valuable resource for researchers and graduate students interested in the regularity theory of PDEs and function spaces, who look for a comprehensive and systematic treatment of the above listed topics. We present new results of our most recent research, including unifying approaches generated while writing the book. The available material is presented in an accessible and adequate way, references for further reading are given at the right places throughout the text. Whenever possible, the book includes figures illustrating connections and main ideas. Hopefully the readers of this book will benefit from the pictures provided for a better understanding, as the students of the author usually do during lectures.

In Chaps. 1 and 2 we start with an introduction and present the relevant function spaces and general concepts needed for our further investigations.

Part I, composed of Chaps. 3–7, is dedicated to studying the regularity of solutions of PDEs in Besov and (fractional) Sobolev spaces. It is formally based on the papers [DS19, DS18, DHSS18a], and [DHSS18b]. The results from [DS19] can be seen as the starting point and the heart of our investigations for parabolic PDEs. In order to round up this manuscript, we extended the results presented in [DS19] from polyhedral cones K to the numerically even more interesting domains of polyhedral type D. This led to essentially new regularity results in Sobolev spaces which are presented in Sect. 5.2.1. We believe that this monograph gains a lot from this generalization. The results for parabolic PDEs from [DS18] rose from discussions when working on [DS19], but are interesting on their own respect. The focus in [DHSS18a] is slightly different. Here, we investigate elliptic problems with a quite general nonlinear term causing some difficulties. Finally, the preprint [DHSS18b] contains all relevant information on Kondratiev spaces. Since our main focus in this book lies in the fractional Sobolev and Besov regularity of the solutions, we only included that material from [DHSS18b] which is needed for our later purposes.

Afterwards, in Part II, we present trace results in various settings assorted in Chaps. 8–10. It consists of the results established in the papers [SV13, SV12, MNS19], and [GS13]. Here our emphasis is on presenting how to obtain trace results in various settings using different methods. For this reason, we do not include any proofs of the homogeneity results from [SV12] which were used for the trace results in [SV13]. Moreover, our findings in [GS13] are only partially included in what follows: To make our presentation clearer, we left out the trace results for vector bundles and focused on the results for function spaces on manifolds instead. In order to unify the concepts and clarify connections and existing relations, we changed the definitions of the atoms used in [MNS19] to fit with the atomic decompositions from [SV13].

All the material is selected and rearranged under the above-described emphasis and restrictions. Hopefully, this focus will make the book accessible to a large audience.

Writing this book would not have been possible without various sources of support. Firstly, I acknowledge the financial support of Deutsche Forschungsgemeinschaft (DFG), Grant Nos. SCHN 1509/1-1 and SCHN 1509/1-2, which helped me to take some time off teaching and focus on my research. I also appreciate the help from Springer Verlag for their cooperation and assistance in publishing the book.

Moreover, this gives me the opportunity to specially thank Stephan Dahlke for the encouragement and support he gave me during the past couple of years. Also my thanks go to Eberhard Bänsch and my colleagues at the chair of Applied Mathematics III in Erlangen for providing a stimulating environment to work in. Furthermore, I appreciate joint work and exchange of ideas on the subject with many colleagues, in particular, Nadine Große, Markus Hansen, Susana Moura, Júlio Neves, Winfried Sickel, and Jan Vybíral.

Finally, my debt of gratitude is owed to my family for their never-ending patience and support. The day after this manuscript was finished, my son and I went to see the Eiffel Tower to have a look beyond Sobolev, Besov ... and Paris!

Nuremberg, Germany   Cornelia Schneider
March 2021

Contents

Quick Guide for Function Spaces

Spaces of Continuous Functions

Lebesgue Spaces

Sobolev Spaces

Besov Spaces (B-Spaces)

Triebel-Lizorkin Spaces (F-Spaces)

Kondratiev Spaces (Weighted Sobolev Spaces)

Smoothness Morrey Spaces

Further Function Spaces

Sequence Spaces

Chapter 1
Introduction

Everything should be made as simple as possible, but not simpler.

—A. Einstein, 1879–1955

Motivation The principal object of this manuscript is to study on the one hand the regularity of solutions of *partial differential equations (=PDEs)* in specific scales of function spaces and traces of such functions on the boundary of domains and submanifolds on the other hand. These kinds of problems are closely linked with each other and arise e.g. when considering an elliptic boundary value problem,

$$\left\{ \begin{aligned} Lu &= f \quad \text{in} \quad \Omega, \\ u\big|_{\partial\Omega} &= g, \end{aligned} \right\} \tag{1.0.1}$$

where $\Omega \subset \mathbb{R}^d$ is a bounded domain with boundary $\partial\Omega$, f a given function defined on Ω, g a function defined on the boundary, and L an elliptic differential operator. In particular, choosing $L = -\Delta$ yields the Poisson problem, which often appears in mechanical engineering and theoretical physics. It describes, for instance, the displacement of a thin membrane fixed at the boundary due to some outer force f or, moreover, the distribution of temperature in a stationary state given an outer heat source f.

The analysis of these problems has a long history. Usually, one is merely interested in the regularity of the solutions u if the initial function f belongs to a Lebesgue space $L_p(\Omega)$, $1 < p < \infty$, or even to a (classical) Sobolev space $W_p^k(\Omega)$, $k \in \mathbb{N}$. Recently, in view of nonlinear approximation and the numerical treatment of these equations, one has also studied questions concerning the exact description of solutions on non-smooth domains Ω, the admittance of generalized differential operators L as well as the further dependence of the regularity and smoothness of the solution u (measured in weighted Sobolev spaces) on the initial data f (and g).

© The Author(s), under exclusive license to Springer Nature Switzerland AG 2021
C. Schneider, *Beyond Sobolev and Besov*, Lecture Notes in Mathematics 2291,
https://doi.org/10.1007/978-3-030-75139-5_1

In Part I of this thesis we address these questions for elliptic, parabolic, and hyperbolic PDEs but restrict our considerations to zero Dirichlet boundary conditions, i.e., we always assume $g = 0$. In particular, we focus on non-smooth domains where it is well-known that singularities of the solutions near the boundary might appear which diminish their Sobolev regularity. Therefore, it is more promising to study the regularity in weighted Sobolev spaces where the weights compensate the blow ups of the solution to a certain extend. Moreover, investigating the regularity of the solutions in special scales of Besov spaces will play a central role. We shall compare the outcome with the fractional Sobolev regularity of the solution, since the smoothness obtained in these two scales is related with the convergence order of adaptive and uniform algorithms, respectively. Since numerical studies clearly indicate that modern adaptive algorithms have a lot of potential, we want to justify the use of adaptive schemes with our results. We explain these relations in more detail below.

On the other hand, when seeking a solution of the boundary value problem (1.0.1), the knowledge of traces (i.e., what can we say about the smoothness of the restriction of the solution u to the boundary $\partial\Omega$) becomes crucial. This question can be rephrased as asking for an adequate function space that g should belong to for the boundary value problem (1.0.1) to be well-defined. This will be our focus in Part II of this manuscript. Also in this context the Besov spaces play a central role: classical results on smooth domains imply that the traces of functions u from the Sobolev space $W_p^k(\Omega)$ belong to the Besov space $\mathbf{B}_{p,p}^{k-\frac{1}{p}}(\partial\Omega)$. We will study corresponding problems on non-smooth domains but also in other settings such as submanifolds and for different kinds of (generalized) function spaces.

From what we outlined above it is clear that the considerations in Parts I and II are both indispensable for a satisfactory treatment of boundary value problems of type (1.0.1). So far the results from both parts have been established separately from each other and are clearly interesting in their own respect. However, we decided to present them now in a joint context and unify the concepts. This might pave the way for combining the results and maybe study the regularity of boundary value problems with nonzero Dirichlet or even von Neumann and Robin conditions in the future.

Throughout this thesis we will mostly be concerned with functions belonging to the scale of Besov spaces as well as closely linked spaces—such as weighted and fractional Sobolev and Triebel-Lizorkin spaces—and certain generalizations of these spaces (smoothness Morrey spaces). An overview of the appearing spaces and their relations is presented in the mindmap on page viii. As for Besov spaces, we deal with the following two approaches. Let $0 < p, q \leq \infty$ in what follows. In Part II of this manuscript we consider the *Fourier-analytical approach* defining spaces $B_{p,q}^s(\mathbb{R}^d)$ as the set of all tempered distributions $f \in \mathcal{S}'(\mathbb{R}^d)$ such that

$$\left\| f \mid B_{p,q}^s(\mathbb{R}^d) \right\| = \left\| \left\{ 2^{js} \left\| \mathcal{F}^{-1}(\varphi_j \mathcal{F} f) \mid L_p(\mathbb{R}^d) \right\| \right\}_{j \in \mathbb{N}_0} \mid \ell_q \right\|$$

is finite, where $s \in \mathbb{R}$ and $\{\varphi_j\}_j$ is a smooth dyadic resolution of unity. However, when studying traces in Part II we are also concerned with the *classical approach* which introduces $\mathbf{B}_{p,q}^s(\mathbb{R}^d)$ as those subspaces of $L_p(\mathbb{R}^d)$ such that

$$\|f|\mathbf{B}_{p,q}^s(\mathbb{R}^d)\| = \|f|L_p(\mathbb{R}^d)\| + \left(\int_0^1 t^{-sq}\omega_r(f,t)_p^q \frac{dt}{t}\right)^{1/q}$$

is finite, where $s > 0$, $r \in \mathbb{N}$ with $r > s$, and $\omega_r(f,t)_p$ is the usual r-th modulus of smoothness of $f \in L_p(\mathbb{R}^d)$. Under certain restrictions on the parameters these approaches coincide. In particular, we have

$$B_{p,q}^s(\mathbb{R}^d) = \mathbf{B}_{p,q}^s(\mathbb{R}^d), \qquad s > n\max\left(\frac{1}{p} - 1, 0\right),$$

(in terms of equivalent norms). We describe in more detail the main scopes of Parts I and II and provide some background material below.

Part I of this book is concerned with regularity estimates of PDEs of elliptic, parabolic, and hyperbolic type. Linear as well as nonlinear equations will be considered and special attention will be paid to PDEs of parabolic type (for elliptic problems a lot of results in this direction are already established).

As a prototype of our investigations one may consider the heat equation on a bounded non-smooth (Lipschitz) domain $\Omega \subset \mathbb{R}^d$, which describes the distribution of heat (or variation in temperature) in the given region Ω over time t,

$$\frac{\partial u}{\partial t}(t, x) - \Delta u(t, x) = f(t, x), \qquad (t, x) \in [0, T] \times \Omega. \tag{1.0.2}$$

We study the spatial regularity of the unknown solution u in specific nonstandard smoothness spaces, i.e., the so-called adaptivity scale of Besov spaces

$$B_{\tau,\tau}^\alpha(\Omega), \qquad \frac{1}{\tau} = \frac{\alpha}{d} + \frac{1}{p}, \qquad \alpha > 0, \tag{1.0.3}$$

where α stands for the smoothness of the solution and τ displays its integrability, cf. [DeV98]. The motivation for these kinds of studies can be explained as follows. PDEs are very often used in science and engineering since they model all kinds of phenomena. Very often, results concerning existence and uniqueness of solutions are known (although for nonlinear equations even these questions might become delicate, see [DM07] for details). However, in most cases, an analytic expression of the solution is not available, so that numerical algorithms for the constructive approximation of the unknown solution up to a given tolerance are needed. In this case *adaptive* schemes are often unavoidable to increase efficiency. In an adaptive strategy, the choice of the underlying degrees of freedom is not a priori fixed but depends on the shape of the unknown solution. In particular, additional degrees of

Fig. 1.1 Uniform mesh

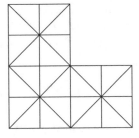

Fig. 1.2 Adaptive mesh

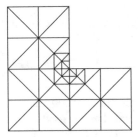

freedom are only spent in regions where the numerical approximation is still 'far away' from the exact solution.

The situation is illustrated for an L-shaped domain Ω. If one considers the Poisson equation with zero Dirichlet boundary conditions, i.e.,

$$
\left\{
\begin{aligned}
-\Delta u &= f && \text{in} \quad \Omega, \\
u\big|_{\partial\Omega} &= 0,
\end{aligned}
\right\}
$$

the solution u has a singular behaviour near the corner of the domain, which results in a finer adaptive mesh in the neighbourhood of the corner (Figs. 1.1 and 1.2).

Although the basic idea is convincing, adaptive algorithms are hard to implement (compared to nonadaptive ones), so that beforehand a rigorous mathematical analysis to justify their use is highly desirable. The guideline to achieve this goal can be described as follows. Given an adaptive algorithm based on a dictionary for the solution spaces of the PDE, the best one can expect is an optimal performance in the sense that it realizes the convergence rate of best N-term approximation schemes, which serves as a benchmark in this context. Given a dictionary $\Psi = \{\psi_\lambda\}_{\lambda \in \Lambda}$ of functions in a Banach space X, the error of best N-term approximation is defined as

$$
\sigma_N\big(u; X\big) = \inf_{\Gamma \subset \Lambda : \#\Gamma \leq N} \inf_{c_\lambda} \left\| u - \sum_{\lambda \in \Gamma} c_\lambda \psi_\lambda \,\big|\, X \right\|, \tag{1.0.4}
$$

i.e., as the name suggests we consider the best approximation by linear combinations of the basis functions consisting of at most N terms. For many dictionaries, in particular for wavelet bases and frames, it has been shown that the convergence

Fig. 1.3 DeVore-Triebel
diagram

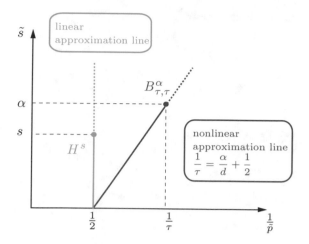

order of best N-term approximations, and therefore the achievable order of adaptive algorithms, depends on the regularity of the target function in the specific scale of Besov spaces (1.0.3), where $p > 1$ is fixed and indicates the L_p-norm in which the error is measured. We refer, e.g. to [CDD01, DDD97, DD97, DJP92, DNS06, Ste09] for details. Quite recently, it has also turned out that similar relations hold for finite element approximations, see [GM14].

On the other hand, it is the regularity of the solution in the scale of Sobolev spaces $H^s_p(D)$, where now s indicates the smoothness of the solution u, which encodes information on the convergence order for nonadaptive (uniform) methods, see, e.g. Hackbusch [Hac92] and [DDD97] for details. Therefore, the use of adaptivity is justified if the Besov smoothness of the exact solution of a given PDE within the scale (1.0.3) is high enough compared to the Sobolev smoothness.

The situation is displayed in the following DeVore-Triebel diagram for $p = 2$, which also shows the relations among the different parameters α, τ, s, and d, where the latter stands for the dimension of the underlying domain Ω (Fig. 1.3). Nonadaptive method (blue):

$$u \in H^s_p(\Omega) \curvearrowright \|u - u_N\|_{L_p(\Omega)} = \mathcal{O}(N^{-s/d})$$

for linear approximation

Adaptive method (red):

$$u \in B^\alpha_{\tau,\tau}(\Omega) \curvearrowright \|u - u_N\|_{L_p(\Omega)} = \mathcal{O}(N^{-\alpha/d})$$

for nonlinear approximation

By rule of thumb, adaptivity pays if $u \in B^\alpha_{\tau,\tau}(\Omega)$ for some

$$\alpha > \sup\{s > 0 : u \in H^s_p(\Omega)\}$$

Fig. 1.4 Fichera corner

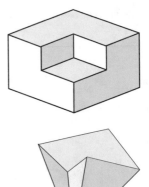

Fig. 1.5 Polyhedral cone

For the case of elliptic PDEs, a lot of positive results in this direction are already established. We refer to [Dah98, Dah99a, Dah99b, Dah02, DDHSW16]. It is well-known that if the domain under consideration, the right-hand side and the coefficients are sufficiently smooth, then the problem is completely regular [ADN59] and there is no reason why the Besov smoothness should be higher than the Sobolev regularity. In particular, for the Poisson equation we have $u \in H^{s+2}(\Omega)$ for $f \in H^s(\Omega)$. However, on non-smooth (Lipschitz) domains having edges and corners (e.g. polygons and polyhedra) the situation changes dramatically.

Typical examples of practically important Lipschitz domains are the Fichera corner and the polyhedral cone illustrated in Figs. 1.4 and 1.5 as well as the L-shaped domain on page 4.

Here singularities at the boundary may occur that diminish the Sobolev regularity s of the solution significantly [Gri92, Gri11]. Moreover, the famous result by Jerison and Kenig [JK95] tells us that in the worst case one has $s \leq 3/2$ for the solution of the Poisson equation, even for smooth right hand side f. Therefore, the $H^{3/2}$-Theorem implies that the optimal rate of convergence for nonadaptive methods of approximation is just $3/2d$ as long as we do not impose further restrictions on Ω. Similar relationships also hold for more specific domains such as domains of polyhedral or polygonal type. We refer e.g. to [Dau88, Gri85, Gri92] in this context.

On the other hand, it turns out that the boundary singularities do not influence the Besov regularity α too much [DD97]. Regularity estimates in quasi-Banach spaces according to (1.0.3) have only been developed quite recently. For linear elliptic operator equations a lot of positive results in this direction already exist, see, e.g. [Dah99b, DDD97, DD97, Han15] (this list is clearly not complete). First studies for nonlinear equations have been reported in [DDHSW16, DS13]. Concerning parabolic PDEs not much is known in this respect. A first step in this direction was done by H. Aimar, I. Gómez and B. Iaffei in [AGI08] and later on continued in [AGI10, AG12], where the authors extended the elliptic setting from [DD97] to the parabolic case (but for a slightly different scale of Besov spaces).

Unfortunately, their studies are limited to the heat equation with zero Dirichlet boundary conditions. Moreover, looking for the generalization of the $H^{3/2}$-Theorem to parabolic equations very little information is available in the literature. In [Gri92] it is shown that for the heat equation on a polygonal domain $\Omega \subset \mathbb{R}^2$ with right hand side $f \in L_2((0, \infty) \times \Omega)$ there is a solution $u \in L_2((0, \infty), H^s(\Omega))$ with $s < \frac{\pi}{\max_j \omega_j} + 1$, where ω_j denote the inner angles of the polygon. Thus, for $\omega_j \to 2\pi$ we see that $s < \frac{3}{2}$ is optimal. The 3-dimensional case is not treated there at all. Nevertheless, the above studies indicate that the use of adaptive algorithms for the heat equation under investigation is completely justified. As for hyperbolic equations, to the best of our knowledge there are no corresponding results in the literature so far. We only mention in this context the seminal paper of DeVore and Lucier [DL90] which is concerned with concervation laws in 1D.

Our contribution to this field of research in Part I is as follows.

(1) Concerning elliptic PDEs, in Chap. 4 we will be concerned with Besov regularity estimates of the solutions to semilinear elliptic partial differential equations of the form

$$-\nabla\big(A(x)\cdot\nabla u(x)\big)+g(x,u(x)) = f(x) \quad \text{in} \quad D, \qquad u\big|_{\partial D} = 0, \qquad (1.0.5)$$

where $D \subset \mathbb{R}^d$, $d = 2, 3$, is a bounded Lipschitz domain of polyhedral type, $A = (a_{i,j})_{i,j=1}^d$ is symmetric and its coefficients satisfy certain smoothness and growth conditions, respectively. Moreover, the nonlinear term $g(x, \xi)$ is assumed to satisfy a quite general growth condition allowing for the important special case $g(x, u(x)) = u^{2n+1}(x)$ to be included in our considerations. Our results are related to [DS13], but we generalize and modify the analysis presented there in the following sense. First of all, in [DS13] only semilinear versions of the Poisson equation have been studied, whereas here much more general elliptic operators are treated. Secondly, in [DS13] only nonlinear terms $g(x, \xi)$ that satisfy a growth condition of the form $|g(x, \xi)| \leq a + b|\xi|^\delta$, $\delta \leq 1$ have been considered. In our investigations this condition is removed to the greatest possible extent.

(2) Concerning parabolic problems we investigate in Chap. 5 several different settings:

(2a) In Problems I, II we study the spacial Besov regularity of linear ($\varepsilon = 0$) and nonlinear ($\varepsilon > 0$) equations of the form

$$\frac{\partial}{\partial t}u + (-1)^m L(t, x, D_x)u + \varepsilon u^M = f \quad \text{in} \quad [0, T] \times D, \qquad (1.0.6)$$

with zero initial and Dirichlet boundary conditions, where $m, M \in \mathbb{N}$, $D \subset \mathbb{R}^d$ is a domain of polyhedral type, and L denotes a uniformly elliptic operator of order $2m$ with sufficiently smooth coefficients. Our results show in the linear case $\varepsilon = 0$ that for every $t \in [0, T]$ the spatial Besov smoothness of the solution to (1.0.6) is always larger than $2m$, provided that some technical conditions on the operator pencils are satisfied, see Theorems 5.4.1 and 5.4.5. The reader should observe that the results are independent of the shape of the polyhedral domain, and that the classical Sobolev smoothness is usually limited by m, see [LL15]. Therefore, for every t, the spatial Besov regularity is more than twice as high as the Sobolev smoothness, which of course justifies the use of (spatial) adaptive algorithms. Moreover, for smooth domains and right-hand sides in L_2, the best one could expect would be smoothness order $2m$ in the classical Sobolev scale. So, the Besov smoothness on polyhedral type domains is at least as high as the Sobolev smoothness on smooth domains.

Afterwards, we generalize this result to nonlinear parabolic equations of the form (1.0.6). We show that in a sufficiently small ball containing the solution of the corresponding linear equation, there exists a unique solution to (1.0.6) possessing the same Besov smoothness in the scale (1.0.3). The proof is performed by a technically quite involved application of the Banach fixed point theorem. The final result is stated in Theorem 5.4.6.

The next natural step is to also study the regularity in time direction. For the linear parabolic problem (1.0.6) with $\varepsilon = 0$ we show that the mapping $t \mapsto u(t, \cdot)$ is in fact a C^l-map into the adaptivity scale of Besov spaces, precisely,

$$u \in C^{l, \frac{1}{2}}((0, T), B^{\alpha}_{\tau, \infty}(\Omega)),$$

see Theorem 5.4.8.

(2b) In Problem III we investigate the fractional Sobolev regularity of the parabolic problem

$$\mathcal{L}u = f \quad \text{in} \quad \mathcal{K} \times \mathbb{R}, \qquad u\big|_{\partial \mathcal{K} \times \mathbb{R}} = 0,$$

where the underlying domain $\mathcal{K} \subset \mathbb{R}^d$ is a generalized wedge and the differential operator $\mathcal{L}$ is of second order given in non-divergence form. Our results now can be seen as a first step to generalize the $H^{3/2}$-Theorem to other parabolic problems. However, we are not as general as [JK95], since the domains we consider are restricted to wedges (or cones) instead of general Lipschitz domains. Our results are in good agreement with the corresponding theory of elliptic equations. Our main findings are stated in Theorem 5.3.1. In particular, for the heat equation on a smooth cone $K \subset \mathbb{R}^3$ we obtain that the fractional Sobolev smoothness of the solution u is bounded from above by $s < \min\left(\frac{3}{2} + \lambda_1^+, 2\right)$, for right hand side $f \in L_2(K)$, where $\lambda_1^+ = -\frac{1}{2} + \sqrt{\Lambda_1 + \frac{1}{4}}$ and Λ_1 is the first eigenvalue of the Dirichlet problem of the Laplace-Beltrami operator in $\Omega = K \cap S^2$.

(2c) In Problem IV we consider the second order parabolic PDEs in non-divergence form on general bounded Lipschitz domains $\mathcal{O} \subset \mathbb{R}^d$,

$$\begin{cases} \frac{\partial}{\partial t} u & = \sum_{i,j=1}^{d} a_{ij} \frac{\partial^2}{\partial x_i \partial x_j} u + \sum_{i=1}^{d} b_i \frac{\partial}{\partial x_i} u + cu + f \text{ on } [0,T] \times \mathcal{O}, \\ u(0,\cdot) = u_0 & \text{on } \mathcal{O}, \end{cases}$$

(1.0.7)

where the coefficients are assumed to satisfy several assumptions. Since in this case all points on the boundary $\partial \mathcal{O}$ are equally bad, the singularities of the solution induced by the boundary have a much stronger influence. Therefore, our regularity results for the linear case of (1.0.6) when $\varepsilon = 0$ on polyhedral type domains are much stronger. To our surprise, it turns out that the spatial regularity results in this general setting are more or less the same as for the case of SPDEs that was already studied in [Cio13] based on [Kim04, Kim12].

(3) In Chap. 7 we are finally concerned with hyperbolic problems of the form

$$\frac{\partial^2}{\partial t^2} u + L(x, t, D_x) u = f \quad \text{in } \Omega \times (0, T),$$

(1.0.8)

with zero initial and Dirichlet boundary conditions in specific Lipschitz domains $\Omega \subset \mathbb{R}^d$, $d > 2$, where L denotes a uniformly elliptic operator of order 2 with sufficiently smooth coefficients. Our main results concerning the Besov regularity of the solution u are stated in Theorem 6.3.1.

It turns out that in all the above studied cases (1)–(3), the Besov regularity is high enough to justify the use of adaptive algorithms! The main ingredients to prove our results are regularity estimates in weighted Sobolev spaces, so-called *Kondratiev spaces* $\mathcal{K}_{p,a}^m(\Omega)$ defined as the collection of all measurable functions which admit m weak derivatives satisfying

$$\|u|\mathcal{K}_{p,a}^m(\Omega)\|^p = \sum_{|\alpha| \le m} \int_\Omega |\rho(x)^{|\alpha|-a} \partial^\alpha u(x)|^p \, dx < \infty.$$

(1.0.9)

Therein, the weight function $\rho : \Omega \longrightarrow [0,1]$ is the smooth distance to the singular set of Ω. The study of solutions to PDEs in Kondratiev spaces has already quite a long history. We refer, e.g. to [KO83, Gri92, JK95, MR10] (this list is clearly not complete). As can be seen from (1.0.9) the basic idea is the following: Since on non-smooth domains the solutions to PDEs as well as their derivatives might become highly singular as one approaches the boundary, their strong growth can to some extent be compensated by means of the specific weights ρ reflecting the regularized distance to the singular set of the domain to some power. Recent studies have also shown that these Kondratiev spaces are very much related with Besov spaces in the adaptivity scale (1.0.3) in the sense that powerful embedding results exist, see, e.g. [Han15]. This means that Besov regularity results can be established by first studying the equation under consideration in Kondratiev spaces and then using known (or deriving new) embeddings into Besov spaces.

Part II has a different focus and is concerned with traces in several function spaces such as Besov- and Triebel-Lizorkin spaces but also quite general smoothness Morrey spaces. We study traces in various settings, in particular, on smooth and non-smooth boundaries $\partial\Omega$ as well as on submanifolds. As outlined above, a clarification of this problem is of crucial interest for solving boundary value problems (1.0.1). However, a lot of our considerations are interesting on their own and grew out of independent interests. In particular, the established non-smooth atomic decompositions for Besov spaces $\mathbf{B}_{p,q}^s(\mathbb{R}^d)$ are surely helpful when tackling other problems related with these kinds of function spaces. Several applications of the obtained results with different scopes than regularity theory of PDEs show the importance of the traces on their own. Thus, on the one hand the intention in this part of the manuscript is to pave the way for future research in order to study boundary value problems (1.0.1) with boundary conditions different from zero Dirichlet. On the other hand we want to present several methods that can be used when studying traces in various settings in a unified and coherent way. We remark that even the understanding of the considered trace operators is not the same in all situations we deal with. This is why we merely collect a selection of the available trace results: We want to link the different cases in a comprehensible way and at the same time shed light on the abstract approaches in order to review things from another perspective. The totality of all these pieces together form the idea we want to present.

As a starting point for questions concerning definitions and characteristics of functions on the boundary $\partial\Omega$ of a domain Ω, one usually considers hyperplanes $\mathbb{R}^{d-1} = \{x = (x_1, \ldots, x_d) \in \mathbb{R}^d : x_d = 0\}$. Restricting ourselves to smooth functions φ, taking the pointwise trace $\mathrm{Tr}\,\varphi = \varphi|_{\mathbb{R}^{d-1}}$ makes sense. But when dealing with functions $f \in L_p(\mathbb{R}^d)$—only explicitly defined up to a set of Lebesgue measure zero—the understanding of the trace operator is not so clear. The possibility of approximating functions via smooth functions (which have boundary values) and defining the trace operator $\mathrm{Tr}\,f$ by completion paves the way and makes sense e.g. in the context of Besov spaces.

The idea of studying traces has attracted a lot of attention over the years, especially in connection with Sobolev spaces. Gagliardo established in [Gag57]— before Besov spaces were fully developed—that the traces of functions belonging to $W_p^k(\mathbb{R}^d)$, $k \in \mathbb{N}$, $1 < p < \infty$, on the hyperplane $\mathbb{R}^{d-1}$ could be defined using descriptions via differences, which in modern terms correspond to the classical Besov spaces with parameters $p = q$ and $s = k - \frac{1}{p}$, i.e., $\mathrm{Tr}\,W_p^k(\mathbb{R}^d) = \mathbf{B}_{p,p}^{k-\frac{1}{p}}(\mathbb{R}^{d-1})$. This already is an instance of the phenomenon, that passing from functions in some source space with integrability p to their traces on hyperplanes of codimension 1 results in a loss of smoothness corresponding to $\frac{1}{p}$ of a derivative. These results extend to more general hyperplanes $\mathbb{R}^m$, $d > m \in \mathbb{N}$.

Furthermore, traces on the boundary $\partial\Omega$ when Ω is a C^k domain were investigated in [Sch11b], where it was shown that Tr is a bounded linear operator from $\mathbf{B}_{p,q}^s(\Omega)$ onto $\mathbf{B}_{p,q}^{s-\frac{1}{p}}(\partial\Omega)$,

$$\mathrm{Tr}\,\mathbf{B}_{p,q}^s(\Omega) = \mathbf{B}_{p,q}^{s-\frac{1}{p}}(\partial\Omega), \tag{1.0.10}$$

where $n \geq 2$, $0 < p, q \leq \infty$ and $s > \frac{1}{p}$ and the additional assumption that $k > s$, i.e., the smoothness of the domain k is larger than the smoothness s of the source space. In this context, a bounded (but nonlinear) extension operator Ex from the trace space into the source space is constructed. Corresponding forerunners for the Besov spaces $B_{p,q}^s(\Omega)$ and Triebel-Lizorkin spaces $F_{p,q}^s(\Omega)$ on smooth domains can be found in [Tri83, Sect. 3.3.3]. Moreover, these results were generalized in [Skr90] to Besov and Triebel-Lizorkin spaces on manifolds—subject to several rigorous restrictions on the considered manifolds and submanifolds.

We build upon these achievements and extend, generalize, and improve them in several directions. Our main investigations can be summarized as follows:

(4) In Chap 8 we compute the trace of Besov spaces $\mathbf{B}_{p,q}^s(\Omega)$ on the boundary $\partial\Omega$ of bounded Lipschitz domains with smoothness s restricted to $0 < s < 1$. Our main result reads as

$$\mathrm{Tr}\,\mathbf{B}_{p,q}^{s+\frac{1}{p}}(\Omega) = \mathbf{B}_{p,q}^s(\partial\Omega),$$

where $n \geq 2$, $0 < s < 1$, and $0 < p, q \leq \infty$, cf. Theorem 8.3.1. The limiting case $s = 0$ is also considered in Corollary 8.3.3. In the range $0 < s < 1$, our results are optimal in the sense that there are no further restrictions on the parameters p, q. The fact that we cover traces in Besov spaces $\mathbf{B}_{p,q}^s(\mathbb{R}^d)$ with $p < 1$ is of particular interest in nonlinear approximation theory, cf. [DP88].
The methods we use in order to prove our results are completely different compared to [Sch11b] and [Tri83, Sect. 3.3.3]. In particular, we develop nonsmooth atomic decompositions for Besov spaces $\mathbf{B}_{p,q}^s(\mathbb{R}^d)$, cf. Theorem 2.3.23 and Corollary 2.3.25. For this we rely on equivalent characterizations for the Besov spaces via smooth atomic decompositions, which allow us to characterize $\mathbf{B}_{p,q}^s(\mathbb{R}^d)$ as the space of those $f \in L_p(\mathbb{R}^d)$ which can be represented as

$$f(x) = \sum_{j=0}^{\infty} \sum_{m \in \mathbb{Z}^d} \lambda_{j,m} a_{j,m}(x), \quad x \in \mathbb{R}^d, \tag{1.0.11}$$

with the sequence of coefficients $\lambda = \{\lambda_{j,m} \in \mathbb{C} : j \in \mathbb{N}_0, m \in \mathbb{Z}^d\}$ belonging to some appropriate sequence space $b_{p,q}^s$, where $s > 0$, $0 < p, q \leq \infty$, and with smooth atoms $a_{j,m}(x)$. We show that one can relax the assumptions on the smoothness of the atoms $a_{j,m}$ used in the representation (1.0.11) and, thus, replace these atoms with more general ones without loosing any crucial information compared to smooth atomic decompositions for functions $f \in \mathbf{B}_{p,q}^s(\mathbb{R}^d)$.

There are only few forerunners dealing with non-smooth atomic decompositions in function spaces so far. We refer to the papers [Tri03, MPP07], and [CL09], all mainly considering the different Fourier-analytical approach for Besov spaces and having in common that they restrict themselves to the technically simpler case when $p = q$. Our approach generalizes and extends these results and seems to be the first one covering the full range of indices $0 < p, q \le \infty$.

The additional freedom we gain in the choice of suitable non-smooth atoms $a_{j,m}$ for the atomic decompositions of $f \in \mathbf{B}_{p,q}^s(\mathbb{R}^d)$ makes this approach well suited to further investigate Besov spaces $\mathbf{B}_{p,q}^s(\Omega)$ on non-smooth domains Ω and their boundaries. In particular, for bounded Lipschitz domains we obtain some interesting new properties concerning interpolation and equivalent quasi-norms for these spaces as well as an atomic decomposition for Besov spaces $\mathbf{B}_{p,q}^s(\partial\Omega)$.

But the main goal is to demonstrate the strength of the newly developed non-smooth atomic decompositions in view of trace results. Moreover, as a by-product we obtain corresponding trace results on Lipschitz domains for Triebel-Lizorkin spaces, defined via atomic decompositions.

The same question for $s \ge 1$ was studied in [JW84]. It turns out that in this case the function spaces on the boundary look very different and also the extension operator must be changed. Moreover, based on the seminal work [JK95], traces on Lipschitz domains were studied in [May05, Th. 1.1.3] for the Fourier-analytical Besov spaces $B_{p,q}^s(\Omega)$ with the natural restrictions $(n - 1)\max\left(\frac{1}{p} - 1, 0\right) < s < 1$ and $\frac{n-1}{n} < p$. Our Theorem 8.3.1 actually covers and extends [May05, Th. 1.1.3], as for the considered range of parameters the Besov spaces defined by differences coincide with the Fourier-analytical Besov spaces. In contrast to Mayboroda we make use of the classical Whitney extension operator and the cone property of Lipschitz domains in order to establish our results instead of potential layers and interpolation. Moreover, the extension operator we construct is not linear—and in fact cannot be whenever $s < (n - 1)\max(\frac{1}{p} - 1, 0)$—compared to the extension operator in [May05, Th. 1.1.3]. Let us recall that the importance of nonlinear extension operators is known in the theory of differentiable spaces since the pioneering work of Gagliardo [Gag57], cf. also [Bur98, Chapter 5].

Finally, as an application we shall use the non-smooth atomic decompositions again to deal with pointwise multipliers in the respective function spaces.

(5) In Chap. 9 we consider a more general scale of function spaces, so called smoothness Morrey spaces and extend (1.0.10) and the counterpart from [Tri83, Sect. 3.3.3] to these spaces. In particular, we study traces of generalized Besov-Morrey spaces $\mathcal{N}_{\mathcal{M}_{p,q}^{\varphi}}^s(\Omega)$ on the boundary of C^k domains Ω with parameters $s \in \mathbb{R}, 0 < p < \infty, 0 < q \le \infty$, and a function parameter $\varphi : (0, \infty) \to$

$(0, \infty)$. The concept of these spaces was developed in [NNS16], where the authors studied several properties and obtained first results concerning traces on hyperplanes. One immediate and obvious advantage of these generalized scales of spaces is that, for different choices of the function parameter φ, one recovers a lot of well-known (scales of) spaces as special cases for which the obtained results immediately follow. In particular, by choosing $\varphi(t) = t^{d/u}$ we get the usual Besov-Morrey spaces, i.e.,

$$\mathcal{N}^s_{u,p,q}(\mathbb{R}^d) = \mathcal{N}^s_{\mathcal{M}^\varphi_{p,q}}(\mathbb{R}^d), \qquad \varphi(t) = t^{d/u},$$

and, furthermore, the classical Besov spaces by choosing $u = p$, i.e.,

$$B^s_{p,q}(\mathbb{R}^d) = \mathcal{N}^s_{p,p,q}(\mathbb{R}^d),$$

as special cases. Similar for the corresponding Triebel-Lizorkin-Morrey spaces. For a detailed overview we refer to the discussion in [NNS16, Appendix A].

Our main result concerning traces on the boundary can be formulated as follows. Let $n \geq 2$,

$$s > \frac{1}{p} + (n-1)\left(\frac{1}{\min(1,p)} - 1\right), \tag{1.0.12}$$

and put $\varphi^*(t) := \varphi(t)\, t^{-1/p}$. Then, subject to some further restriction on φ, we have

$$\mathrm{Tr}\,\mathcal{N}^s_{\mathcal{M}^\varphi_{p,q}}(\Omega) = \mathcal{N}^{s-\frac{1}{p}}_{\mathcal{M}^{\varphi^*}_{p,q}}(\partial\Omega), \tag{1.0.13}$$

where we require the smoothness k of the domain Ω to be large enough, cf. Theorem 9.3.2. Similar for the generalized Triebel-Lizorkin Morrey spaces. Let us remark that the proof of this trace result is completely different from the methods used in Chap. 8 as briefly outlined beforehand. Here a delicate construction of a lift and an extension operator is needed.

Additionally, to complete our investigation, we also study traces of Besov-type spaces $B^{s,\tau}_{p,q}(\Omega)$, which by are related but not covered by our scale of generalized Besov-Morrey spaces $\mathcal{N}^s_{\mathcal{M}^\varphi_{p,q}}(\Omega)$ and show in Theorem 9.3.9 that for $0 \leq \tau \leq \frac{1}{p}$, $n \geq 2$, and (1.0.12), we have

$$\mathrm{Tr}\, B^{s,\tau}_{p,q}(\Omega) = B^{s-\frac{1}{p}, \frac{n\tau}{n-1}}_{p,q}(\partial\Omega), \tag{1.0.14}$$

where again the smoothness k of the domain Ω has to be large enough. For the proof of (1.0.14) we develop in Theorem 2.5.23 a quarkonial decomposition for the spaces $B^{s,\tau}_{p,q}(\mathbb{R}^d)$, which itself is of independent interest. Finally, we apply the trace results for the Besov-type spaces and obtain some *a priori* estimates for solutions of elliptic boundary value problems, extending results from [ElB05].

(6) Finally, we adress the problem to what extend results from classical anaysis on Euclidean space carry over to the setting of Riemannian manifolds— without making unnecessary assumptions about the manifold. Here we are particularly interested in noncompact manifolds since the compact case presents no difficulties and is well understood. To be precise, we generalize (1.0.10) to Sobolev and Besov spaces $H_p^s(M)$ and $B_{p,q}^s(M)$, respectively, defined on noncompact Riemannian manifolds M with bounded geometry by considering their traces on submanifolds N.

Usually, spaces on manifolds of bounded geometry are defined via localization and pull-back using geodesic normal coordinates and corresponding function spaces on $\mathbb{R}^d$, cf. [Tri92, Sect. 7.2.2, 7.4.5] and also [Skr98, Def. 1]. Unfortunately, for some applications the choice of geodesic normal coordinates is not convenient as can be seen from [Skr90, Thm. 1], where traces on manifolds were studied using geodesic normal coordinates. Since these coordinates in general do not take into account the structure of the underlying submanifold where the trace is taken, in this setting one is limited to so-called *geodesic* submanifolds. This is highly restrictive, since geodesic submanifolds are very exceptional. Therefore, we do not wish to limit ourselves to these coordinates and pursue another idea for establishing our results: We consider a more general definition of the function spaces on manifolds subject to different local coordinates and give sufficient conditions on the corresponding coordinates resulting in equivalent norms. We speak about *admissible trivializations* in this context and mean all suitable local coordinates subordinate to locally finite coverings and partitions of unity of the manifold yielding basically the same spaces as the geodesic coordinates. Then choosing coordinates that are more adapted to the situation will immediately enable us to compute the trace on a much larger class of submanifolds compared to [Skr90]. The coordinates of choice for proving the trace results are Fermi coordinates. We show in Theorem 10.1.9 that for a certain cover with Fermi coordinates our spaces coincide with the ones defined via geodesic coordinates. The main trace result is then stated in Theorem 10.2.1, where we prove that if M is a manifold of dimension $n \geq 2$, N a submanifold of dimension $k < n$, and (M, N) of bounded geometry, we have for $s > \frac{n-k}{p}$,

$$\operatorname{Tr}_N H_p^s(M) = B_{p,p}^{s-\frac{n-k}{p}}(N). \tag{1.0.15}$$

The spaces on the right hand side of (1.0.15) are Besov spaces obtained via real interpolation of the fractional Sobolev spaces $H_p^s(M)$. Moreover, we extend (1.0.15) also to traces in Besov and Triebel-Lizorkin spaces defined on M. We believe that the method using admissible trivializations presented in Chap. 10 is very well suited in order to tackle the trace problem on manifolds. Alternatively, one could also think of computing traces using atomic decompositions of the spaces $H_p^s(M)$ as established in [Skr98], as we did before in Chap. 8 when dealing with traces of Besov spaces on bounded Lipschitz domains. But on (sub-)manifolds it should be complicated (if not impossible) to obtain a linear and continuous extension operator from the trace space into the source space—which by our method follows immediately from corresponding results on $\mathbb{R}^d$. Ultimately, we also provide another application of the concept of admissible trivializations and use them in the context of spaces with symmetries.

This outlines the main goals pursued in this manuscript. Further references are given at the corresponding places.

Structure of this book The structure of this book is as follows. At the beginning in Chap. 2 we present the relevant function spaces and general concepts needed for our further investigations. Part I composed of Chaps. 3–7 is dedicated to studying the regularity of solutions of PDEs in Besov and (fractional) Sobolev spaces. Afterwards in Part II we present trace results in various settings assorted in Chaps. 8–10.

Part I of this book is formally based on the papers [DS19, DS18, DHSS18a], and the recent preprint [DHSS18b]. The results from [DS19] can be seen as the starting point and the heart of our investigations for parabolic PDEs. In order to round up this manuscript we extended the results presented in [DS19] from polyhedral cones K to the numerically even more interesting domains of polyhedral type D. This lead to essentially new regularity results in Sobolev spaces which are presented in Sect. 5.2.1. We believe that this thesis gains a lot from this generalization. In particular, these results are not published elsewhere so far. The results for parabolic PDEs from [DS18] arouse from discussions when working on [DS19], but are interesting on their own respect. The focus in [DHSS18a] is slightly different. Here we investigate elliptic problems with a quite general nonlinear term causing some difficulties. Finally, the preprint [DHSS18b] contains all relevant information on Kondratiev spaces. Since our main focus in this manuscript lies in the fractional Sobolev and Besov regularity of the solutions we only included that material from [DHSS18b], which is needed for our later purposes.

Part II consists of the results established in the papers [SV13, SV12, MNS19], and [GS13]. Here our emphasis is on presenting how to obtain trace results in various settings using different methods. For this reason we do not include any proofs of the homogeneity results from [SV12] which were used for the trace results in [SV13]. Moreover, our findings in [GS13] are only partially included in what follows: To make our presentation clearer we left out the trace results for vector bundles and focused on the results for function spaces on manifolds instead. In order to unify the concepts and clarify connections and existing relations we changed the definitions of the atoms used in [MNS19] to fit with the atomic decompositions from [SV13].

All the material is selected and rearranged under the above described emphasis and restrictions. Hopefully, this focus will make the manuscript accessible to a large audience interested in these topics.

Chapter 2
Function Spaces and General Concepts

Function spaces will play a major role in our later investigations in Parts I and II of this thesis. It is our intention in Chap. 2 to systematically study all relevant function spaces needed later on, which are of interest for their own sake. The material is presented according to our needs. In some parts we merely collect what is needed avoiding detailed proofs but providing references for further reading. At other places we develop the existing theory further and present new properties and tools for the function spaces. In this case, of course, we provide detailed proofs of our new results in order to shed light on the basic concepts and methods we pursue.

We wish to demonstrate that the theory of function spaces is an indispensable tool for the study of ordinary and partial differential equations as well as in other areas in analysis. Therefore, we try to present a unified theory with special emphasis on the relations between the spaces studied. Presenting this material before the main results about to come in Parts I and II, hopefully prevents an overload of information and facilitates the reading of this manuscript.

2.1 Preliminaries

We start by collecting some general notation used throughout the thesis and recall the definition and some basic properties of Lipschitz domains, since the concept will be central for our studies.

2.1.1 Notation

As usual, we denote by $\mathbb{N}$ the set of all natural numbers, $\mathbb{N}_0 = \mathbb{N} \cup \{0\}$, $\mathbb{C}$ the complex plane, and $\mathbb{R}^d$, $d \in \mathbb{N}$, the d-dimensional real Euclidean space. Moreover,

© The Author(s), under exclusive license to Springer Nature Switzerland AG 2021
C. Schneider, *Beyond Sobolev and Besov*, Lecture Notes in Mathematics 2291,
https://doi.org/10.1007/978-3-030-75139-5_2

$\mathbb{R}^d_+ = \{x = (x', x_d) \in \mathbb{R}^d, x' \in \mathbb{R}^{d-1}, x_d > 0\}$ stands for the half-space. By $\mathbb{Z}^d$ we denote the lattice of all points in $\mathbb{R}^d$ with integer components. For convenience, let both dx and $|\cdot|$ stand for the d-dimensional Lebesgue measure in the sequel, and $\langle x \rangle := \sqrt{1 + |x|^2}$.

If $a \in \mathbb{R}$, then $a_+ := \max(a, 0)$ and $[a]$ denotes the integer part of a. For later use we introduce the numbers

$$\sigma_p = d\left(\frac{1}{p} - 1\right)_+ \quad \text{and} \quad \sigma_{p,q} = d\left(\frac{1}{\min(p, q)} - 1\right)_+, \tag{2.1.1}$$

where $0 < p, q \leq \infty$. Moreover, for $p \in (0, \infty]$, the number p' is defined by $1/p' := (1 - 1/p)_+$ with the convention that $1/\infty = 0$.

If E is a measurable subset of $\mathbb{R}^d$, we denote by χ_E its characteristic function.

For $x \in \mathbb{R}^d$ and $r \in (0, \infty)$ we denote by $Q(x, r)$ the compact cube centred at x with side length r, whose sides are parallel to the axes of coordinates. We write simply $Q(r) = Q(0, r)$ when $x = 0$. For each cube $Q \subset \mathbb{R}^d$ we denote its centre by x_Q, its side length by $\ell(Q)$, and, for $r > 0$, we denote by rQ the cube concentric with Q having the side length $r\ell(Q)$. Our convention for dyadic cubes is as follows: Let $Q_{j,m}$ with $j \in \mathbb{N}_0$ and $m \in \mathbb{Z}^d$ denote a cube in $\mathbb{R}^d$ with sides parallel to the axes of coordinates, centered at $2^{-j}m$, and with side length 2^{-j+1}. Furthermore, $\chi_{j,m}$ stands for the characteristic function of $Q_{j,m}$. Let $\mathcal{Q} := \{Q_{j,m} : j \in \mathbb{Z}, m \in \mathbb{Z}^d\}$, $\mathcal{Q}^* := \{Q \in \mathcal{Q} : \ell(Q) \leq 1\}$ and $j_Q := -\log_2 \ell(Q)$ for all $Q \in \mathcal{Q}$.

Given two quasi-Banach spaces X and Y, we write $X \hookrightarrow Y$ if $X \subset Y$ and the natural embedding is bounded. Moreover, by $\mathcal{L}(X, Y)$ we denote the space of all linear continuous functions from X to Y. If X is a normed linear space, the dual space which contains all continuous linear functionals on X is denoted by $X' = \mathcal{L}(X, \mathbb{R})$. It is a Banach-space normed by

$$\|x'|X'\| := \sup_{0 \neq x \in X} \frac{|x'(x)|}{\|x|X\|}.$$

We denote the duality pairing by $\langle \cdot, \cdot \rangle_{X \times X'}$ and write for the application of x' to x

$$\langle x, x' \rangle_{X \times X'} = \langle x', x \rangle_{X' \times X} = x'(x). \tag{2.1.2}$$

If X is a Hilbert space it follows that $X = X'$ and we write $(\cdot, \cdot)_X$ for the inner product.

By supp f we denote the support of the function f. Let $\Omega \subset \mathbb{R}^d$ be a domain, where domain always stands for connected open set. The boundary of Ω is denoted by $\Gamma = \partial\Omega$. Then $L_p(\Omega)$, $0 < p \leq \infty$, stands for the Lebesgue spaces, which contain all measurable functions $f : \Omega \to \mathbb{R}$ such that the quasi-norm

$$\|f|L_p(\Omega)\| = \left(\int_\Omega |f(x)|dx\right)^{1/p} < \infty$$

(with the usual modification if $p = \infty$). Moreover, $L_1^{\text{loc}}(\Omega)$ denotes the space of locally integrable functions on Ω and more general $L_p^{\text{loc}}(\Omega)$, $1 \leq p \leq \infty$ stands for the set of all p-locally integrable functions, i.e., all measurable functions : $\Omega \to \mathbb{C}$, whose restriction $f|_K$ belongs to $L_p(K)$ for any compact subset K of Ω.

Let $C(\Omega)$ be the space of bounded continuous functions on Ω, equipped with the sup-norm as usual. For $k \in \mathbb{N}$ we define $C^k(\Omega) = \{ f : D^{(\alpha)} f \in C(\mathbb{R}^d)$ for all $|\alpha| \leq k \}$. Here $\alpha = (\alpha_1, \ldots, \alpha_d) \in \mathbb{N}_0^d$ stands for some multi-index with $|\alpha| := \alpha_1 + \ldots + \alpha_d = k$, $k \in \mathbb{N}_0$, and for a k-times differentiable function $u : \Omega \to \mathbb{R}$, we write

$$D^{(\alpha)} u = \frac{\partial^{|\alpha|}}{\partial x_1^{\alpha_1} \ldots \partial x_d^{\alpha_d}} u$$

for the corresponding classical partial derivative as well as $u^{(k)} := D^{(k)} u$ in the one-dimensional case. Hence, the space $C^k(\Omega)$ is normed by

$$\| u | C^k(\Omega) \| := \max_{|\alpha| \leq k} \sup_{x \in \Omega} |D^{(\alpha)} u(x)| < \infty. \tag{2.1.3}$$

Additionally, $C^\infty(\Omega)$ contains the set of smooth and bounded functions on Ω, i.e.,

$$C^\infty(\Omega) := \bigcap_{k \in \mathbb{N}_0} C^k(\Omega).$$

Moreover, we put $x^\alpha = (x_1)^{\alpha_1} \cdots (x_d)^{\alpha_d}$. If $u \in C(\Omega)$ is bounded and uniformly continuous on Ω, then it possesses a unique, bounded continuous extension to the closure $\overline{\Omega}$ of Ω. We define the vector space $C^k(\overline{\Omega})$ to consist of all those functions $u \in C^k(\Omega)$ for which $D^{(\alpha)} u$ is bounded and uniformly continuous on Ω for $0 \leq |\alpha| \leq k$ (this convenient abuse of notation leads to ambiguities if Ω is unbounded, e.g. $C^k(\mathbb{R}^d) \neq C^k(\mathbb{R}^d)$ even though $\overline{\mathbb{R}^d} = \mathbb{R}^d$). $C^k(\overline{\Omega})$ is a closed subspace of $C^k(\Omega)$, and therefore also a Banach space with the same norm (2.1.3). Let $\mathcal{D}(\Omega)$ or $C_0^\infty(\Omega)$ stand for the set of test functions, i.e., the collection of all infinitely differentiable functions with support compactly contained in Ω. Given $\alpha \in (0, 1)$, we denote by $\mathcal{C}^\alpha(\Omega)$ the Hölder space containing all $f \in C(\Omega)$ such that

$$\| f | \mathcal{C}^\alpha(\Omega) \| := \| f | C(\Omega) \| + \sup_{\substack{x, y \in \Omega, \\ x \neq y}} \frac{|f(x) - f(y)|}{|x - y|^\alpha} < \infty. \tag{2.1.4}$$

The space $\mathcal{C}^\alpha(\overline{\Omega})$ consequently contains all $f \in C(\overline{\Omega})$ for which (2.1.4) is finite. Furthermore, for $\alpha = 1$ in (2.1.4) we recover the space of Lipschitz functions $\text{Lip}(\Omega)$. More general, the Hölder space $\mathcal{C}^{k,\alpha}(\Omega)$, $k \in \mathbb{N}_0$, $\alpha \in (0, 1]$, contains all $f \in C^k(\Omega)$, such that $D^{(\alpha)} f$ with $|\alpha| = k$ satisfies (2.1.4).

Let $\mathcal{S}(\mathbb{R}^d)$ stand for the Schwartz space of rapidly decreasing functions. The set of distributions on Ω will be denoted by $\mathcal{D}'(\Omega)$, whereas $\mathcal{S}'(\mathbb{R}^d)$ denotes the set of tempered distributions on $\mathbb{R}^d$. Furthermore, let $\mathcal{F}$ stand for the Fourier-transform on $\mathcal{S}'(\mathbb{R}^d)$ with inverse $\mathcal{F}^{-1}$. The terms *distribution* and *generalized function* will be used synonymously. For the application of a distribution $u \in \mathcal{D}'(\Omega)$ to a test function $\varphi \in \mathcal{D}(\Omega)$ we write $\langle u, \varphi \rangle$ in good agreement with (2.1.2). Similar if $u \in \mathcal{S}'(\mathbb{R}^d)$ and $\varphi \in \mathcal{S}(\mathbb{R}^d)$. For $u \in \mathcal{D}'(\Omega)$ and a multi-index $\alpha = (\alpha_1, \ldots, \alpha_d) \in \mathbb{N}_0^d$, we write $D^\alpha u$ for the α-th *generalized* or *distributional derivative* of u with respect to $x = (x_1, \ldots, x_d) \in \Omega$, i.e., $D^\alpha u$ is a distribution on Ω, uniquely determined by the formula

$$\langle D^\alpha u, \varphi \rangle := (-1)^{|\alpha|} \langle u, D^{(\alpha)} \varphi \rangle, \qquad \varphi \in \mathcal{D}(\Omega).$$

In particular, if $u \in L_1^{\mathrm{loc}}(\Omega)$ and there exists a function $v \in L_1^{\mathrm{loc}}(\Omega)$ such that

$$\int_\Omega v(x)\varphi(x)dx = (-1)^{|\alpha|} \int_\Omega u(x)D^{(\alpha)}\varphi(x)dx \qquad \text{for all} \qquad \varphi \in \mathcal{D}(\Omega),$$

we say that v is the α-*th weak derivative* of u and write $D^\alpha u = v$. We also use the notation $\partial^\alpha u := D^\alpha u$ as well as $\frac{\partial^k}{\partial x_j^k} u := D^\beta u$ and $\partial_{x_j^k} := D^\beta u$ for some multi-index $\beta = (0, \ldots, k, \ldots, 0)$ with $\beta_j = k$, $k \in \mathbb{N}$. Furthermore, for $m \in \mathbb{N}_0$, we write $D^m u$ for any (generalized) m-th order derivative of u, where $D^0 u := u$ and $Du := D^1 u$. Sometimes we shall use subscripts such as D_x^m, ∂_x^α or D_t to emphasize that we only take derivatives with respect to $x = (x_1, \ldots, x_d) \in \Omega$ or $t \in \mathbb{R}$.

Notations Concerning Manifolds When investigating function spaces and traces on manifolds in Sect. 2.6 and Chap. 10, respectively, we will use slightly different notations in order to avoid confusions and make our considerations clearer. In contrast to the other chapters the standard coordinates on $\mathbb{R}^d$ are denoted by $x = (x^1, x^2, \ldots, x^d)$ in order to avoid confusion with other coordinates with lower indices (from the tangent space). Then the partial derivative operators in direction of the coordinates are denoted by $\partial_i = \partial/\partial x^i$ for $1 \le i \le d$. For multi-indices we use the notation $\mathfrak{a} = (\mathfrak{a}_1, \ldots, \mathfrak{a}_d)$, $\mathfrak{a}_i \in \mathbb{N}_0$, $i = 1, \ldots, d$, with $D^{\mathfrak{a}} f = \partial_1^{\mathfrak{a}_1} \ldots \partial_d^{\mathfrak{a}_d} f = \frac{\partial^{|\mathfrak{a}|} f}{(\partial x^1)^{\mathfrak{a}_1} \ldots (\partial x^d)^{\mathfrak{a}_d}}$, having its usual meaning for a function f on $\mathbb{R}^d$ (as α will be used as index for the local charts on a manifold in this setting). Moreover, we use the Einstein's sum convention and the symbol $\sqcup$ instead of $\cup$ to emphasize when a union of sets is disjoint.

2.1.2 Lipschitz Domains and Their Boundaries

Lipschitz domains will play a major role in our investigations. We are going to study the regularity of solutions of PDEs on these domains and specific subclasses as well as traces on their boundary. Therefore, we recall the basic notion of a Lipschitz domain and some further types of related domains, which we deal with in the sequel.

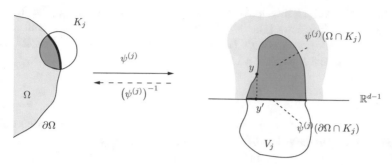

Fig. 2.1 Lipschitz diffeomorphism $\psi^{(j)}$

A one-to-one mapping $\Phi : \mathbb{R}^d \mapsto \mathbb{R}^d$, is called a *Lipschitz diffeomorphism*, if the components $\Phi_k(x)$ of $\Phi(x) = (\Phi_1(x), \ldots, \Phi_d(x))$ are Lipschitz functions on $\mathbb{R}^d$ and

$$|\Phi(x) - \Phi(y)| \sim |x - y|, \quad x, y \in \mathbb{R}^d, \quad |x - y| \le 1,$$

where the equivalence constants are independent of x and y. Of course the inverse of Φ^{-1} is also a Lipschitz diffeomorphism on $\mathbb{R}^d$ (Fig. 2.1).

Definition 2.1.1 (Lipschitz Domain) Let Ω be a bounded domain in $\mathbb{R}^d$ with boundary $\Gamma = \partial\Omega$. Then Ω is said to be a *Lipschitz domain*, if there exist N open balls $K_1, \ldots, K_N$ such that $\bigcup_{j=1}^{N} K_j \supset \Gamma$ and $K_j \cap \Gamma \ne \emptyset$ if $j = 1, \ldots, N$, with the following property: for every ball K_j there are Lipschitz diffeomorphisms $\psi^{(j)}$ such that $\psi^{(j)} : K_j \longrightarrow V_j$ for $j = 1, \ldots, N$, where $V_j := \psi^{(j)}(K_j)$ and

$$\psi^{(j)}(K_j \cap \Omega) \subset \mathbb{R}^d_+, \qquad \psi^{(j)}(K_j \cap \Gamma) \subset \mathbb{R}^{d-1}.$$

Remark 2.1.2

(i) We recover the definition of a bounded C^k *domain*, if we assume $\psi^{(j)}$ to be a *k-diffeomorphism* (C^k*-diffeomorphism*) in Definition 2.1.1, which means that all components are C^k functions for some $k \in \mathbb{N} \cup \{\infty\}$. Note that our understanding of a k-diffeomorphism implies that the inverse $\left(\psi^{(j)}\right)^{-1}$ is also a C^k map. In [NNS16, Sect. 5.3] a diffeomorphism with this property was called *regular*. We do not make this distinction. Later on we will sometimes speak of smooth domains when we deal with C^k domains, $k \in \mathbb{N} \cup \{\infty\}$, and non-smooth domains when we think of Lipschitz domains.
Furthermore, the maps $\psi^{(j)}$ can be extended outside K_j in such a way that the extended vector functions (denoted by $\psi^{(j)}$ as well) yield diffeomorphic mappings from $\mathbb{R}^d$ onto itself (Lipschitz or k-diffeomorphisms, respectively).

(ii) There are several equivalent definitions of Lipschitz domains in the literature. Our approach follows [Dac04]. Another version as can be found in [Ste70], which defines first a *simple (unbounded) Lipschitz domain* Ω in $\mathbb{R}^d$ as the domain above the graph of a Lipschitz function $h : \mathbb{R}^{d-1} \longrightarrow \mathbb{R}$, i.e.,

$$\Omega = \{(x', x_d) : h(x') < x_d\}.$$

Then a *bounded Lipschitz domain* Ω in $\mathbb{R}^d$ is defined as a bounded domain where the boundary $\Gamma = \partial\Omega$ can be covered by finitely many open balls B_j in $\mathbb{R}^d$ with $j = 1, \ldots, J$, centered at Γ such that

$$B_j \cap \Omega = B_j \cap \Omega_j \qquad \text{for } j = 1, \ldots, J,$$

where Ω_j are rotations of suitable simple Lipschitz domains in $\mathbb{R}^d$. We shall occasionally use this alternative definition, in particular, since it usually suffices to consider simple Lipschitz domains in our proofs (the related covering involves only finitely many balls), simplifying the notation considerably.

Examples 2.1.3 Two examples of Lipschitz domains are the circle and the segment of a circle with opening angle $0 < \alpha < 2\pi$. In particular, the circle is also a C^∞ domain whereas the segment of a circle is not (not even C^k for any $k \in \mathbb{N}$) because of its corner (Figs. 2.2 and 2.3).

The two domains with the cusp in the figures below are not Lipschitz domains (let locally $\Omega \subset \mathbb{R}^2$ be below or above the curve $|x_1|^\alpha$, $0 < \alpha < 1$, respectively) (Figs. 2.4 and 2.5).

Fig. 2.2 Circle

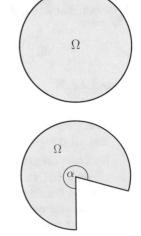

Fig. 2.3 Segment of circle

Fig. 2.4 Cusp 1

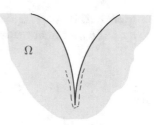

Fig. 2.5 Cusp 2

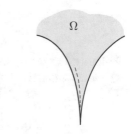

Fig. 2.6 Cover of domain Ω

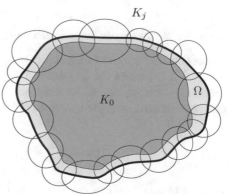

Remark 2.1.4 (Resolution of Unity) Let K_j with $j = 1, \ldots, N$ be the same balls as in Definition 2.1.1 and let K_0 be an inner domain with $\overline{K_0} \subset \Omega$, as indicated in Fig. 2.6; hence

$$\partial \Omega \subset \bigcup_{j=1}^{N} K_j$$

and

$$\Omega \subset K_0 \cup \left(\bigcup_{j=1}^{N} K_j \right).$$

Moreover, let $\{\varphi_j\}_{j=0}^N$ denote a related resolution of unity of $\overline{\Omega}$ subject to this cover, i.e., φ_j are nonnegative functions with

$$\varphi_j \in D(K_j), \quad j = 0, \ldots, N, \tag{2.1.5}$$

and

$$\sum_{j=0}^N \varphi_j(x) = 1 \quad \text{if } x \in \overline{\Omega}. \tag{2.1.6}$$

Obviously, the restriction of φ_j to $\partial\Omega$ is also a resolution of unity with respect to $\partial\Omega$. Now we can decompose $f \in L_p(\Omega)$ such that

$$f(x) = \varphi_0(x)f(x) + \sum_{j=1}^N \varphi_j(x)f(x), \qquad x \in \Omega,$$

where the term $\varphi_0 f$ can be extended outside of Ω by zero.

Spaces on Boundaries of Lipschitz and C^k Domains The boundary $\partial\Omega = \Gamma$ of a bounded Lipschitz (or C^k) domain Ω will be furnished in the usual way with a surface measure $d\sigma$. If we consider subspaces of $L_p(\mathbb{R}^d)$, the corresponding complex-valued Lebesgue spaces $L_p(\Gamma), 0 < p \le \infty$, are normed by

$$\|g|L_p(\Gamma)\| = \left(\int_\Gamma |g(\gamma)|^p d\sigma(\gamma) \right)^{1/p}$$

(with obvious modifications if $p = \infty$). Later on we require the introduction of several function spaces on Γ. We rely on the resolution of unity according to (2.1.6) and the local Lipschitz (or k-) diffeomorphisms $\psi^{(j)}$ mapping $\Gamma_j = \Gamma \cap K_j$ onto $W_j = \psi^{(j)}(\Gamma_j)$, recall Definition 2.1.1. We define

$$g_j(y) := (\varphi_j f) \circ (\psi^{(j)})^{-1}(y), \qquad j = 1, \ldots, N, \tag{2.1.7}$$

which restricted to $y = (y', 0) \in W_j$,

$$g_j(y') = (\varphi_j f) \circ (\psi^{(j)})^{-1}(y'), \qquad j = 1, \ldots, N, \quad f \in L_p(\Gamma), \tag{2.1.8}$$

makes sense. This results in functions $g_j \in L_p(W_j)$ with compact supports in the $(d-1)$-dimensional Lipschitz (or C^k) domain W_j. We do not distinguish notationally between g_j and $(\psi^{(j)})^{-1}$ as functions of $(y', 0)$ and of y'.

Our constructions enable us to transport function spaces naturally from $\mathbb{R}^{d-1}$ to the boundary Γ of a bounded Lipschitz (or C^k) domain via pull-back and a partition of unity (Fig. 2.7).

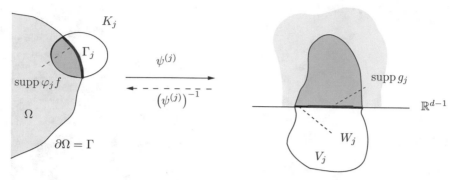

Fig. 2.7 Local pull-back of boundary Γ onto $\mathbb{R}^{d-1}$

Remark 2.1.5 If we consider subspaces of $\mathcal{S}'(\mathbb{R}^d)$ rather than $L_p(\mathbb{R}^d)$ we have to change the above setting in the following way: Let $\mathcal{D}'(\Gamma)$ stand for the distributions on the compact C^k manifold Γ. Then we replace (2.1.8) above by

$$g_j(y') = (\varphi_j f) \circ (\psi^{(j)})^{-1}(y'), \qquad j = 1, \ldots, N, \quad f \in \mathcal{D}'(\Gamma),$$

which results in distributions $g_j \in \mathcal{D}'(W_j)$ with compact supports in the $(d-1)$-dimensional C^k domain W_j.

Special Lipschitz Domains We provide some more examples of specific Lipschitz domains, which will be important for our investigations later on.

When studying the fractional Sobolev regularity of parabolic PDEs in Chap. 5 we deal with generalized wedges, which are unbounded Lipschitz domains. The precise definition of these kinds of domains is given below.

Definition 2.1.6 (Generalized Wedge) Let $2 \leq m \leq d$ and consider a smooth cone in $\mathbb{R}^m$ defined as

$$K := \{x \in \mathbb{R}^m : x/|x| = \omega \in \Omega\},$$

where we assume that $\Omega = K \cap S^{m-1}$ is of class $C^{1,1}$. Then a *generalized wedge* in $\mathbb{R}^d$ is given by

$$\mathcal{K} = K \times \mathbb{R}^{d-m}.$$

Remark 2.1.7 Note that in our considerations the case $\mathcal{K} = K$, i.e., $m = d$, is not excluded. Two important examples for domains which are covered by Definition 2.1.6 are given in Figs. 2.8 and 2.9.

In particular, the domain in Fig. 2.8 is also called a *dihedron* denoted by $\mathcal{D} \subset \mathbb{R}^3$, which in polar coordintates can be described as

$$\mathcal{D} = K \times \mathbb{R}, \qquad K = \{(x_1, x_2) : 0 < r < \infty, \; -\theta/2 < \varphi < \theta/2\},$$

Fig. 2.8 $m = 2 < 3 = d$,
$\mathcal{K} = K \times \mathbb{R}$

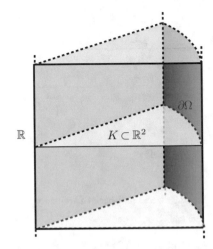

$\mathbb{R}$ $K \subset \mathbb{R}^2$ $\partial\Omega$

Fig. 2.9 $m = 3 = d, \mathcal{K} = K$

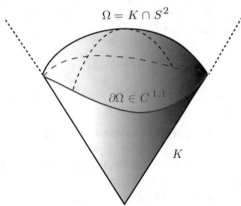

$\Omega = K \cap S^2$

$\partial\Omega \in C^{1,1}$

K

where the opening angle θ of the 2-dimensional wedge K satisfies $0 < \theta \le 2\pi$. Figure 2.9 illustrates a smooth cone $K \subset \mathbb{R}^3$. Later on we will also deal with the *truncated cone*

$$K_0 = \{x \in K \; : \; |x| < r_0\} \subset \mathbb{R}^m, \tag{2.1.9}$$

and the *bounded wedge*

$$\mathcal{K}_0 := K_0 \times B_{r_0}^{d-m}(0). \tag{2.1.10}$$

The following domains will be considered in Chap. 6, when we study the regularity of hyperbolic problems.

Definition 2.1.8 (Special Lipschitz Domain) Let $\Omega \subset \mathbb{R}^d$, $d > 2$, be a Lipschitz domain, whose boundary $\partial\Omega$ consists of two (smooth) surfaces Γ_1 and Γ_2 intersecting along a manifold l_0. We assume that in a neighbourhood of each point of l_0 the

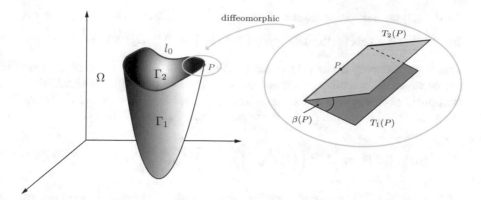

Fig. 2.10 Special Lipschitz domain Ω

set $\overline{\Omega}$ is diffeomorphic to a dihedral angle. For any $P \in l_0$ define $T_1(P)$ and $T_2(P)$ as (part of) the tangent spaces in P w.r.t. Γ_1 and Γ_2 (Fig. 2.10).

2.2   Classical Function Spaces

In this section we briefly collect the basics concerning Sobolev and fractional Sobolev spaces needed later on. Moreover, when studying the regularity of parabolic PDEs, we also have to deal with Banach-space valued versions of these (and other) spaces. Thus, we also provide all material needed in this context. In particular, a generalized version of Sobolev's embedding theorem for Banach-space valued functions can be found in Theorem 2.2.5.

2.2.1   Sobolev and Fractional Sobolev Spaces

Sobolev Spaces Let $1 \le p \le \infty$ and $m \in \mathbb{N}_0$. The Sobolev space $W_p^m(\Omega)$ contains all complex-valued functions $u(x)$ defined on Ω such that the norm

$$\|u|W_p^m(\Omega)\| = \left(\sum_{|\alpha| \le m} \int_\Omega |D^\alpha u(x)|^p dx \right)^{1/p} < \infty$$

(with obvious modifications if $p = \infty$). In particular, we see that $W_p^0(\Omega) = L_p(\Omega)$. If $p = 2$ the spaces are Hilbert spaces and we write

$$H^m(\Omega) := W_2^m(\Omega).$$

By $\overset{\circ}{W}^m_p(\Omega)$ we denote the closure of $\mathcal{D}(\Omega)$ in $W^m_p(\Omega)$. Moreover, if $1 < p < \infty$, then $W^{-m}_p(\Omega)$ stands for the dual space $\left(\overset{\circ}{W}^m_{p'}(\Omega)\right)'$ of $\overset{\circ}{W}^m_{p'}(\Omega)$, where $\frac{1}{p'} = 1 - \frac{1}{p}$.

Fractional Sobolev Spaces In order to fill the gaps in the scale of Sobolev spaces, where the smoothness is given by $m \in \mathbb{N}_0$, one possibility is to introduce fractional Sobolev spaces as follows. For $s \in \mathbb{R}$ and $1 < p < \infty$ we define the fractional Sobolev spaces $H^s_p(\mathbb{R}^d)$ as the collection of all $u \in \mathcal{S}'(\mathbb{R}^d)$ such that

$$\|u|H^s_p(\mathbb{R}^d)\| := \left\| \mathcal{F}^{-1}\left(\left(1 + |\xi|^2\right)^{s/2} \mathcal{F}u \right) |L_p(\mathbb{R}^d) \right\| < \infty, \qquad (2.2.1)$$

cf. [Tri92, Ch. 1]. These spaces partially coincide with the classical Sobolev spaces, i.e., we have $H^s_p(\mathbb{R}^d) = W^m_p(\mathbb{R}^d)$ for $s = m$ with $m \in \mathbb{N}_0$. Corresponding spaces on domains can be defined via restrictions of functions from $H^s_p(\mathbb{R}^d)$ equipped with the norm

$$\|u|H^s_p(\Omega)\| := \inf\left\{ \|g|H^s_p(\mathbb{R}^d)\| : \ g \in H^s_p(\mathbb{R}^d),\ g|_\Omega = u \right\}. \qquad (2.2.2)$$

If $p = 2$ we abbreviate $H^s(\Omega) = H^s_2(\Omega)$.

Remark 2.2.1

(i) It is also possible to characterize fractional Sobolev spaces with the help of the Laplace operator via

$$H^s_p = (\mathrm{Id} - \Delta)^{-s/2}L_p,$$

   cf. [Tri92, Ch. 1].
(ii) For $p = 2$, $0 \le s \notin \mathbb{N}$, and $s = k + \lambda$ with $k \in \mathbb{N}_0$ and $\lambda \in (0, 1)$, an alternative norm for $H^s_p(\Omega)$ comes from the Slobodeckij spaces and is given by

$$\|u|H^s(\Omega)\| = \|u|H^k(\Omega)\| + \sum_{|\alpha|=k} \left(\int_{\Omega\times\Omega} \frac{|D^\alpha u(x) - D^\alpha u(y)|^2}{|x - y|^{d+2\lambda}} dx dy \right)^{1/2} < \infty.$$
$$(2.2.3)$$

Properties For later use we recall the following lemma concerning pointwise multipliers and diffeomorphisms in the fractional Sobolev scale. In particular, this is a special case of Theorem 2.5.19, where the situation for the generalized smoothness Morrey spaces is treated. For a proof in the more general setting of Triebel-Lizorkin spaces we refer to [Tri92, Sect. 4.2, 4.3].

Lemma 2.2.2 *Let $s \in \mathbb{R}$, $1 < p < \infty$ and $k > s > 0$ with $k \in \mathbb{N}$.*

(i) *Let $g \in C^k(\mathbb{R}^d)$. Then $f \to gf$ is a linear and bounded operator from $H_p^s(\mathbb{R}^d)$ into itself, i.e., there exists a positive constant $C(k)$ such that*

$$\|gf|H_p^s(\mathbb{R}^d)\| \leq C(k)\|g|C^k(\mathbb{R}^d)\| \cdot \|f|H_p^s(\mathbb{R}^d)\|.$$

(ii) *Let ψ be a k-diffeomorphism. Then $f \to f \circ \psi$ is a linear and bounded operator from $H_p^s(\mathbb{R}^d)$ into itself. In particular, we have for some positive constant $C(\psi)$,*

$$\|f \circ \psi|H_p^s(\mathbb{R}^d)\| \leq C(\psi)\|f|H_p^s(\mathbb{R}^d)\|.$$

Remark 2.2.3 In Chap. 10 we deal with local charts and therefore use a local version of Lemma 2.2.2(ii): There we consider functions $f \in H_p^s(\mathbb{R}^d)$ with supp $f \subset U \subset \mathbb{R}^d$ for U open and k-diffeomorphisms $\psi : V \subset \mathbb{R}^d \to U \subset \mathbb{R}^d$. This causes no problems since the k-diffeomorphisms can be extended to k-diffeomorphisms on $\mathbb{R}^d$.

2.2.2 Banach-Space Valued Hölder, Lebesgue, and Sobolev Spaces

The Hölder Spaces $C^{k,\alpha}(I, X)$ Consider a Banach space X and an interval $I = [0, T] \subset \mathbb{R}$ (or $I = (0, T)$) with $T < \infty$. We write $C(I, X)$ for the space consisting of all bounded and (uniformly) continuous functions $u : I \to X$ normed by

$$\|u|C(I, X)\| := \max_{t \in I} \|u(t)|X\|.$$

Moreover, we say that $u \in C^k(I, X)$, $k \in \mathbb{N}_0$, if u has a Taylor expansion

$$u(t + h) = u(t) + u'(t)h + \frac{1}{2}u''(t)h^2 + \ldots + \frac{1}{k!}u^{(k)}(t)h^k + r_k(t, h)$$

for all $t + h, t \in I$ such that

- $u^{(j)}(t)$ depends continuously on t for all $j = 0, \ldots, k$,
- $\lim_{|h| \to 0} \dfrac{\|r_k(t, h)|X\|}{|h|^k} = 0$.

The space $C^k(I, X)$ is then equipped with the following norm

$$\|u|C^k(I, X)\| := \sum_{j=0}^{k} \|u^{(j)}|C(I, X)\|.$$

Given $\alpha \in (0, 1)$, we denote by $\mathcal{C}^\alpha(I, X)$ the Hölder space containing all $u \in C(I, X)$ such that

$$\|u|\mathcal{C}^\alpha(I, X)\| := \|u|C(I, X)\| + |u|_{\mathcal{C}^\alpha(I,X)}$$

$$= \|u|C(I, X)\| + \sup_{\substack{t,s \in I \\ t \neq s}} \frac{\|u(t) - u(s)|X\|}{|t - s|^\alpha} < \infty.$$

Consequently, $\mathcal{C}^{k,\alpha}(I, X)$ contains all functions $u \in C(I, X)$ such that

$$\|u|\mathcal{C}^{k,\alpha}(I, X)\| := \|u|C^k(I, X)\| + |u^{(k)}|_{\mathcal{C}^\alpha(I,X)} < \infty.$$

Remark 2.2.4 Let Y be some (quasi-)Banach space such that $X \hookrightarrow Y$. Then it follows that

$$C^k(I, X) \hookrightarrow C^k(I, Y) \qquad \text{and} \qquad \mathcal{C}^{k,\alpha}(I, X) \hookrightarrow \mathcal{C}^{k,\alpha}(I, Y).$$

This is an immediate consequence of the definition of the spaces. Let $u : I \to X \in C^k(I, X)$ with Taylor expansion

$$u(t + h) = u(t) + u'(t)h + \frac{1}{2}u''(t)h^2 + \ldots + \frac{1}{k!}u^{(k)}(t)h^k + r_k(t, h).$$

Then also $u : I \to Y$ and we have

- $u^{(j)}(t)$ depends continuously on t for all $j = 0, \ldots, k$, since

$$|t - t_0| < \delta \implies \|u^{(j)}(t) - u^{(j)}(t_0)|Y\| \leq c\|u^{(j)}(t) - u^{(j)}(t_0)|X\| < \varepsilon,$$

- $$\lim_{|h|\to 0} \frac{\|r_k(t, h)|Y\|}{|h|^k} \leq \lim_{|h|\to 0} \frac{\|r_k(t, h)|X\|}{|h|^k} = 0,$$

from which we deduce that $u \in C^k(I, Y)$. Moreover, concerning the generalized Hölder spaces we now observe that

$$\|u|\mathcal{C}^{k,\alpha}(I, Y)\| = \|u|C^k(I, Y)\| + \sup_{\substack{t,s \in I \\ t \neq s}} \frac{\|u(t) - u(s)|Y\|}{|t - s|^\alpha}$$

$$\leq \|u|C^k(I, X)\| + \sup_{\substack{t,s \in I \\ t \neq s}} \frac{\|u(t) - u(s)|X\|}{|t - s|^\alpha} = \|u|\mathcal{C}^{k,\alpha}(I, X)\|,$$

which gives the desired result.

Lebesgue Spaces $L_p(I, X)$ **and Sobolev Spaces** $W_p^m(I, X)$ Let us briefly recall
the definition of Lebesgue and Sobolev spaces for functions with values in a Banach
space X. We denote by $L_p(I, X)$, $1 \leq p \leq \infty$, the space of (equivalence classes
of) measurable functions $u : I \to X$ such that the mapping $t \mapsto \|u(t)\|_X$ belongs
to $L_p(I)$, which is endowed with the norm

$$\|u|L_p(I, X)\| = \begin{cases} \left(\int_I \|u(t)|X\|^p dt \right)^{1/p} & \text{if } p < \infty, \\ \underset{t \in I}{\text{ess sup}} \, \|u(t)|X\| & \text{if } p = \infty. \end{cases}$$

The definition of weak derivatives of Banach-space valued distributions is a natural
generalization of the one for real-valued distributions. We refer to [Cap14, Part I,
Sect. 3] in this context. Let $\mathcal{D}'(I, X) := \mathcal{L}(\mathcal{D}(I), X)$ be the space of X-valued
distributions, where $\mathcal{L}(U, V)$ denotes the space of all linear continuous functions
from U to V. For the application of a distribution $u \in \mathcal{D}'(I, X)$ to a test function
$\varphi \in \mathcal{D}(I)$, we use the notation $\langle u, \varphi \rangle$. For $u \in \mathcal{D}'(I, X)$ and $k \in \mathbb{N}$, the k-th
generalized or *distributional derivative* $\partial_{t^k} u$ is defined as an X-valued distribution
satisfying

$$\langle \partial_{t^k} u, \varphi \rangle := (-1)^k \langle u, \partial_{t^k} \varphi \rangle, \qquad \varphi \in \mathcal{D}(I).$$

In particular, if $u : I \to X$ is an integrable function and there exists an integrable
function $v : I \to X$ satisfying

$$\int_I v(t)\varphi(t)dt = (-1)^k \int_I u(t)\partial_{t^k}\varphi(t)dt \qquad \text{for all} \quad \varphi \in \mathcal{D}(I),$$

where the integrals above are Bochner integrals, cf. [Coh13], we say that v is the k-
th weak derivative of u and write $\partial_{t^k} u = v$. For $m \in \mathbb{N}_0$ we denote by $W_p^m(I, X)$ the
space of all functions $u \in L_p(I, X)$, whose weak derivatives of order $0 \leq k \leq m$
belong to $L_p(I, X)$, normed by

$$\|u|W_p^m(I, X)\| = \begin{cases} \left(\sum_{k=0}^m \|\partial_{t^k} u | L_p(I, X)\|^p \right)^{1/p} & \text{if } p < \infty, \\ \underset{0 \leq k \leq m}{\max} \, \|\partial_{t^k} u | L_\infty(I, X)\| & \text{if } p = \infty. \end{cases}$$

$L_p(I, X)$ and $W_p^m(I, X)$ are Banach spaces. Moreover, if $p = 2$ we write
$H^m(I, X) := W_2^m(I, X)$.

We will also need a version of Sobolev's embedding theorem for Banach-space
valued functions.

Theorem 2.2.5 (Generalized Sobolev's Embedding Theorem) *Let* $1 < p < \infty$ *and* $m \in \mathbb{N}$. *Then*

$$W_p^m(I, X) \hookrightarrow \mathcal{C}^{m-1, 1-\frac{1}{p}}(I, X). \tag{2.2.4}$$

Proof Note that the theorem of Meyers-Serrin extends to the spaces $W_p^m(I, X)$, cf. [Kre15, Thm. 4.11]. Hence, $C^\infty(I, X)$ is dense in $W_p^m(I, X)$. It is also shown in [Kre15, Prop. 4.3] that in this case weak derivatives coincide with normal derivatives. Since $k = m - 1$, using [CH98, Thm. 1.4.35] together with Bochner's Theorem and Hölder's inequality gives for $u \in C^\infty(I, X)$,

$$
\begin{aligned}
\left\| u^{(k)}(t + h) - u^{(k)}(t) | X \right\| &= \left\| \int_t^{t+h} u^{(k+1)}(s) ds | X \right\| \\
&\leq \int_t^{t+h} \left\| u^{(m)} | X \right\| ds \\
&\leq h^{1-\frac{1}{p}} \left(\int_t^{t+h} \left\| u^{(m)} | X \right\|^p ds \right)^{\frac{1}{p}} \\
&\leq h^\alpha \| u | W_p^m(I, X) \|.
\end{aligned}
$$

Hence, we see that id is a linear and bounded operator from the dense subset $C^\infty(I, X)$ of $W_p^m(I, X)$ into the Banach-space $\mathcal{C}^{k,\alpha}(I, X)$ admitting an extension $\widetilde{\text{id}}$ onto $W_p^m(I, X)$ with equal norm. This completes the proof. □

Remark 2.2.6 In particular, we have $W_p^1(I, X) \hookrightarrow \mathcal{C}^{0, 1-\frac{1}{p}}(I, X)$ for $m = 1$. This was proven in [CH98, Thm. 1.4.38]. Theorem 2.2.5 can also be found in [Sim90, Cor. 26].

The Spaces $L_p(\Omega_T)$ and $H^{m,l*}(\Omega_T)$ We collect some more notation for specific Banach-space valued Lebesgue and Sobolev spaces, which will be used when studying the regularity of solutions of parabolic PDEs in Chap. 5.

Let $\Omega_T := [0, T] \times \Omega$. Then we abbreviate

$$L_p(\Omega_T) := L_p([0, T], L_p(\Omega)).$$

Moreover, we put

$$H^{m,l*}(\Omega_T) := H^{l-1}([0, T], \mathring{H}^m(\Omega)) \cap H^l([0, T], H^{-m}(\Omega)).$$

normed by

$$\| u | H^{m,l*}(\Omega_T) \| = \| u | H^{l-1}([0, T], \mathring{H}^m(\Omega)) \| + \| u | H^l([0, T], H^{-m}(\Omega)) \|.$$

2.3 Besov and Triebel-Lizorkin Spaces

In this section we introduce Besov spaces and Triebel-Lizorkin spaces—sometimes briefly denoted as B- and F-spaces in the sequel—and present some basic properties and concepts. For a detailed study together with historical remarks we refer to Triebel [Tri83, Tri92, Tri06] and the references given there. In particular, these two scales of spaces cover many well-known spaces of functions and distributions such as Hölder-Zygmund spaces, (fractional) Sobolev spaces, Bessel-potential spaces and Hardy spaces.

Our main focus are the Besov spaces, which have been investigated for several decades, resulting, for instance, from the study of partial differential equations, interpolation theory, approximation theory, harmonic analysis. There are several definitions of B-spaces to be found in the literature. For our purposes we present two of the most prominent approaches, which are the *Fourier-analytic approach* using Fourier transforms on the one hand (Besov spaces $B_{p,q}^s$) and the *classical approach* via higher order differences involving the modulus of smoothness on the other (Besov spaces $\mathbf{B}_{p,q}^s$). These two definitions are equivalent only with certain restrictions on the parameters, but may otherwise share similar properties.

We also deal with Triebel-Lizorkin spaces, since their theory is strongly linked with the theory of Besov spaces. They were introduced independently by Triebel and Lizorkin in the early 1970s. One usually gives priority to the Besov spaces but sometimes working with F-spaces has some advantages. We briefly explain where this happens in our context:

- When studying regularity of PDEs in Part II of this thesis we will heavily rely on embeddings between Kondratiev and B- or F-spaces, respectively. In this situation it turns out that the embedding into the F-scale in Theorem 2.4.16 is better compared to the corresponding embedding into the B-scale in Theorem 2.4.10— no additional knowledge of the smoothness in the space $B_{p,\infty}^s$ is required in the first case. This is the reason why we prefer to work with Theorem 2.4.16 in Chap. 4. In particular, in Chap. 4 we first establish regularity of the solution of our PDE in specific scales of F-spaces and afterwards make use of an embedding into the B-scale.

- Particularly with regard to the fractional Sobolev spaces the F-scale has the advantage that the whole Sobolev scale is included as a special case. We make use of this relation in Chap. 5, Theorem 5.3.1, when studying the fractional Sobolev regularity of Problem III.

- Finally, when we consider B- and F-spaces on manifolds, the F-spaces are easier to hande due to the localization principle.

2.3.1 Spaces of Type $B^s_{p,q}$, $F^s_{p,q}$

Fourier-Analytical Approach We now introduce Besov and Triebel-Lizorkin spaces via the Fourier-analytical approach in terms of dyadic Littlewood-Paley decompositions. For this, we recall briefly the basic ingredients needed in this context.

We start with the notion of a *smooth dyadic resolution of unity*. Let

$$A_l = \{x \in \mathbb{R}^d : 2^{l-1} < |x| < 2^{l+1}\}, \qquad l \in \mathbb{N},$$

complemented by

$$A_0 = \{x \in \mathbb{R}^d : |x| < 2\}$$

be the usual dyadic annuli in $\mathbb{R}^d$. Let $\{\varphi_j\}_{j=0}^\infty$ be a sequence of C_0^∞ functions satisfying the following conditions:

(i) $\operatorname{supp} \varphi_j \subset \overline{A_j}, \quad j \in \mathbb{N}_0$,

(ii) for any multi-index $\gamma = (\gamma_1, \ldots, \gamma_d) \in \mathbb{N}_0^d$ there exists a positive constant c_γ such that

$$2^{j|\gamma|}|D^\gamma \varphi_j(x)| \le c_\gamma \quad \text{for all} \quad x \in \mathbb{R}^d, \quad |\gamma| = \gamma_1 + \cdots + \gamma_d,$$

(iii) $\displaystyle\sum_{j=0}^\infty \varphi_j(x) = 1, \quad x \in \mathbb{R}^d.$

Then $\{\varphi_j\}_{j=0}^\infty$ is said to be a *smooth dyadic resolution of unity*. Such a dyadic resolution of unity can be constructed as follows: Let $\varphi_0 = \varphi \in \mathcal{S}(\mathbb{R}^d)$ be such that

$$\operatorname{supp} \varphi \subset \left\{y \in \mathbb{R}^d : |y| < 2\right\} \quad \text{and} \quad \varphi(x) = 1 \quad \text{if} \quad |x| \le 1, \tag{2.3.1}$$

and for each $j \in \mathbb{N}$ let

$$\varphi_j(x) = \varphi(2^{-j}x) - \varphi(2^{-j+1}x).$$

Then $\{\varphi_j\}_{j=0}^\infty$ forms a smooth dyadic resolution of unity (Fig. 2.11).

Fig. 2.11 Smooth dyadic resolution of unity

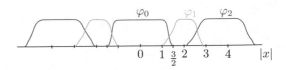

Remark 2.3.1 Sometimes a less restrictive definition is used: instead of (iii) the sum of all the function values at a particular point is only required to be positive, rather that 1, for each point in the space. However, given such a set of functions, one can obtain a partition of unity in the strict sense by dividing every function by the sum of all functions (which is defined, since at any point it has only a finite number of terms). This generalized version of a smooth resolution of unity will be used when introducing the generalized smoothness Morrey spaces and Besov-type spaces in Sect. 2.5.

Based on such resolutions of unity, we can decompose every tempered distribution $f \in S'(\mathbb{R}^d)$ into a series of entire analytical functions,

$$f = \sum_{j=0}^{\infty} \mathcal{F}^{-1}(\varphi_j \mathcal{F} f),$$

converging in $S'(\mathbb{R}^d)$.

Definition 2.3.2 (Besov and Triebel-Lizorkin Spaces) Let $s \in \mathbb{R}$, $0 < q \le \infty$, and $\{\varphi_j\}_{j=0}^{\infty}$ a smooth dyadic resolution of unity.

(i) Let $0 < p \le \infty$. The *Besov space* $B_{p,q}^s(\mathbb{R}^d)$ is the set of all distributions $f \in S'(\mathbb{R}^d)$ such that

$$\left\| f \mid B_{p,q}^s(\mathbb{R}^d) \right\| = \left(\sum_{j=0}^{\infty} 2^{jsq} \| \mathcal{F}^{-1}(\varphi_j \mathcal{F} f) | L_p(\mathbb{R}^d) \|^q \right)^{1/q} \qquad (2.3.2)$$

is finite (with the usual modification if $q = \infty$).

(ii) Let $0 < p < \infty$. The *Triebel-Lizorkin space* $F_{p,q}^s(\mathbb{R}^d)$ is the set of all distributions $f \in S'(\mathbb{R}^d)$ such that

$$\left\| f \mid F_{p,q}^s(\mathbb{R}^d) \right\| = \left\| \left(\sum_{j=0}^{\infty} 2^{jsq} | \mathcal{F}^{-1}(\varphi_j \mathcal{F} f)(\cdot) |^q \right)^{1/q} | L_p(\mathbb{R}^d) \right\| \qquad (2.3.3)$$

is finite (with the usual modification if $q = \infty$).

Remark 2.3.3

(i) We sometimes write $A_{p,q}^s$ instead of $B_{p,q}^s$ or $F_{p,q}^s$, when both scales of spaces are meant simultaneously in some context. An important aspect of Triebel-Lizorkin spaces is their close relation to many classical function spaces. In particular, the scale $F_{p,q}^s$ generalizes fractional Sobolev spaces, i.e., we have the coincidences

$$F_{p,2}^s(\mathbb{R}^d) = H_p^s(\mathbb{R}^d) \qquad \text{and} \qquad F_{2,2}^m(\mathbb{R}^d) = H^m(\mathbb{R}^d) = W_2^m(\mathbb{R}^d),$$

(2.3.4)

where $s \in \mathbb{R}$, $m \in \mathbb{N}_0$, and $1 < p < \infty$, cf. [Tri83, p. 51].

(ii) We see from Definition 2.3.2 that Besov and Triebel-Lizorkin spaces on $\mathbb{R}^d$ are basically defined in the same way by interchanging the order in which the ℓ_q- and L_p-norms are taken. In particular, if $p = q$, we have

$$B_{p,p}^s(\mathbb{R}^d) = F_{p,p}^s(\mathbb{R}^d), \qquad 0 < p < \infty,$$

and we extend this to $p = \infty$ by putting $F_{\infty,\infty}^s(\mathbb{R}^d) := B_{\infty,\infty}^s(\mathbb{R}^d)$. The scales $F_{p,q}^s(\mathbb{R}^d)$ and $B_{p,q}^s(\mathbb{R}^d)$ were studied in detail in [Tri83, Tri92], where the reader may also find further references to the literature.

(iii) We mention here that Besov spaces $B_{p,p}^s(\mathbb{R}^d)$ can alternatively be defined via interpolation of Sobolev spaces (this will be our approach for defining Besov spaces on manifolds M, cf. Definition 2.6.14). In particular, let $(\cdot, \cdot)_{\Theta,p}$ stand for the real interpolation method, cf. [Tri92, Sect. 1.6.2]. Then for $s_0, s_1 \in \mathbb{R}$, $1 < p < \infty$, and $0 < \Theta < 1$, we have

$$B_{p,p}^s(\mathbb{R}^d) = \left(H_p^{s_0}(\mathbb{R}^d), H_p^{s_1}(\mathbb{R}^d) \right)_{\Theta,p},$$

where $s = \Theta s_0 + (1 - \Theta)s_1$. Note that $B_{p,p}^s(\mathbb{R}^d)$ does not depend on the choice of s_0, s_1, Θ.

Spaces of Type $B_{p,q}^s$, $F_{p,q}^s$ on Domains We define corresponding B- and F-spaces on domains $\Omega \subset \mathbb{R}^d$ via restriction. Let $A \in \{B, F\}$. Then

$$A_{p,q}^s(\Omega) := \{f \in \mathcal{D}'(\Omega) : \exists g \in A_{p,q}^s(\mathbb{R}^d), \ g|_\Omega = f\},$$

normed by

$$\|f | A_{p,q}^s(\Omega)\| := \inf_{g|_\Omega = f} \|g | A_{p,q}^s(\mathbb{R}^d)\|.$$

Defining spaces on domains this way has the advantage that the embeddings for the spaces on $\mathbb{R}^d$ carry over to the spaces on domains. In particular, it follows that the two scales of function spaces are linked via

$$B_{p,\min(p,q)}^s(\Omega) \hookrightarrow F_{p,q}^s(\Omega) \hookrightarrow B_{p,\max(p,q)}^s(\Omega)$$

(2.3.5)

and they coincide for $p = q$, cf. [Tri83, Prop. 3.2.4].

Properties We recall what is known concerning embeddings results for spaces of type $A_{p,q}^s$.

Proposition 2.3.4 (Embeddings Within the Scales $B_{p,q}^s$, $F_{p,q}^s$) *Let $\Omega \subset \mathbb{R}^d$ be a domain. Furthermore, assume $s \in \mathbb{R}$, $0 < p \leq \infty$ ($p < \infty$ for F-spaces), and $0 < q \leq \infty$.*

(i) Let $\varepsilon > 0$, $0 < u \leq \infty$, and $q \leq v \leq \infty$, then

$$A_{p,q}^s(\Omega) \hookrightarrow A_{p,u}^{s-\varepsilon}(\Omega) \qquad and \qquad A_{p,q}^s(\Omega) \hookrightarrow A_{p,v}^s(\Omega).$$

(ii) (Sobolev-type embedding) Let $\sigma < s$ and $p < \tau$ be such that

$$s - \frac{n}{p} = \sigma - \frac{n}{\tau}, \tag{2.3.6}$$

then

$$A_{p,q}^s(\Omega) \hookrightarrow A_{\tau,r}^\sigma(\Omega), \tag{2.3.7}$$

where $0 < r \leq \infty$ and, additionally, $q \leq r$ if $A = B$ and '=' holds in (2.3.6).
(iii) Additionally, let $\Omega \subset \mathbb{R}^d$ be bounded and $\tau \leq p$. Then we have the embedding

$$A_{p,q}^s(\Omega) \hookrightarrow A_{\tau,q}^s(\Omega), \qquad \tau \leq p. \tag{2.3.8}$$

Remark 2.3.5 The above results can be found in [Tri83, Prop. 2.3.2/2, Thms. 2.7.1, 3.3.1].

In the DeVore-Triebel diagram in Fig. 2.12 we have illustrated the area of possible embeddings of a fixed original space $A_{p,q}^s(\Omega)$ into spaces $A_{\tau_1,r_1}^{\sigma_1}(\Omega)$ and $A_{\tau_2,r_2}^{\sigma_2}(\Omega)$. The lighter shaded area corresponds to the additonal embeddings we have if the underlying domain Ω is bounded.

2.3.2   Wavelet Characterization for $B_{p,q}^s$

Under certain restrictions on the parameters the Besov spaces $B_{p,q}^s$ allow a characterization in terms of wavelet decompositions. In this context we refer e.g. to [Coh03, Mey92]. In particular, this wavelet characterization will turn out to be extremely useful when proving embeddings of weighted Sobolev spaces into Besov spaces from the non-linear approximation scale (1.0.3).

Wavelets are specific orthonormal bases for $L_2(\mathbb{R})$ that are obtained by dilating, translating and scaling one fixed function, the so-called *mother wavelet* ψ. The mother wavelet is usually constructed by means of a so-called *multiresolution analysis,* that is, a sequence $\{V_j\}_{j \in \mathbb{Z}}$ of shift-invariant, closed subspaces of $L_2(\mathbb{R})$

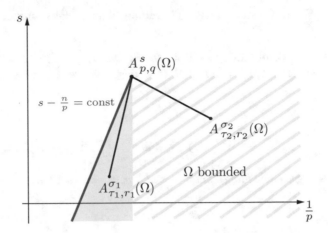

Fig. 2.12 Embeddings within the scales $B_{p,q}^s$, $F_{p,q}^s$

whose union is dense in L_2 while their intersection is zero. Moreover, all the spaces are related via dyadic dilation, and the space V_0 is spanned by the translates of one fixed function ϕ, called the *generator*. In her fundamental work [Dau98, Dau92] Daubechies has shown that there exist families of compactly supported wavelets. By taking tensor products, a compactly supported orthonormal basis for $L_2(\mathbb{R}^d)$ can be constructed which will also be used in this paper.

Let ϕ be a father wavelet of tensor product type on $\mathbb{R}^d$ and let $\Psi' = \{\psi_i : i = 1, \ldots, 2^d - 1\}$ be the set containing the corresponding multivariate mother wavelets such that, for a given $r \in \mathbb{N}$ and some $N > 0$ the following localization, smoothness, and vanishing moment conditions hold. For all $\psi \in \Psi'$,

$$\operatorname{supp} \phi, \ \operatorname{supp} \psi \ \subset \ [-N, N]^d, \tag{2.3.9}$$

$$\phi, \ \psi \ \in \ C^r(\mathbb{R}^d), \tag{2.3.10}$$

$$\int_{\mathbb{R}^d} x^\alpha \psi(x) dx \ = \ 0 \quad \text{for all } \alpha \in \mathbb{N}_0^d \text{ with } |\alpha| \le r. \tag{2.3.11}$$

We refer again to [Dau98, Dau92] for a detailed discussion. Recall that the set of all dyadic cubes $Q_{j,k} \subset \mathbb{R}^d$, $j \in \mathbb{N}_0$, $k \in \mathbb{Z}^d$ with measure at most 1 is denoted by

$$\mathcal{Q}^* := \big\{ I = Q_{j,k} : l(I) \le 1 \big\}$$

and we set $\mathcal{Q}_j := \{I \in \mathcal{Q}^* : |I| = 2^{-jd}\}$. For the dyadic shifts and dilations of the father wavelet and the corresponding wavelets we use the abbreviations

$$\phi_k(x) := \phi(x - k), \quad \psi_I(x) := 2^{jd/2}\psi(2^j x - k) \quad \text{for} \quad j \in \mathbb{N}_0, \ k \in \mathbb{Z}^d, \ \psi \in \Psi'. \tag{2.3.12}$$

It follows that

$$\left\{\phi_k,\ \psi_I\ :\ k \in \mathbb{Z}^d,\ I \in \mathcal{Q}^*,\ \psi \in \Psi'\right\}$$

is an orthonormal basis in $L_2(\mathbb{R}^d)$. Denote by $Q(I)$ some dyadic cube (of minimal size) such that $\operatorname{supp} \psi_I \subset Q(I)$ for every $\psi \in \Psi'$. Then, we clearly have $Q(I) = 2^{-j}k + 2^{-j}Q$ for some dyadic cube Q. Put $\Lambda' = \mathcal{Q}^* \times \Psi'$. Then, every function $f \in L_2(\mathbb{R}^d)$ can be written as

$$f = \sum_{k \in \mathbb{Z}^d} \langle f, \phi_k \rangle \phi_k + \sum_{(I, \psi) \in \Lambda'} \langle f, \psi_I \rangle \psi_I.$$

Later on, it will be convenient to include ϕ into the set Ψ'. We use the notation $\phi_I := 0$ for $|I| < 1$, $\phi_I = \phi(\cdot - k)$ for $I = k + [-1/2, 1/2]^d$, and can simply write

$$f = \sum_{(I, \psi) \in \Lambda} \langle f, \psi_I \rangle \psi_I, \qquad \Lambda = \mathcal{Q}^* \times \Psi, \qquad \Psi = \Psi' \cup \{\phi\}.$$

We describe Besov spaces $B^s_{p,q}(\mathbb{R}^d)$ by decay properties of the wavelet coefficients, if the parameters fulfill certain conditions.

Theorem 2.3.6 (Wavelet Characterization) *Let* $0 < p, q < \infty$ *and* $s > \sigma_p$. *Choose* $r \in \mathbb{N}$ *such that* $r > s$ *and construct a wavelet Riesz basis as described above. Then a function* $f \in L_p(\mathbb{R}^d)$ *belongs to the Besov space* $B^s_{p,q}(\mathbb{R}^d)$ *if, and only if,*

$$f = \sum_{k \in \mathbb{Z}^d} \langle f, \phi_k \rangle \phi_k + \sum_{(I, \psi) \in \Lambda'} \langle f, \psi_I \rangle \psi_I \qquad (2.3.13)$$

(convergence in $\mathcal{S}'(\mathbb{R}^d)$*) with*

$$\|f \mid B^s_{p,q}(\mathbb{R}^d)\| \sim \left(\sum_{k \in \mathbb{Z}^d} |\langle f, \phi_k \rangle|^p \right)^{1/p}$$

$$+ \left(\sum_{j=0}^{\infty} 2^{j\left(s + d\left(\frac{1}{2} - \frac{1}{p}\right)\right)q} \left(\sum_{(I, \psi) \in \mathcal{Q}_j \times \Psi'} |\langle f, \psi_I \rangle|^p \right)^{q/p} \right)^{1/q} < \infty.$$

$$(2.3.14)$$

Remark 2.3.7

(i) For parameter $q = \infty$ we use the usual modification (replacing the outer sum by a supremum), i.e.,

$$\| f \mid B^s_{p,\infty}(\mathbb{R}^d)\| \sim \left(\sum_{k \in \mathbb{Z}^d} |\langle f, \phi_k \rangle|^p \right)^{1/p}$$

$$+ \sup_{j \geq 0} 2^{j\left(s + d\left(\frac{1}{2} - \frac{1}{p}\right)\right)} \left(\sum_{(I,\psi) \in \mathcal{Q}_j \times \Psi'} |\langle f, \psi_I \rangle|^p \right)^{1/p} < \infty.$$

(ii) In particular, for our adaptivity scale (1.0.3), i.e., $B^s_{\tau,\tau}(\mathbb{R}^d)$ with $s = d\left(\frac{1}{\tau} - \frac{1}{p}\right)$, we see that the norm (2.3.14) becomes

$$\| f \mid B^s_{\tau,\tau}(\mathbb{R}^d)\| \sim \left(\sum_{k \in \mathbb{Z}^d} |\langle f, \phi_k \rangle|^\tau \right)^{1/\tau}$$

$$+ \left(\sum_{j=0}^{\infty} 2^{jd\left(\frac{1}{2} - \frac{1}{p}\right)\tau} \sum_{(I,\psi) \in \mathcal{Q}_j \times \Psi'} |\langle f, \psi_I \rangle|^\tau \right)^{1/\tau}.$$

$$\tag{2.3.15}$$

2.3.3 Spaces of Type $\mathbf{B}^s_{p,q}$

In this subsection, we are concerned with the classical Besov spaces $\mathbf{B}^s_{p,q}$ defined via higher order differences as can be found in [Tri83, Sect. 2.5.12]. Under certain restrictions on the parameters the spaces $\mathbf{B}^s_{p,q}$ coincide with the spaces $B^s_{p,q}$ defined via the Fourier-analytical approach, cf. formula (2.3.17) below. The reason why we prefer to stick with two different notations for Besov spaces is, that the approach via differences and its characterization via smooth and non-smooth atomic decompositions will be the main tool for our studies of traces on Lipschitz domains in Chap. 8. In this context we will particularly deal with ranges of parameters where the spaces $\mathbf{B}^s_{p,q}$ differ from the spaces $B^s_{p,q}$.

If f is an arbitrary function on $\mathbb{R}^d$, $h \in \mathbb{R}^d$ and $r \in \mathbb{N}$, then

$$(\Delta^1_h f)(x) = f(x + h) - f(x) \quad \text{and} \quad (\Delta^{r+1}_h f)(x) = \Delta^1_h (\Delta^r_h f)(x)$$

are the usual iterated differences. Given a function $f \in L_p(\mathbb{R}^d)$ the *r-th modulus of smoothness* is defined by

$$\omega_r(f, t)_p = \sup_{|h| \leq t} \|\Delta_h^r f \mid L_p(\mathbb{R}^d)\|, \quad t > 0, \quad 0 < p \leq \infty.$$

Definition 2.3.8 (Besov Spaces) Let $0 < p, q \leq \infty$, $s > 0$, and $r \in \mathbb{N}$ such that $r > s$. Then the *Besov space* $\mathbf{B}_{p,q}^s(\mathbb{R}^d)$ contains all $f \in L_p(\mathbb{R}^d)$ such that

$$\|f \mid \mathbf{B}_{p,q}^s(\mathbb{R}^d)\| = \|f \mid L_p(\mathbb{R}^d)\| + \left(\int_0^1 t^{-sq} \omega_r(f, t)_p^q \frac{dt}{t} \right)^{1/q} < \infty.$$

(with the usual modification if $q = \infty$).

Remark 2.3.9 Definition 2.3.8 is independent of r, meaning that different values of $r > s$ result in quasi-norms which are equivalent. Furthermore, the spaces are quasi-Banach spaces (Banach spaces if $p, q \geq 1$). Note that we deal with subspaces of $L_p(\mathbb{R}^d)$, in particular, for $s > 0$ and $0 < q \leq \infty$, we have the embedding

$$\mathbf{B}_{p,q}^s(\mathbb{R}^d) \hookrightarrow L_p(\mathbb{R}^d), \quad 0 < p \leq \infty.$$

This scale of Besov spaces contains many well-known function spaces. For example, if $p = q = \infty$, one recovers the Hölder-Zygmund spaces $\mathcal{C}^s(\mathbb{R}^d)$, i.e.,

$$\mathbf{B}_{\infty,\infty}^s(\mathbb{R}^d) = \mathcal{C}^s(\mathbb{R}^d), \quad s > 0. \tag{2.3.16}$$

Moreover, under certain restrictions on the parameters the different approaches for Besov spaces coincide, i.e., we have

$$\mathbf{B}_{p,q}^s(\mathbb{R}^d) = B_{p,q}^s(\mathbb{R}^d) \quad \text{if} \quad s > \sigma_p, \quad 0 < p, q \leq \infty, \tag{2.3.17}$$

in the sense of equivalent quasi-norms. In particular, for the adaptivity scale of Besov spaces $B_{\tau,\tau}^\alpha(\mathbb{R}^d)$ from (1.0.3) we deal with when studying the regularity of solutions of PDEs later on, we are always in the case where $\alpha > \sigma_p$, hence, the approaches coincide. This follows immediately from the restriction on the parameters $\frac{1}{\tau} = \frac{\alpha}{d} + \frac{1}{p}$, which together with the assumptions $\tau \leq p$, i.e., $\frac{1}{\tau} - \frac{1}{p} \geq 0$ and $p \geq 1$, yields

$$\alpha = d \left(\frac{1}{\tau} - \frac{1}{p} \right) \geq d \left(\frac{1}{\tau} - 1 \right)_+.$$

Later on we will need the following homogeneity estimate proved in [SV12, Thm. 2] based on [CLT07].

Theorem 2.3.10 (Homogeneity Result) *Let* $0 < \lambda \leq 1$ *and* $f \in \mathbf{B}^s_{p,q}(\mathbb{R}^d)$ *with* supp $f \subset B_\lambda(0)$. *Then*

$$\|f(\lambda \cdot)|\mathbf{B}^s_{p,q}(\mathbb{R}^d)\| \sim \lambda^{s-d/p}\|f|\mathbf{B}^s_{p,q}(\mathbb{R}^d)\|. \tag{2.3.18}$$

Spaces of Type $\mathbf{B}^s_{p,q}$ **on Domains** We are now going to define the spaces $\mathbf{B}^s_{p,q}(\Omega)$ on domains via higher order differences.

Let us introduce an adapted notation of generalized differences

$$\Delta^r_h f(x, \Omega) := \begin{cases} \Delta^r_h f(x), & x, x+h, \ldots, x+rh \in \Omega, \\ 0, & \text{otherwise,} \end{cases} \tag{2.3.19}$$

where $x, h \in \mathbb{R}^d$ and $r \in \mathbb{N}$. The modulus of smoothness of order r of a function $f \in L_p(\Omega)$ is then

$$\omega_r(f, t, \Omega)_p := \sup_{|h| \leq t} \|\Delta^r_h f(\cdot, \Omega)|L_p(\Omega)\|, \qquad t > 0.$$

Definition 2.3.11 (Besov Spaces on Domains) Let $0 < p, q \leq \infty$, $s > 0$, and $r \in \mathbb{N}$ such that $r > s$. Then the Besov space $\mathbf{B}^s_{p,q}(\Omega)$ contains all $f \in L_p(\Omega)$ such that

$$\|f|\mathbf{B}^s_{p,q}(\Omega)\| = \|f|L_p(\Omega)\| + \left(\int_0^1 t^{-sq}\omega_r(f, t, \Omega)^q_p \frac{dt}{t}\right)^{1/q} \tag{2.3.20}$$

(with the usual modification if $q = \infty$) is finite.

Remark 2.3.12 The above definition is intrinsic as it only relies on values of the function f on points inside of the domain Ω. The question arises whether it is possible to obtain the same spaces via restriction of the corresponding spaces $\mathbf{B}^s_{p,q}(\mathbb{R}^d)$ on $\mathbb{R}^d$, since then a lot of results known for $\mathbf{B}^s_{p,q}(\mathbb{R}^d)$ (e.g. embedding results) carry over to $\mathbf{B}^s_{p,q}(\Omega)$. This might not always be the case but is true for so-called (ε, δ)-domains, which play a crucial role concerning questions of extendability. We refer to [DS93, Sect. 5] in this context. In particular, on these domains it is possible to define an extension operator Ex, which extends functions in $\mathbf{B}^s_{p,q}(\Omega)$ to all of $\mathbb{R}^d$, cf. [DS93, Thm. 6.1]. Bounded Lipschitz domains and also the half space $\mathbb{R}^d_+$ are examples of (ε, δ)-domains. For these domains the spaces $\mathbf{B}^s_{p,q}(\Omega)$ can as well be regarded as restrictions of the corresponding spaces on $\mathbb{R}^d$ in the usual interpretation, i.e.,

$$\mathbf{B}^s_{p,q}(\Omega) = \{f \in L_p(\Omega) : \quad \text{there exists} \quad g \in \mathbf{B}^s_{p,q}(\mathbb{R}^d) \quad \text{with} \quad g|_\Omega = f\},$$

furnished with the norm

$$\|f|\mathbf{B}_{p,q}^s(\Omega)\| = \inf\left\{\|g|\mathbf{B}_{p,q}^s(\mathbb{R}^d)\| \quad \text{with} \quad g\big|_\Omega = f\right\},$$

where $g\big|_\Omega = f$ denotes the restriction of g to Ω. Therefore, well-known embedding results for spaces defined on $\mathbb{R}^d$ carry over to those defined on domains Ω. Let $s > 0, \varepsilon > 0, 0 < q, u \leq \infty$, and $q \leq v \leq \infty$. Then we have

$$\mathbf{B}_{p,u}^{s+\varepsilon}(\Omega) \hookrightarrow \mathbf{B}_{p,q}^s(\Omega) \qquad \text{and} \qquad \mathbf{B}_{p,q}^s(\Omega) \hookrightarrow \mathbf{B}_{p,u}^s(\Omega),$$

cf. [HS09, Thm. 1.15], where also further embeddings for the Besov spaces may be found.

2.3.4 Smooth and Non-smooth Atomic Decompositions for $\mathbf{B}_{p,q}^s$

We give equivalent characterizations for the Besov spaces $\mathbf{B}_{p,q}^s(\Omega)$ through their decomposition properties. This provides a constructive approach expanding functions f via smooth and non-smooth atoms (excluding any moment conditions) and suitable coefficients, where the latter belong to certain sequence spaces denoted by $b_{p,q}^s(\Omega)$ defined below.

Let $G \subset \mathbb{R}^d$ and $j \in \mathbb{N}_0$. We use the abbreviation

$$\sum_{m \in \mathbb{Z}^d}^{G,j} = \sum_{m \in \mathbb{Z}^n, Q_{j,m} \cap G \neq \emptyset}, \tag{2.3.21}$$

where G will usually denote either a domain Ω in $\mathbb{R}^d$ or its boundary Γ.

Smooth Atomic Decomposition We start with the relevant sequence spaces.

Definition 2.3.13 (Sequence Spaces) Let $0 < p, q \leq \infty, s \in \mathbb{R}$. Furthermore, let $\Omega \subset \mathbb{R}^d$ and $\lambda = \{\lambda_{j,m} \in \mathbb{C} : j \in \mathbb{N}_0, m \in \mathbb{Z}^d\}$. Then

$$b_{p,q}^s(\Omega) = \left\{\lambda : \|\lambda|b_{p,q}^s(\Omega)\| = \left(\sum_{j=0}^\infty 2^{j(s-\frac{d}{p})q}\left(\sum_{m \in \mathbb{Z}^d}^{\Omega,j}|\lambda_{j,m}|^p\right)^{q/p}\right)^{1/q} < \infty\right\}$$

(with the usual modification if $p = \infty$ and/or $q = \infty$).

Remark 2.3.14 If $\Omega = \mathbb{R}^d$, we simply write $b_{p,q}^s$ and $\sum_m$ instead of $b_{p,q}^s(\Omega)$ and $\sum_m^{\Omega,j}$, respectively.

Now we define the smooth atoms.

Definition 2.3.15 (Smooth Atoms) Let $K \in \mathbb{N}_0$ and $\tilde{d} > 1$. A K-times continuously differentiable complex-valued function a on $\mathbb{R}^d$ (continuous if $K = 0$) is called a K-atom if for some $j \in \mathbb{N}_0$,

$$\operatorname{supp} a \subset \tilde{d} Q_{j,m} \quad \text{for some} \quad m \in \mathbb{Z}^d, \tag{2.3.22}$$

and

$$|D^\alpha a(x)| \leq 2^{|\alpha|j} \quad \text{for} \quad |\alpha| \leq K. \tag{2.3.23}$$

It is convenient to write $a_{j,m}(x)$ instead of $a(x)$ if this atom is located at $Q_{j,m}$ according to (2.3.22). Furthermore, K denotes the smoothness of the atom, cf. (2.3.23).

We can characterize Besov spaces $\mathbf{B}^s_{p,q}(\Omega)$ using atomic decompositions as follows.

Theorem 2.3.16 (Smooth Atomic Decomposition) *Let $\Omega \subset \mathbb{R}^d$ be an bounded Lipschitz domain, $s > 0$, and $0 < p, q \leq \infty$. Let $\tilde{d} > 1$ and $K \in \mathbb{N}_0$ with*

$$K \geq (1 + [s])$$

be fixed. Then $f \in L_p(\Omega)$ belongs to $\mathbf{B}^s_{p,q}(\Omega)$ if, and only if, it can be represented as

$$f(x) = \sum_{j=0}^{\infty} \sum_{m \in \mathbb{Z}^d}^{\Omega, j} \lambda_{j,m} a_{j,m}(x), \tag{2.3.24}$$

where the $a_{j,m}$ are K-atoms ($j \in \mathbb{N}_0$) with

$$\operatorname{supp} a_{j,m} \subset \tilde{d} Q_{j,m}, \qquad j \in \mathbb{N}_0, \quad m \in \mathbb{Z}^d,$$

and $\lambda \in b^s_{p,q}(\Omega)$, convergence being in $L_p(\Omega)$. Furthermore,

$$\|f|\mathbf{B}^s_{p,q}(\Omega)\| := \inf \|\lambda|b^s_{p,q}(\Omega)\|, \tag{2.3.25}$$

where the infimum is taken over all admissible representations (2.3.24).

Remark 2.3.17 According to [Tri06], based on [HN07], the space $\mathbf{B}^s_{p,q}(\Omega)$ is independent of $\tilde{d}$ and K. This may justify our omission of K and $\tilde{d}$ in (2.3.25).

Since the atoms $a_{j,m}$ used in Theorem 2.3.16 are defined also outside of Ω, the spaces $\mathbf{B}^s_{p,q}(\Omega)$ characterized in Theorem 2.3.16 can as well be regarded as restrictions of the corresponding spaces on $\mathbb{R}^d$. Thus, in view of Remark 2.3.12,

Theorem 2.3.16 in general holds for (ε, δ)-domains Ω and, in particular, also for the half-space $\mathbb{R}^d_+$.

There are also atomic decompositions available for the spaces $B^s_{p,q}$, cf. [Tri97, Thm. 13.8]. Note that the atoms used in this context have to satisfy moment conditions up to a certain order for the range of parameters where the two approaches $\mathbf{B}^s_{p,q}$ and $B^s_{p,q}$ do not coincide.

Non-smooth Atomic Decomposition   Our aim is to provide a non-smooth atomic characterization of Besov spaces $\mathbf{B}^s_{p,q}(\mathbb{R}^d)$ by relaxing the assumptions about the smoothness of the atoms $a_{j,m}$ in Definition 2.3.15.

The idea is as follows. Note that condition (2.3.23) is equivalent to

$$\|a(2^{-j}\cdot)|C^K(\mathbb{R}^d)\| \leq 1. \qquad (2.3.26)$$

We replace the C^K-norm with $K > s$ by a Besov quasi-norm $\mathbf{B}^\sigma_{p,p}(\mathbb{R}^d)$ with $\sigma > s$ or in case of $0 < s < 1$ by a norm in the space of Lipschitz functions $\mathrm{Lip}(\mathbb{R}^d)$. This concept leads to the following non-smooth atoms, which were introduced in [Tri02]. They will be very adequate when considering (non-smooth) atomic decompositions of spaces defined on Lipschitz domains (or as we shall see later on the boundary of a Lipschitz domain, respectively).

Definition 2.3.18 (Non-smooth Atom)   We say that $a \in \mathrm{Lip}(\mathbb{R}^d)$ is a Lip-atom, if for some $j \in \mathbb{N}_0$,

$$\mathrm{supp}\, a \subset \tilde{d} Q_{j,m}, \quad m \in \mathbb{Z}^d, \quad \tilde{d} > 1, \qquad (2.3.27)$$

and

$$|a(x)| \leq 1, \qquad |a(x) - a(y)| \leq 2^j |x - y|. \qquad (2.3.28)$$

Remark 2.3.19   One might use alternatively in (2.3.28) that

$$\|a(2^{-j}\cdot)|\mathrm{Lip}(\mathbb{R}^d)\| \leq 1. \qquad (2.3.29)$$

We use the abbreviation

$$\mathbf{B}^s_p(\mathbb{R}^d) = \mathbf{B}^s_{p,p}(\mathbb{R}^d) \quad \text{with} \quad 0 < p \leq \infty, \quad s > 0.$$

In particular, in view of (2.3.16),

$$C^s(\mathbb{R}^d) = \mathbf{B}^s_\infty(\mathbb{R}^d), \quad s > 0,$$

are the Hölder-Zygmund spaces.

Definition 2.3.20 (Non-smooth Atom) Let $0 < p \le \infty$, $\sigma > 0$, and $\tilde{d} > 1$. Then $a \in \mathbf{B}_p^\sigma(\mathbb{R}^d)$ is called a (σ, p)-atom if for some $j \in \mathbb{N}_0$,

$$\operatorname{supp} a \subset \tilde{d} Q_{j,m} \qquad \text{for some} \quad m \in \mathbb{Z}^d, \tag{2.3.30}$$

and

$$\|a(2^{-j}\cdot)|\mathbf{B}_p^\sigma(\mathbb{R}^d)\| \le 1. \tag{2.3.31}$$

Remark 2.3.21 Note that if $\sigma < \frac{d}{p}$ then (σ, p)-atoms might be unbounded. Roughly speaking, they arise by dilating $\mathbf{B}_p^\sigma$-normalized functions. Obviously, the condition (2.3.31) is a straightforward modification of (2.3.26) and (2.3.29).

In general, as for smooth atoms, it is convenient to write $a_{j,m}(x)$ instead of $a(x)$ if the atoms are located at $Q_{j,m}$ according to (2.3.27) and (2.3.30), respectively. Furthermore, σ denotes the 'non-smoothness' of the atom, cf. (2.3.23).

We remark that the non-smooth atoms we consider in Definition 2.3.20, are renormalized versions of the non-smooth $(s, p)^\sigma$-atoms considered in [Tri03] and [TW96], where (2.3.31) is replaced by

$$a \in B_p^\sigma(\mathbb{R}^d) \qquad \text{with} \qquad \|a(2^{-j}\cdot)|B_p^\sigma(\mathbb{R}^d)\| \le 2^{j(\sigma-s)},$$

where $B_p^\sigma(\mathbb{R}^d) = B_{p,p}^\sigma(\mathbb{R}^d)$ are the Besov spaces defined via the Fourier-analytical approach. The spaces $\mathbf{B}_p^\sigma$ we consider now are different from B_p^σ for certain ranges of parameters σ and p, cf. (2.3.17). Moreover, the different normalization of the atoms results in corresponding changes in the definition of the sequence spaces $b_{p,q}^s$ used for the atomic decomposition. For our purposes (studying traces later on) it is more convenient to shift the factors $2^{j(s-\frac{d}{p})}$ to the sequence spaces.

The use of atoms with limited smoothness (i.e. finite element functions or splines) was studied already in [Osw94], where the author deals with spline approximation (and traces) in Besov spaces.

We wish to compare these atoms with the smooth atoms in Definition 2.3.15.

Proposition 2.3.22 *Let* $0 < p \le \infty$ *and* $0 < \sigma < K$. *Furthermore, let* $\tilde{d} > 1$, $j \in \mathbb{N}_0$, *and* $m \in \mathbb{Z}^d$. *Then any* K-*atom* $a_{j,m}$ *is a* (σ, p)-*atom.*

Proof Since the functions $a_{j,m}(2^{-j}\cdot)$ have compact support, we obtain

$$\|a_{j,m}(2^{-j}\cdot)|\mathbf{B}_p^\sigma(\mathbb{R}^d)\| \lesssim \|a_{j,m}(2^{-j}\cdot)|C^K(\mathbb{R}^d)\| \le 1,$$

with constants independent of j, giving the desired result for non-smooth atoms from Definition 2.3.20. $\qquad\qquad\qquad\qquad\qquad\qquad\qquad\qquad\qquad\qquad\qquad\qquad\square$

The following theorem contains the main result of this subsection. It gives the counterpart of Theorem 2.3.16 and provides a non-smooth atomic decomposition of the spaces $\mathbf{B}^s_{p,q}(\mathbb{R}^d)$.

Theorem 2.3.23 (Non-smooth Atomic Decomposition) *Let* $0 < p, q \leq \infty$, $0 < s < \sigma$, *and* $\tilde{d} > 1$. *Then* $f \in L_p(\mathbb{R}^d)$ *belongs to* $\mathbf{B}^s_{p,q}(\mathbb{R}^d)$ *if, and only if, it can be represented as*

$$f = \sum_{j=0}^{\infty} \sum_{m \in \mathbb{Z}^d} \lambda_{j,m} a_{j,m}, \tag{2.3.32}$$

where the $a_{j,m}$ *are* (σ, p)-*atoms with* $\operatorname{supp} a_{j,m} \subset \tilde{d} Q_{j,m}$, $j \in \mathbb{N}_0$, $m \in \mathbb{Z}^d$, *and* $\lambda \in b^s_{p,q}$, *convergence being in* $L_p(\mathbb{R}^d)$. *Furthermore,*

$$\|f|\mathbf{B}^s_{p,q}(\mathbb{R}^d)\| = \inf \|\lambda|b^s_{p,q}\|, \tag{2.3.33}$$

where the infimum is taken over all admissible representations (2.3.32).

Proof We have the atomic decomposition based on smooth K-atoms according to Theorem 2.3.16. By Proposition 2.3.22 classical K-atoms are special (σ, p)-atoms. Hence, it is enough to prove that

$$\|f|\mathbf{B}^s_{p,q}(\mathbb{R}^d)\| \lesssim \left(\sum_{k=0}^{\infty} 2^{k(s - \frac{d}{p})q} \left(\sum_{l \in \mathbb{Z}^d} |\lambda_{k,l}|^p \right)^{q/p} \right)^{1/q} \tag{2.3.34}$$

for any atomic decomposition

$$f = \sum_{k=0}^{\infty} \sum_{l \in \mathbb{Z}^d} \lambda_{k,l} a^{k,l}, \tag{2.3.35}$$

where $a^{k,l}$ are (σ, p)-atoms according to Definition 2.3.20.

For this purpose we expand each function $a^{k,l}(2^{-k} \cdot)$ optimally in $\mathbf{B}^{\sigma}_p(\mathbb{R}^d)$ with respect to classical K-atoms $b^{j,w}_{k,l}$ where $\sigma < K$,

$$a^{k,l}(2^{-k} x) = \sum_{j=0}^{\infty} \sum_{w \in \mathbb{Z}^d} \eta^{k,l}_{j,w} b^{j,w}_{k,l}(x), \quad x \in \mathbb{R}^d, \tag{2.3.36}$$

with

$$\operatorname{supp} b^{j,w}_{k,l} \subset Q_{j,w}, \qquad \left| D^{\alpha} b^{j,w}_{k,l}(x) \right| \leq 2^{|\alpha|j}, \quad |\alpha| \leq K, \tag{2.3.37}$$

and

$$\left(\sum_{j=0}^{\infty} 2^{j(\sigma - \frac{d}{p})p} \sum_{w \in \mathbb{Z}^d} |\eta_{j,w}^{k,l}|^p \right)^{\frac{1}{p}} = \|\eta^{k,l}|b_{p,p}^{\sigma}\| \sim \|a^{k,l}(2^{-k}\cdot)|\mathbf{B}_p^{\sigma}(\mathbb{R}^d)\| \lesssim 1.$$

$$(2.3.38)$$

Hence,

$$a^{k,l}(x) = \sum_{j=0}^{\infty} \sum_{w \in \mathbb{Z}^d} \eta_{j,w}^{k,l} b_{k,l}^{j,w}(2^k x),$$

where the functions $b_{k,l}^{j,w}(2^k \cdot)$ are supported by cubes with side lengths $\sim 2^{-k-j}$. By (2.3.37) we have

$$\left| D^{\alpha} b_{k,l}^{j,w}(2^k x) \right| = 2^{k|\alpha|} \left| (D^{\alpha} b_{k,l}^{j,w})(2^k x) \right| \le 2^{(j+k)|\alpha|}.$$

Replacing $j + k$ by j and putting $d_{k,l}^{j,w}(x) := b_{k,l}^{j-k,w}(2^k x)$, we obtain that

$$a^{k,l}(x) = \sum_{j=k}^{\infty} \sum_{w \in \mathbb{Z}^d} \eta_{j-k,w}^{k,l} d_{k,l}^{j,w}(x), \tag{2.3.39}$$

where $d_{k,l}^{j,w}$ are classical K-atoms supported by cubes with side lengths $\sim 2^{-j}$. We insert (2.3.39) into the expansion (2.3.35). We fix $j \in \mathbb{N}_0$ and $w \in \mathbb{Z}^d$, and collect all non-vanishing terms $d_{k,l}^{j,w}$ in the expansions (2.3.39). We have $k \le j$. Furthermore, multiplying (2.3.36) if necessary with suitable cut-off functions it follows that there is a natural number N such that for fixed k only at most N points $l \in \mathbb{Z}^d$ contribute to $d_{k,l}^{j,w}$. We denote this set by (j, w, k). Hence its cardinality is at most N, where N is independent of j, w, k. Then

$$d^{j,w}(x) = \frac{\sum_{k \le j} \sum_{l \in (j,w,k)} \eta_{j-k,w}^{k,l} \cdot \lambda_{k,l} \cdot d_{k,l}^{j,w}(x)}{\sum_{k \le j} \sum_{l \in (j,w,k)} |\eta_{j-k,w}^{k,l}| \cdot |\lambda_{k,l}|}$$

are correctly normalized smooth K-atoms located in cubes with side lengths $\sim 2^{-j}$ and centered at $2^{-j} w$. Let

$$v_{j,w} = \sum_{k \le j} \sum_{l \in (j,w,k)} |\eta_{j-k,w}^{k,l}| \cdot |\lambda_{k,l}|. \tag{2.3.40}$$

Then we obtain a classical atomic decomposition in the sense of Theorem 2.3.16,

$$f = \sum_j \sum_w v_{j,w} d^{j,w}(x),$$

where $d^{j,w}$ are K-atoms and

$$\|f|\mathbf{B}^s_{p,q}(\mathbb{R}^d)\| \lesssim \|v|b^s_{p,q}\|.$$

Therefore, in order to prove (2.3.34), it is enough to show, that

$$\|v|b^s_{p,q}\| \lesssim \|\lambda|b^s_{p,q}\| \tag{2.3.41}$$

if (2.3.38) holds.

Let $0 < \varepsilon < \sigma - s$. Then we obtain by (2.3.40) that (assuming $p < \infty$)

$$|v_{j,w}|^p \lesssim \sum_{k \leq j} \sum_{l \in (j,w,k)} 2^{(j-k)p\varepsilon} |\eta^{k,l}_{j-k,w}|^p |\lambda_{k,l}|^p, \tag{2.3.42}$$

where we used the bounded cardinality of the sets (j, w, k).

This gives for $q/p \leq 1$

$$\|v|b^s_{p,q}\|^q = \sum_{j=0}^{\infty} 2^{j(s-d/p)q} \left(\sum_{w \in \mathbb{Z}^d} |v_{j,w}|^p \right)^{q/p}$$

$$\lesssim \sum_{j=0}^{\infty} 2^{j(s-d/p)q} \left(\sum_{w \in \mathbb{Z}^d} \sum_{k=0}^{j} \sum_{l \in (j,w,k)} 2^{(j-k)p\varepsilon} |\eta^{k,l}_{j-k,w}|^p |\lambda_{k,l}|^p \right)^{q/p}$$

$$\leq \sum_{j=0}^{\infty} 2^{j(s-d/p)q} \sum_{k=0}^{j} \left(\sum_{w \in \mathbb{Z}^d} \sum_{l \in (j,w,k)} 2^{(j-k)p\varepsilon} |\eta^{k,l}_{j-k,w}|^p |\lambda_{k,l}|^p \right)^{q/p}$$

$$= \sum_{k=0}^{\infty} \sum_{j=k}^{\infty} 2^{j(s-d/p)q} \left(\sum_{w \in \mathbb{Z}^d} \sum_{l \in (j,w,k)} 2^{(j-k)p\varepsilon} |\eta^{k,l}_{j-k,w}|^p |\lambda_{k,l}|^p \right)^{q/p}$$

$$= \sum_{k=0}^{\infty} \sum_{j=0}^{\infty} 2^{(j+k)(s-d/p)q} \left(\sum_{w \in \mathbb{Z}^d} \sum_{l \in (j+k,w,k)} 2^{jp\varepsilon} |\eta^{k,l}_{j,w}|^p |\lambda_{k,l}|^p \right)^{q/p}$$

$$= \sum_{k=0}^{\infty} 2^{k(s-d/p)q} \sum_{j=0}^{\infty} 2^{j(s-\sigma+\varepsilon)q}$$

$$\times \left(\sum_{w \in \mathbb{Z}^d} \sum_{l \in (j+k,w,k)} 2^{j(\sigma-d/p)p} |\eta^{k,l}_{j,w}|^p |\lambda_{k,l}|^p \right)^{q/p}$$

$$\lesssim \sum_{k=0}^{\infty} 2^{k(s-d/p)q} \left(\sum_{j=0}^{\infty} \sum_{w \in \mathbb{Z}^d} \sum_{l \in (j+k,w,k)} 2^{j(\sigma-d/p)p} |\eta_{j,w}^{k,l}|^p |\lambda_{k,l}|^p \right)^{q/p}$$

$$\leq \sum_{k=0}^{\infty} 2^{k(s-d/p)q} \left(\sum_{j=0}^{\infty} \sum_{w \in \mathbb{Z}^d} \sum_{l \in \mathbb{Z}^d} 2^{j(\sigma-d/p)p} |\eta_{j,w}^{k,l}|^p |\lambda_{k,l}|^p \right)^{q/p}$$

$$= \sum_{k=0}^{\infty} 2^{k(s-d/p)q} \left(\sum_{l \in \mathbb{Z}^d} |\lambda_{k,l}|^p \sum_{j=0}^{\infty} \sum_{w \in \mathbb{Z}^d} 2^{j(\sigma-d/p)p} |\eta_{j,w}^{k,l}|^p \right)^{q/p}$$

$$\lesssim \sum_{k=0}^{\infty} 2^{k(s-d/p)q} \left(\sum_{l \in \mathbb{Z}^d} |\lambda_{k,l}|^p \right)^{q/p} = \|\lambda|b_{p,q}^s\|^q.$$

We have used (2.3.38) in the last inequality.

If $q/p > 1$, we shall use the following inequality, which holds for every non-negative sequence $\{\gamma_{j,k}\}_{0 \leq k \leq j < \infty}$, every $\alpha \geq 1$ and every $\varepsilon > 0$:

$$\sum_{j=0}^{\infty} \left(\sum_{k=0}^{j} 2^{-(j-k)\varepsilon} \gamma_{j,k} \right)^{\alpha} \leq c_{\alpha,\varepsilon} \sum_{k=0}^{\infty} \left(\sum_{j=k}^{\infty} \gamma_{j,k} \right)^{\alpha}. \tag{2.3.43}$$

If $\alpha = \infty$, (2.3.43) has to be modifed appropriately. To prove (2.3.43) for $\alpha < \infty$, we use Hölder's inequality and the embedding $\ell_1 \hookrightarrow \ell_\alpha$

$$\sum_{j=0}^{\infty} \left(\sum_{k=0}^{j} 2^{-(j-k)\varepsilon} \gamma_{j,k} \right)^{\alpha} \leq \sum_{j=0}^{\infty} \left(\sum_{k=0}^{j} 2^{-(j-k)\varepsilon\alpha'} \right)^{\alpha/\alpha'} \left(\sum_{k=0}^{j} \gamma_{j,k}^{\alpha} \right)^{\alpha/\alpha}$$

$$\lesssim \sum_{j=0}^{\infty} \sum_{k=0}^{j} \gamma_{j,k}^{\alpha} = \sum_{k=0}^{\infty} \sum_{j=k}^{\infty} \gamma_{j,k}^{\alpha} \leq \sum_{k=0}^{\infty} \left(\sum_{j=k}^{\infty} \gamma_{j,k} \right)^{\alpha}.$$

We use (2.3.42) and (2.3.43) with $p(\sigma - s - \varepsilon)$ instead of ε and $\alpha = q/p > 1$,

$$\|v|b_{p,q}^s\|^q \lesssim \sum_{j=0}^{\infty} 2^{j(\sigma-\frac{d}{p})q} \left(\sum_{w \in \mathbb{Z}^d} \sum_{k=0}^{j} \sum_{l \in (j,w,k)} 2^{(j-k)p\varepsilon} |\eta_{j-k,w}^{k,l}|^p |\lambda_{k,l}|^p \right)^{q/p}$$

$$= \sum_{j=0}^{\infty} \left(\sum_{k=0}^{j} 2^{-(j-k)p(\sigma-s-\varepsilon)} \right.$$

$$\times \sum_{w \in \mathbb{Z}^d} \sum_{l \in (j,w,k)} 2^{k(s-\frac{d}{p})p} 2^{(j-k)(\sigma-\frac{d}{p})p} |\eta_{j-k,w}^{k,l}|^p |\lambda_{k,l}|^p \right)^{q/p}$$

$$\lesssim \sum_{k=0}^{\infty} \left(\sum_{j=k}^{\infty} \sum_{w \in \mathbb{Z}^d} \sum_{l \in (j,w,k)} 2^{k(s-\frac{d}{p})p} 2^{(j-k)(\sigma-\frac{d}{p})p} |\eta_{j-k,w}^{k,l}|^p |\lambda_{k,l}|^p \right)^{q/p}$$

$$= \sum_{k=0}^{\infty} 2^{k(s-\frac{d}{p})q} \left(\sum_{j=0}^{\infty} \sum_{w \in \mathbb{Z}^d} \sum_{l \in (j+k,w,k)} 2^{j(\sigma-\frac{d}{p})p} |\eta_{j,w}^{k,l}|^p |\lambda_{k,l}|^p \right)^{q/p}$$

$$= \sum_{k=0}^{\infty} 2^{k(s-\frac{d}{p})q} \left(\sum_{l \in \mathbb{Z}^d} \sum_{j=0}^{\infty} \sum_{w \in \mathbb{Z}^d : l \in (j+k,w,k)} 2^{j(\sigma-\frac{d}{p})p} |\eta_{j,w}^{k,l}|^p |\lambda_{k,l}|^p \right)^{q/p}$$

$$\lesssim \sum_{k=0}^{\infty} 2^{k(s-\frac{d}{p})q} \left(\sum_{l \in \mathbb{Z}^d} |\lambda_{k,l}|^p \sum_{j=0}^{\infty} \sum_{w \in \mathbb{Z}^d} 2^{j(\sigma-\frac{d}{p})p} |\eta_{j,w}^{k,l}|^p \right)^{q/p}$$

$$\leq \sum_{k=0}^{\infty} 2^{k(s-\frac{d}{p})q} \left(\sum_{l \in \mathbb{Z}^d} |\lambda_{k,l}|^p \right)^{q/p} = \||\lambda|b_{p,q}^s\|^q.$$

The proof of (2.3.41) is finished. We again used (2.3.38) in the last inequality. If p and/or q are equal to infinity, only notational changes are necessary. □

Remark 2.3.24 Our results generalize [Tri03, Thm. 2] and [TW96, Thm. 2.3], where non-smooth atomic decompositions for spaces $B_{p,p}^s(\mathbb{R}^d) = \mathbf{B}_{p,p}^s(\mathbb{R}^d)$ with $s > \sigma_p$ can be found, to $\mathbf{B}_{p,q}^s(\mathbb{R}^d)$ with no restrictions on the parameters. In particular, the case when $p \neq q$ is completely new.

Using the Lip-atoms from Definition 2.3.18, as a Corollary we now obtain the following non-smooth atomic decomposition for Besov spaces with smoothness $0 < s < 1$.

Corollary 2.3.25 (Non-smooth Atomic Decomposition) *Let $0 < p, q \leq \infty$, $0 < s < 1$, and $\tilde{d} > 1$. Then $f \in L_p(\mathbb{R}^d)$ belongs to $\mathbf{B}_{p,q}^s(\mathbb{R}^d)$ if, and only if, it can be represented as*

$$f = \sum_{j=0}^{\infty} \sum_{m \in \mathbb{Z}^d} \lambda_{j,m} a_{j,m}, \tag{2.3.44}$$

where the $a_{j,m}$ are Lip-atoms with $\operatorname{supp} a_{j,m} \subset \tilde{d} Q_{j,m}$, $j \in \mathbb{N}_0$, $m \in \mathbb{Z}^d$, and $\lambda \in b_{p,q}^s$, convergence being in $L_p(\mathbb{R}^d)$. Furthermore,

$$\|f|\mathbf{B}_{p,q}^s(\mathbb{R}^d)\| = \inf \||\lambda|b_{p,q}^s\|, \tag{2.3.45}$$

where the infimum is taken over all admissible representations (2.3.44).

Proof We use the embedding

$$\mathrm{Lip}(\mathbb{R}^d) \hookrightarrow B^1_\infty(\mathbb{R}^d)$$

from [Tri83, p. 89/90] and assume that $0 < s < \sigma < 1$. Then for functions with compact support (e.g. atoms since $\mathrm{supp}\, a_{j,m} \subset \tilde{d} Q_{j,m}$ is bounded), we have the following chain of embeddings:

$$C^1 \hookrightarrow \mathrm{Lip} \hookrightarrow B^1_\infty = \mathbf{B}^1_\infty \hookrightarrow \mathbf{B}^1_{p,\infty} \hookrightarrow \mathbf{B}^\sigma_p, \tag{2.3.46}$$

where the third embedding follows since on bounded domains Ω we have $L_\infty(\Omega) \hookrightarrow L_p(\Omega)$. From Theorems 2.3.16 and 2.3.23 we see that the spaces $\mathbf{B}^s_{p,q}(\mathbb{R}^d)$ with $s < 1$ can either be decomposed with smooth C^1-atoms or non-smooth (σ, p)-atoms, respectively. Therefore, by (2.3.46), we deduce that the spaces can also be decomposed using Lip-atoms. $\square$

2.3.5 Besov Spaces $\mathbf{B}^s_{p,q}$ on Boundaries of Lipschitz Domains

In Sect. 2.1.2 on page 24 we explained how to define function spaces on the boundary Γ of bounded Lipschitz domains via pull-back and a partition of unity using corresponding spaces on $\mathbb{R}^{d-1}$. This leads to the following precise definition for Besov spaces $\mathbf{B}^s_{p,q}(\Gamma)$ on the boundary of a Lipschitz domain. Concerning the notation we refer to the construction (2.1.7) and the explanations given there.

Definition 2.3.26 (Besov Spaces on the Boundary) Let $d \geq 2$, and let Ω be a bounded Lipschitz domain in $\mathbb{R}^d$ with boundary Γ, and $\varphi_j, \psi^{(j)}, W_j$ be as given on page 24. Assume $0 < s < 1$ and $0 < p, q \leq \infty$. Then we introduce

$$\mathbf{B}^s_{p,q}(\Gamma) = \{ f \in L_p(\Gamma) : g_j \in \mathbf{B}^s_{p,q}(W_j),\ j = 1, \ldots, N \},$$

equipped with the quasi-norm $\| f | \mathbf{B}^s_{p,q}(\Gamma) \| := \sum_{j=1}^N \| g_j | \mathbf{B}^s_{p,q}(W_j) \|.$

Remark 2.3.27 The spaces $\mathbf{B}^s_{p,q}(\Gamma)$ turn out to be independent of the particular choice of the resolution of unity $\{\varphi_j\}_{j=1}^N$ and the local diffeomorphisms $\psi^{(j)}$ (the proof is similar to the proof of [Tri83, Prop. 3.2.3(ii)], making use of Propositions 2.3.35 and 2.3.36 below). We furnish $\mathbf{B}^s_{p,q}(W_j)$ with the $(d-1)$-dimensional norms according to Theorem 2.3.16. Note that we could furthermore replace W_j in the definition of the norm above by $\mathbb{R}^{d-1}$ if we extend g_j outside W_j with zero, i.e.,

$$\| f | \mathbf{B}^s_{p,q}(\Gamma) \| \sim \sum_{j=1}^N \| g_j | \mathbf{B}^s_{p,q}(\mathbb{R}^{d-1}) \|. \tag{2.3.47}$$

In particular, the equivalence (2.3.47) yields that characterizations for B-spaces defined on $\mathbb{R}^{d-1}$ can be generalized to B-spaces defined on Γ. This will be done in Theorem 2.3.31 for non-smooth atomic decompositions and is very likely to work as well for characterizations in terms of differences.

Non-smooth Atomic Decompositions for $\mathbf{B}^s_{p,q}(\Gamma)$ Similar to the non-smooth atomic decompositions constructed for $\mathbf{B}^s_{p,q}$ we now establish corresponding atomic decompositions for Besov spaces defined on Lipschitz boundaries. They will be very useful when investigating traces on Lipschitz domains in Chap. 10.

The relevant sequence spaces and Lipschitz-atoms on the boundary Γ we shall define next are closely related to the sequence spaces $b^s_{p,q}(\Omega)$ and Lip-atoms used for the non-smooth atomic decompositions in Corollary 2.3.25.

Definition 2.3.28 (Sequence Spaces) Let $0 < p, q \leq \infty$, $s \in \mathbb{R}$. Furthermore, let Γ be the boundary of a bounded Lipschitz domain $\Omega \subset \mathbb{R}^d$, and $\lambda = \{\lambda_{j,m} \in \mathbb{C} : j \in \mathbb{N}_0, m \in \mathbb{Z}^d\}$. Then

$$b^s_{p,q}(\Gamma) = \left\{ \lambda : \|\lambda | b^s_{p,q}(\Gamma)\| = \left(\sum_{j=0}^{\infty} 2^{j(s-\frac{d-1}{p})q} \left(\sum_{m \in \mathbb{Z}^d}^{\Gamma,j} |\lambda_{j,m}|^p \right)^{q/p} \right)^{1/q} < \infty \right\}$$

(with the usual modification if $p = \infty$ and/or $q = \infty$).

Definition 2.3.29 (Non-smooth Boundary Atoms) Let $j \in \mathbb{N}_0$, $m \in \mathbb{Z}^d$, $\tilde{d} > 1$, and let Γ be the boundary of a bounded Lipschitz domain $\Omega \subset \mathbb{R}^d$. Put $Q^{\Gamma}_{j,m} := \tilde{d} Q_{j,m} \cap \Gamma \neq \emptyset$. A function $a \in \text{Lip}(\Gamma)$ is a Lip^Γ-atom, if

$$\text{supp}\, a \subset Q^{\Gamma}_{j,m}, \qquad \tilde{d} > 1,$$

$$\|a|L_\infty(\Gamma)\| \leq 1 \quad \text{and} \quad \sup_{\substack{x,y \in \Gamma, \\ x \neq y}} \frac{|a(x) - a(y)|}{|x - y|} \leq 2^j. \qquad (2.3.48)$$

Remark 2.3.30 Note that if we put $2^j \Gamma := \{2^j x : x \in \Gamma\}$, we can state (2.3.48) like $\|a(2^{-j}\cdot)|\text{Lip}(2^j \Gamma)\| \leq 1$.

The theorem below provides non-smooth atomic decompositions for the spaces $\mathbf{B}^s_{p,q}(\Gamma)$.

Theorem 2.3.31 (Non-smooth Atomic Decomposition) *Let $\Omega \subset \mathbb{R}^d$ be a bounded Lipschitz domain with boundary Γ. Furthermore, let $0 < s < 1$ and $0 < p, q \leq \infty$. Then $f \in L_p(\Gamma)$ belongs to $\mathbf{B}^s_{p,q}(\Gamma)$ if, and only if,*

$$f = \sum_{j,m} \lambda_{j,m} a_{j,m},$$

where $a_{j,m}$ are Lip$^\Gamma$-*atoms with* supp $a_{j,m} \subset Q_{j,m}^\Gamma$ *and* $\lambda \in b_{p,q}^s(\Gamma)$, *convergence being in* $L_p(\Gamma)$. *Furthermore,*

$$\|f|\mathbf{B}_{p,q}^s(\Gamma)\| = \inf \|\lambda|b_{p,q}^s(\Gamma)\|,$$

where the infimum is taken over all possible representations.

Proof

Step 1.   Fix $f \in \mathbf{B}_{p,q}^s(\Gamma)$. For simplicity, we suppose that supp $f \subset \{x \in \Gamma :$ $\varphi_l(x) = 1\}$ for some $l \in \{1, 2, \ldots, N\}$. If this is not the case the arguments have to be slightly modified to incorporate the decomposition of unity (2.1.6). To simplify the notation we write φ instead of φ_l and ψ instead of $\psi^{(l)}$. Then we obtain

$$\|f|\mathbf{B}_{p,q}^s(\Gamma)\| = \|f \circ \psi^{-1}|\mathbf{B}_{p,q}^s(\mathbb{R}^{d-1})\|.$$

We use Corollary 2.3.25 with d replaced by $d-1$ to obtain an optimal atomic decomposition

$$f \circ \psi^{-1} = \sum_{j,m} \lambda_{j,m} a_{j,m} \quad \text{where} \quad \|f \circ \psi^{-1}|\mathbf{B}_{p,q}^s(\mathbb{R}^{d-1})\| \sim \|\lambda|b_{p,q}^s(\mathbb{R}^{d-1})\|.$$

$$(2.3.49)$$

For $j \in \mathbb{N}_0$ and $m \in \mathbb{Z}^{d-1}$ fixed, we consider the function $a_{j,m}(\psi(x))$. Due to the Lipschitz properties of ψ, this function is supported in $Q_{j,l}^\Gamma$ for some $l \in \mathbb{Z}^n$ and we denote it by $a_{j,l}^\Gamma(x)$. Furthermore, we set $\lambda_{j,l}' = \lambda_{j,m}$. This leads to the decomposition

$$f = \sum_{j,l} \lambda_{j,l}' a_{j,l}^\Gamma. \qquad (2.3.50)$$

It is straightforward to verify that $a_{j,l}^\Gamma$ are Lip$^\Gamma$-atoms since $\|a_{j,l}^\Gamma|L_\infty(\Gamma)\| \lesssim$ $\|a_{j,m}|L_\infty(W_l)\| \lesssim 1$ and

$$\frac{|a_{j,l}^\Gamma(x) - a_{j,l}^\Gamma(y)|}{|x - y|} = \frac{|a_{j,m}(x') - a_{j,m}(y')|}{|\psi^{-1}(x') - \psi^{-1}(y')|}$$

$$\sim \frac{|a_{j,m}(x') - a_{j,m}(y')|}{|x' - y'|} \lesssim 2^j, \quad x, y \in \Gamma.$$

Furthermore, we have the estimate

$$\|f|\mathbf{B}_{p,q}^s(\Gamma)\| = \|f \circ \psi^{-1}|\mathbf{B}_{p,q}^s(\mathbb{R}^{d-1})\| \sim \|\lambda|b_{p,q}^s(\mathbb{R}^{d-1})\| = \|\lambda'|b_{p,q}^s(\Gamma)\|.$$

Step 2.   The proof of the opposite direction follows along the same lines. If f on Γ is given by

$$f = \sum_{j,l} \lambda'_{j,l} a^\Gamma_{j,l},$$

then $f \circ \psi^{-1} = \sum_{j,m} \lambda_{j,m} a_{j,m}$, where $a_{j,m}(x) = a^\Gamma_{j,l}(\psi^{-1}(x))$ and $\lambda_{j,m} = \lambda'_{j,l}$ for suitable $m \in \mathbb{Z}^{d-1}$. Again it follows that $a_{j,m}$ are Lip-atoms on $\mathbb{R}^{d-1}$ and

$$\|f|\mathbf{B}^s_{p,q}(\Gamma)\| = \|f \circ \psi^{-1}|\mathbf{B}^s_{p,q}(\mathbb{R}^{d-1})\| \lesssim \|\lambda|b^s_{p,q}(\mathbb{R}^{d-1})\| = \|\lambda'|b^s_{p,q}(\Gamma)\|.$$

Step 3.   The convergence in $L_p(\Gamma)$ of the representation $f = \sum_{j,m}^{j,\Gamma} \lambda_{j,m} a^\Gamma_{j,m}$, follows for $p \leq 1$ by

$$\left\|\sum_{j,m}^{j,\Gamma} \lambda_{j,m} a^\Gamma_{j,m}|L_p(\Gamma)\right\|^p \leq \sum_{j,m}^{j,\Gamma} |\lambda_{j,m}|^p \|a^\Gamma_{j,m}|L_p(\Gamma)\|^p$$

$$\lesssim \sum_j 2^{-j(d-1)} \sum_m^{j,\Gamma} |\lambda_{j,m}|^p$$

$$= \|\lambda|b^0_{p,p}(\Gamma)\|^p \lesssim \|\lambda|b^s_{p,q}(\Gamma)\|^p \qquad (2.3.51)$$

and using

$$\left\|\sum_{j,m}^{j,\Gamma} \lambda_{j,m} a^\Gamma_{j,m}|L_p(\Gamma)\right\| \leq \sum_j \left\|\sum_m^{j,\Gamma} \lambda_{j,m} a^\Gamma_{j,m}|L_p(\Gamma)\right\|$$

$$\lesssim \sum_j 2^{-j(d-1)/p} \left(\sum_m^{j,\Gamma} |\lambda_{j,m}|^p\right)^{1/p}$$

$$= \|\lambda|b^0_{p,1}(\Gamma)\| \lesssim \|\lambda|b^s_{p,q}(\Gamma)\| \qquad (2.3.52)$$

for $p > 1$.

$\square$

2.3.6 Properties of Besov Spaces $\mathbf{B}^s_{p,q}$ on Lipschitz Domains

We establish several properties for Besov spaces on Lipschitz domains and their boundaries using non-smooth atomic decompositions of $f \in \mathbf{B}^s_{p,q}(\Omega)$. The results are essentially new, which is why we mostly give detailed proofs below.

Interpolation Results Interpolation results for $\mathbf{B}^s_{p,q}(\mathbb{R}^d)$ as obtained in [DP88, Cor. 6.2, 6.3] carry over to the spaces $\mathbf{B}^s_{p,q}(\Gamma)$, which follows immediately from their definition and properties of real interpolation.

Theorem 2.3.32 *Let $\Omega \subset \mathbb{R}^d$ be a bounded Lipschitz domain with boundary Γ.*

(i) Let $0 < p, q, q_0, q_1 \leq \infty$, $s_0 \neq s_1$, and $0 < s_i < 1$. Then

$$\left(\mathbf{B}^{s_0}_{p,q_0}(\Gamma), \mathbf{B}^{s_1}_{p,q_1}(\Gamma)\right)_{\theta,q} = \mathbf{B}^s_{p,q}(\Gamma),$$

where $0 < \theta < 1$ and $s = (1-\theta)s_0 + \theta s_1$.

(ii) Let $0 < p_i, q_i \leq \infty$, $s_0 \neq s_1$ and $0 < s_i < 1$. Then for each $0 < \theta < 1$, $s = (1-\theta)s_0 + \theta s_1$, $\frac{1}{p} = \frac{1-\theta}{p_0} + \frac{\theta}{p_1}$, and for $\frac{1}{q} = \frac{1-\theta}{q_0} + \frac{\theta}{q_1}$, we have

$$\left(\mathbf{B}^{s_0}_{p_0,q_0}(\Gamma), \mathbf{B}^{s_1}_{p_1,q_1}(\Gamma)\right)_{\theta,q} = \mathbf{B}^s_{p,q}(\Gamma),$$

provided $p = q$.

Proof By definition of the spaces $\mathbf{B}^s_{p,q}(\Gamma)$ we can construct a well-defined and bounded linear operator

$$E : \mathbf{B}^s_{p,q}(\Gamma) \longrightarrow \oplus_{1 \leq j \leq N} \mathbf{B}^s_{p,q}(\mathbb{R}^{d-1}),$$

$$(Ef)_j := (\varphi_j f) \circ \psi^{(j)^{-1}} \quad \text{on } \mathbb{R}^{d-1}, \quad 1 \leq j \leq N,$$

which has a bounded and linear left inverse given by

$$R : \oplus_{1 \leq j \leq N} \mathbf{B}^s_{p,q}(\mathbb{R}^{d-1}) \longrightarrow \mathbf{B}^s_{p,q}(\Gamma)$$

$$R\left((g_j)_{1 \leq j \leq N}\right) := \sum_{j=1}^{N} \Psi_j \left(g_j \circ \psi_j\right) \quad \text{on } \Gamma,$$

where $\Psi_j \in C_0^\infty(\mathbb{R}^d)$, supp $\Psi_j \subset K_j$, $\Psi \equiv 1$ in a neighbourhood of supp φ_j (concerning the notation we refer to Remark 2.1.4).

Straightforward calculation shows for $f \in \mathbf{B}_{p,q}^s(\Gamma)$ that

$$(R \circ E)f = R(Ef) = R\left(\left((\varphi_j f) \circ \psi^{(j)-1}\right)_{1 \leq j \leq N}\right) = \sum_{j=1}^N \Psi_j \varphi_j f = \sum_{j=1}^N \varphi_j f = f,$$

i.e.,

$$R \circ E = I, \quad \text{the identity operator on } \mathbf{B}_{p,q}^s(\Gamma).$$

One arrives at a standard situation in interpolation theory. Hence, by the method of retraction-coretraction, cf. [Tri78, Sect. 1.2.4, 1.17.1], the results for $\mathbf{B}_{p,q}^s(\mathbb{R}^{d-1})$ carry over to the spaces $\mathbf{B}_{p,q}^s(\Gamma)$. Therefore, (i) and (ii) are a consequence of [DP88, Cor. 6.2, 6.3].   □

Furthermore, we briefly show that the interpolation results for Besov spaces $\mathbf{B}_{p,q}^s(\mathbb{R}^d)$ also hold for spaces on domains $\mathbf{B}_{p,q}^s(\Omega)$. This is not automatically clear in our context since the extension operator

$$\mathrm{Ex} : \mathbf{B}_{p,q}^s(\Omega) \longrightarrow \mathbf{B}_{p,q}^s(\mathbb{R}^d)$$

constructed in [DS93] is not linear. The situation is different for spaces $B_{p,q}^s(\Omega)$. Here Rychkov's (linear) extension operator, cf. [Ryc00], automatically yields interpolation results for B-spaces on domains.

Theorem 2.3.33 *Let $\Omega \subset \mathbb{R}^d$ be a bounded Lipschitz domain.*

(i) Let $0 < p, q, q_0, q_1 \leq \infty$, $s_0 \neq s_1$, and $0 < s_i < 1$. Then

$$\left(\mathbf{B}_{p,q_0}^{s_0}(\Omega), \mathbf{B}_{p,q_1}^{s_1}(\Omega)\right)_{\theta,q} = \mathbf{B}_{p,q}^s(\Omega),$$

where $0 < \theta < 1$ and $s = (1-\theta)s_0 + \theta s_1$.

(ii) Let $0 < p_i, q_i \leq \infty$, $s_0 \neq s_1$ and $0 < s_i < 1$. Then for each $0 < \theta < 1$, $s = (1-\theta)s_0 + \theta s_1$, $\frac{1}{p} = \frac{1-\theta}{p_0} + \frac{\theta}{p_1}$, and for $\frac{1}{q} = \frac{1-\theta}{q_0} + \frac{\theta}{q_1}$, we have

$$\left(\mathbf{B}_{p_0,q_0}^{s_0}(\Omega), \mathbf{B}_{p_1,q_1}^{s_1}(\Omega)\right)_{\theta,q} = \mathbf{B}_{p,q}^s(\Omega),$$

provided $p = q$.

Proof In spite of our remarks before the theorem, we can nevertheless use the extension operator

$$\mathrm{Ex} : \mathbf{B}_{p,q}^s(\Omega) \longrightarrow \mathbf{B}_{p,q}^s(\mathbb{R}^d)$$

constructed in [DS93] to show that interpolation results for spaces $\mathbf{B}^s_{p,q}(\mathbb{R}^d)$ carry over to spaces $\mathbf{B}^s_{p,q}(\Omega)$. Let $X_i(\Omega) := \mathbf{B}^{s_i}_{p_i,q_i}(\Omega)$. By the explanations given in [DS93, p. 859] we have the estimate

$$K(f, t, X_0(\Omega), X_1(\Omega)) \sim K(\mathrm{Ex}\, f, t, X_0(\mathbb{R}^d), X_1(\mathbb{R}^d)) \qquad (2.3.53)$$

although the operator Ex is not linear. Let $\mathbf{B}^\theta(\Omega) := \left(\mathbf{B}^{s_0}_{p_0,q_0}(\Omega), \mathbf{B}^{s_1}_{p_1,q_1}(\Omega)\right)_{\theta,q}$ with the given restrictions on the parameters given in (i) and (ii), respectively. We have to prove that

$$\mathbf{B}^\theta(\Omega) = \mathbf{B}^s_{p,q}(\Omega),$$

but this follows immediately from [DP88, Cor. 6.2,6.3] using (2.3.53), since

$$\|f|\mathbf{B}^\theta(\Omega)\| \sim \|\mathrm{Ex}\, f|\mathbf{B}^\theta(\mathbb{R}^d)\| \sim \|\mathrm{Ex}\, f|\mathbf{B}^s_{p,q}(\mathbb{R}^d)\| \sim \|f|\mathbf{B}^s_{p,q}(\Omega)\|.$$

$\square$

Moreover, we use the following simple variant of the Gagliardo-Nirenberg inequality, cf. [Pee76, Ch. 5] (for $p = q$ this is a consequence of Theorem 2.3.33(ii)).

Lemma 2.3.34 (Interpolation Inequality) *Let* $0 < s_0, s_1 < \infty$, $0 < p_0, p_1, q_0, q_1 \le \infty$, *and* $0 < \theta < 1$. *Put*

$$s = (1-\theta)s_0 + \theta s_1, \quad \frac{1}{p} = \frac{1-\theta}{p_0} + \frac{\theta}{p_1}, \quad \frac{1}{q} = \frac{1-\theta}{q_0} + \frac{\theta}{q_1}. \qquad (2.3.54)$$

Then

$$\|f|\mathbf{B}^s_{p,q}(\Omega)\| \lesssim \|f|\mathbf{B}^{s_0}_{p_0,q_0}(\Omega)\|^{1-\theta} \cdot \|f|\mathbf{B}^{s_1}_{p_1,q_1}(\Omega)\|^\theta \qquad (2.3.55)$$

for all $f \in \mathbf{B}^{s_0}_{p_0,q_0}(\Omega) \cap \mathbf{B}^{s_1}_{p_1,q_1}(\Omega)$.

Proof The straightforward proof uses the definition of B-spaces through differences and Hölder's inequality. $\square$

Diffeomorphisms and Pointwise Multipliers The non-smooth atomic decomposition enables us to generalize [Sch10, Prop. 2.5] and obtain new results concerning diffeomorphisms and pointwise multipliers in $\mathbf{B}^s_{p,q}(\mathbb{R}^d)$ in the following way. For related matters we also refer to [May05, Thm. 3.3.3].

Proposition 2.3.35 *Let* $0 < p, q \le \infty$, $0 < s < 1$, *and* $\sigma > s$.

(i) *Let* ψ *be a Lipschitz diffeomorphism. Then* $f \longrightarrow f \circ \psi$ *is a linear and bounded operator from* $\mathbf{B}^s_{p,q}(\mathbb{R}^d)$ *onto itself.*

(ii) *Let* $h \in \mathcal{C}^\sigma(\mathbb{R}^d)$. *Then* $f \longrightarrow hf$ *is a linear and bounded operator from* $\mathbf{B}^s_{p,q}(\mathbb{R}^d)$ *into itself.*

Proof Concerning (i), we make use of the atomic decomposition as in (2.3.44) with the Lip-atoms from Definition 2.3.18. Then we have

$$f \circ \psi = \sum_{j=0}^{\infty} \sum_{m \in \mathbb{Z}^d} \lambda_{j,m} a_{j,m} \circ \psi$$

and $a \circ \psi$ is a Lip-atom based on a new cube, and multiplied with a constant depending on ψ, since

$$|(a_{j,m} \circ \psi)(x) - (a_{j,m} \circ \psi)(y)| \leq 2^j |\psi(x) - \psi(y)| \lesssim 2^j |x - y|$$

To prove (ii) we argue as follows. First, we may suppose that $0 < s < \sigma < 1$. Furthermore, we choose a real parameter σ' with $s < \sigma' < \sigma$. We take the smooth atomic decomposition (2.3.24) with K-atoms $a_{j,m}$, where $K = 1$. Multiplied with $h \in C^\sigma$, it gives a new (non-smooth) atomic decomposition of hf. Its convergence in $L_p(\mathbb{R}^d)$ follows from the convergence of (2.3.24) in $L_p(\mathbb{R}^d)$ and the boundedness of h.

It remains to verify, that $ha_{j,m}$ are non-smooth (σ', p)-atoms. The support property follows immediately from the support property of $a_{j,m}$. We use the bounded support of $(ha_{j,m})(2^{-j}\cdot)$ and the multiplier assertion for $\mathbf{B}^\sigma_\infty(\mathbb{R}^d)$ as presented in [RS96, Sect. 4.6.1, Thm. 2] to get

$$\|(ha_{j,m})(2^{-j}\cdot)|\mathbf{B}^{\sigma'}_p(\mathbb{R}^d)\| \leq \|(ha_{j,m})(2^{-j}\cdot)|\mathbf{B}^\sigma_\infty(\mathbb{R}^d)\|$$

$$= \|h(2^{-j}\cdot) \cdot a_{j,m}(2^{-j}\cdot)|\mathbf{B}^\sigma_\infty(\mathbb{R}^d)\|$$

$$\lesssim \|h(2^{-j}\cdot)|\mathbf{B}^\sigma_\infty(\mathbb{R}^d)\| \cdot \|a_{j,m}(2^{-j}\cdot)|\mathbf{B}^\sigma_\infty(\mathbb{R}^d)\|.$$

The last product is bounded by a constant due to the inequality

$$\|h(2^{-j}\cdot)|\mathbf{B}^\sigma_\infty(\mathbb{R}^d)\| \lesssim \|h|\mathbf{B}^\sigma_\infty(\mathbb{R}^d)\|, \quad j \in \mathbb{N}_0,$$

which may be verified directly (or found in [Bou83, Sect. 1.7] or [ET96, Sect. 2.3.1]), combined with the fact that $a_{j,m}$ are K-atoms for $K = 1$. $\square$

An Equivalent Quasi-norm We establish an equivalent quasi-norm for $\mathbf{B}^s_{p,q}(\Omega)$.

Proposition 2.3.36 *Let* $0 < p, q \leq \infty$, $0 < s < 1$, *and* $\Omega \subset \mathbb{R}^d$ *be a bounded Lipschitz domain. Then*

$$\|\varphi_0 f|\mathbf{B}^s_{p,q}(\mathbb{R}^d)\| + \sum_{j=1}^{N} \|(\varphi_j f)(\psi^{(j)}(\cdot))^{-1}|\mathbf{B}^s_{p,q}(\mathbb{R}^d_+)\| \tag{2.3.56}$$

is an equivalent quasi-norm in $\mathbf{B}^s_{p,q}(\Omega)$.

Proof Let Ω_1 be a bounded domain with

$$\overline{\Omega}_1 \subset \left\{ x \in \mathbb{R}^d : \sum_{j=0}^{N} \varphi_j(x) = 1 \right\}$$

and $\overline{\Omega} \subset \Omega_1$, where $\{\varphi_j\}_{j=0}^{N}$ is the resolution of unity from Remark 2.1.4. Let $f \in \mathbf{B}_{p,q}^s(\Omega)$. If we restrict the infimum in (2.3.12) to $g \in \mathbf{B}_{p,q}^s(\mathbb{R}^d)$ with

$$g|_{\Omega} = f \quad \text{and} \quad \operatorname{supp} g \subset \Omega_1, \tag{2.3.57}$$

then we obtain a new equivalent quasi-norm in $\mathbf{B}_{p,q}^s(\Omega)$. This follows from Proposition 2.3.35(ii) if one multiplies an arbitrary element $g \in \mathbf{B}_{p,q}^s(\mathbb{R}^d)$ with a fixed infinitely differentiable function $\varkappa(x)$ with

$$\varkappa(x) = 1 \quad \text{if} \quad x \in \Omega \quad \text{and} \quad \operatorname{supp} \varkappa \subset \Omega_1.$$

For elements $g \in \mathbf{B}_{p,q}^s(\mathbb{R}^d)$ with (2.3.57),

$$\sum_{k=0}^{N} \|\varphi_k g|\mathbf{B}_{p,q}^s(\mathbb{R}^d)\|$$

is an equivalent quasi-norm. This is also a consequence of Proposition 2.3.35(ii). Applying part (i) of that proposition to $g(x) \to g(\psi^{(j)}(x))$, we see that

$$\|\varphi_0 g|\mathbf{B}_{p,q}^s(\mathbb{R}^d)\| + \sum_{k=1}^{N} \|(\varphi_k g)(\psi^{(k)}(\cdot))^{-1}|\mathbf{B}_{p,q}^s(\mathbb{R}^d)\|$$

is an equivalent quasi-norm for all $g \in \mathbf{B}_{p,q}^s(\mathbb{R}^d)$ with (2.3.57). But the infimum over all admissible g with (2.3.57) yields (2.3.56).   □

2.3.7   Spaces of Type $\mathbf{F}_{p,q}^s$

In view of Theorem 2.3.16 we could have defined the spaces $\mathbf{B}_{p,q}^s(\Omega)$ directly via atomic decompositions (instead of differences) as containing functions which can be expanded via (smooth) atoms and suitable coefficients belonging to the right sequence spaces $b_{p,q}^s(\Omega)$.

This alternative will serve us now for the definition of corresponding Triebel-Lizorkin spaces $\mathbf{F}_{p,q}^s(\Omega)$, since for these kinds of spaces the approaches via differences and atomic decompositions do not always yield the same spaces for all possible parameters s, p, and q. Since we heavily rely on atomic decompositions when computing the traces in Chap. 8 this is the right choice for our purposes.

The relevant sequence spaces are given below.

Definition 2.3.37 (Sequence Spaces) Let $0 < p < \infty, 0 < q \leq \infty$, and $s \in \mathbb{R}$. Furthermore, let $\Omega \subset \mathbb{R}^d$ and $\lambda = \{\lambda_{j,m} \in \mathbb{C} : j \in \mathbb{N}_0, m \in \mathbb{Z}^d\}$. Then

$$
f_{p,q}^s(\Omega) = \left\{ \lambda : \|\lambda \mid f_{p,q}^s(\Omega)\| \right.
$$

$$
= \left\| \left(\sum_{j=0}^{\infty} \sum_{m \in \mathbb{Z}^d}^{\Omega,j} 2^{jsq} |\lambda_{j,m}|^q \chi_{j,m}(\cdot) \right)^{1/q} \mid L_p(\mathbb{R}^d) \right\| < \infty \left. \right\}
$$

(with the usual modification if $q = \infty$).

Now we define the corresponding function spaces as follows.

Definition 2.3.38 (Triebel-Lizorkin Spaces) Let $0 < p < \infty, 0 < q \leq \infty$, and $s > 0$. Let $\tilde{d} > 1$ and $K \in \mathbb{N}_0$ with

$$
K \geq (1 + [s])
$$

be fixed. Then $\mathbf{F}_{p,q}^s(\Omega)$ is the collection of all $f \in L_p(\Omega)$ which can be represented as

$$
f(x) = \sum_{j=0}^{\infty} \sum_{m \in \mathbb{Z}^d}^{\Omega,j} \lambda_{j,m} a_{j,m}(x), \tag{2.3.58}
$$

where the $a_{j,m}$ are K-atoms according to Definition 2.3.15 with

$$
\operatorname{supp} a_{j,m} \subset \tilde{d} Q_{j,m}, \qquad j \in \mathbb{N}_0, \quad m \in \mathbb{Z}^d,
$$

and $\lambda \in f_{p,q}^s(\Omega)$, convergence being in $L_p(\Omega)$. Furthermore,

$$
\|f \mid \mathbf{F}_{p,q}^s(\Omega)\| := \inf \|\lambda \mid f_{p,q}^s(\Omega)\|,
$$

where the infimum is taken over all admissible representations (2.3.58).

Remark 2.3.39

(i) According to their definition, the spaces $\mathbf{F}_{p,q}^s(\Omega)$ can be regarded as restrictions of the corresponding spaces on $\mathbb{R}^d$, i.e.,

$$
\mathbf{F}_{p,q}^s(\Omega) = \{f \in L_p(\Omega) : \quad \text{there exists} \quad g \in \mathbf{F}_{p,q}^s(\mathbb{R}^d) \quad \text{with} \quad g|_\Omega = f\},
$$

furnished with the norm

$$\|f|\mathbf{F}^s_{p,q}(\Omega)\| = \inf\left\{\|g|\mathbf{F}^s_{p,q}(\mathbb{R}^d)\| \quad \text{with} \quad g\big|_\Omega = f\right\},$$

where $g\big|_\Omega = f$ denotes the restriction of g to Ω.

(ii) Comparing the sequence spaces one can see that $b^s_{p,p}(\Omega) = f^s_{p,p}(\Omega)$, which yields the coincidence

$$\mathbf{B}^s_{p,p}(\Omega) = \mathbf{F}^s_{p,p}(\Omega) \qquad \text{for} \quad s > 0, \quad 0 < p < \infty,$$

and Ω being a bounded Lipschitz domain (or even a more general (ε, δ)-domain).

(iii) Moreover, from [Tri06, Prop. 9.14, Rem 9.15] we have the coincidence

$$\mathbf{F}^s_{p,q}(\Omega) = F^s_{p,q}(\Omega) \quad \text{for} \quad s > \sigma_{p,q}, \quad 0 < p < \infty, \quad 0 < q \leq \infty.$$

Note that the spaces $\mathbf{F}^s_{p,q}$ were denoted by $\mathfrak{F}^s_{p,q}$ in [Tri06].

2.4 Kondratiev Spaces

In this section, we introduce a family of weighted Sobobev spaces, the so-called *Kondratiev spaces*. They play a central role in the regularity theory for elliptic PDEs on domains with piecewise smooth boundary, particularly polygons (2D) and polyhedra (3D).

The relevance of Kondratiev spaces comes from the fact that within this scale we can prove shift theorems on non-smooth domains analogous to those in the usual Sobolev spaces on smooth domains, since the weights compensate for singularities which are known to emerge at the boundary even for smooth data. Regularity estimates in these spaces will be the major tool we use in order to establish regularity results in Besov and fractional Sobolev spaces later on.

Kondratiev spaces have their origin in the midsixties in the pioneering work of Kondratiev [Kon67, Kon77], see also the survey of Kondratiev and Oleinik [KO83]. Later these kinds of spaces, partly more general, have been considered by Kufner, Sändig [KS87], Babuska, Guo [BG97], Maz'ya, Rossmann [KMR97, MR10], Nistor, Mazzucato [MN10], and Costabel, Dauge, Nicaise [CDN10], to mention at least a few contributions in this context.

Whereas in the mentioned references the weight was always chosen to be a power of the distance to the singular set of the boundary, there are also publications dealing with the weight being a power of the distance to the whole boundary. We refer e.g. to Kufner, Sändig [KS87], Triebel [Tri78, Sect. 3.2.3] and Lototsky [Lot00].

2.4.1 Definition and Basic Properties

Definition 2.4.1 (Kondratiev Spaces) Let Ω be a domain of $\mathbb{R}^d$ and let M be a nontrivial closed subset of its boundary $\partial\Omega$. Furthermore, let $1 \leq p \leq \infty$, $m \in \mathbb{N}_0$, and $a \in \mathbb{R}$. We define the space $\mathcal{K}^m_{p,a}(\Omega, M)$ as the collection of all $u \in \mathcal{D}'(\Omega)$, which have m generalized derivatives satisfying

$$\|u|\mathcal{K}^m_{p,a}(\Omega, M)\| := \Big(\sum_{|\alpha|\leq m} \int_\Omega |\rho^{|\alpha|-a}(x)\partial^\alpha u(x)|^p \, dx \Big)^{1/p} < \infty \qquad (2.4.1)$$

if $p < \infty$, modified by

$$\|u|\mathcal{K}^m_{\infty,a}(\Omega, M)\| := \sum_{|\alpha|\leq m} \sup_{x\in\Omega} |\rho^{|\alpha|-a}(x)\partial^\alpha u(x)| < \infty \qquad (2.4.2)$$

if $p = \infty$. Therein, the weight function ρ is defined by

$$\rho(x) := \min\{1, \text{dist}(x, M)\}, \qquad x \in \Omega.$$

The closure of the set $C_0^\infty(\Omega)$ with respect to the norms (2.4.1) and (2.4.2), respectively, is denoted by $\mathring{\mathcal{K}}^m_{p,a}(\Omega, M)$.

Finally, for $1 < p < \infty$ and $m \in \mathbb{N}$ we also define $\mathcal{K}^{-m}_{p,a}(\Omega, M) = \Big(\mathring{\mathcal{K}}^m_{p',-a}(\Omega, M)\Big)'$, the dual space equipped with its usual norm, where $\frac{1}{p'} = 1 - \frac{1}{p}$.

Remark 2.4.2

(i) We will not distinguish spaces which differ by an equivalent norm.
(ii) In our applications the set M will usually be the *singular set* S of the domain Ω, i.e., the set of all points $x \in \partial\Omega$ for which for any $\varepsilon > 0$ the set $\partial\Omega \cap B_\varepsilon(x)$ is not smooth. In this case, we simply abbreviate

$$\mathcal{K}^m_{p,a}(\Omega) := \mathcal{K}^m_{p,a}(\Omega, S).$$

(iii) If Ω is a polygon in $\mathbb{R}^2$ or a polyhedral domain in $\mathbb{R}^3$, then the singular set S consists of the vertices of the polygon or the vertices and edges of the polyhedra, respectively.
(iv) If $\mathcal{O} \subset \mathbb{R}^d$ is a general bounded Lipschitz domain, the whole boundary $\partial\mathcal{O}$ might coincide with the singular set. In this case the Kondratiev spaces $\mathcal{K}^m_{p,a}(\mathcal{O})$ from (2.4.1) are normed by

$$\|u|\mathcal{K}^m_{p,a}(\mathcal{O})\| := \Big(\sum_{|\alpha|\leq m} \int_\mathcal{O} |\rho^{(|\alpha|-a)}(x)\partial^\alpha u(x)|^p dx \Big)^{1/p} < \infty, \qquad (2.4.3)$$

where the weight function $\rho : \mathcal{O} \rightarrow [0, 1]$ stands for the distance to $\partial \mathcal{O}$, i.e., $\rho(x) = \min\{1, \text{dist}(x, \partial \mathcal{O})\}$. For $1 < p < \infty$ these spaces and their generalizations were investigated in [Lot00].

(v) In Sect. 2.4.2 we introduce Kondratiev spaces on domains of polyhedral type, where the set M does not necessarily coincide with the singular set S. Moreover, in Chaps. 5 and 6 we will deal with PDEs defined on specific domains which are not always of polyhedral type (such as generalized wedges and specific Lipschitz domains, respectively) and regularity results in corresponding Kondratiev spaces defined on these domains, where the set M again not necessarily coincides with the singular set of Ω. However, to simplify our notation we usually drop the set M in our considerations. Therefore, in order to avoid any misunderstanding, we shall briefly recall the definition of the Kondratiev spaces and sets M considered in each context right before presenting the regularity results for each problem.

(vi) For functions depending on the time $t \in [0, T]$ with values in some Kondratiev space $\mathcal{K}_{p,a}^m(\Omega)$ we generalize the above concepts. In particular, we consider the *generalized Kondratiev spaces* $L_q([0, T], \mathcal{K}_{p,a}^m(\Omega))$ which contain all functions $u(t, x)$ such that

$$\|u|L_q([0, T], \mathcal{K}_{p,a}^m(\Omega))\|$$

$$:= \left(\int_{[0,T]} \|u(t, \cdot)|\mathcal{K}_{p,a}^m(\Omega)\|^q dt \right)^{1/q}$$

$$= \left(\int_{[0,T]} \left(\sum_{|\alpha| \leq m} \int_\Omega |\varrho(x)|^{p(|\alpha|-a)} |D_x^\alpha u(t, x)|^p dx \right)^{q/p} dt \right)^{1/q} < \infty,$$

$$(2.4.4)$$

with $0 < q \leq \infty$ and parameters a, p, m as above. Similar for the spaces $W_q^l([0, T], \mathcal{K}_{p,a}^m(\Omega))$, $l \in \mathbb{N}_0$, and $C^{k,\alpha}([0, T], \mathcal{K}_{p,a}^m(\Omega))$, $k \in \mathbb{N}_0$ and $\alpha \in (0, 1)$.

Remark 2.4.3 (Basic Properties of Kondratiev Spaces) We collect basic properties of Kondratiev spaces that will be useful in what follows.

(i) $\mathcal{K}_{p,a}^m(\Omega, M)$ is a Banach space, see [KO84, KO86].
(ii) The scale of Kondratiev spaces is monotone in m and a, i.e.,

$$\mathcal{K}_{p,a}^m(\Omega, M) \hookrightarrow \mathcal{K}_{p,a}^{m'}(\Omega, M) \qquad \text{and} \qquad \mathcal{K}_{p,a}^m(\Omega, M) \hookrightarrow \mathcal{K}_{p,a'}^m(\Omega, M)$$

$$(2.4.5)$$

if $m' < m$ and $a' < a$.

(iii) Regularized distance function: There exists a function $\tilde{\varrho} : \overline{\Omega} \to [0, \infty)$, which is infinitely often differentiable in Ω, and positive constants A, B, C_α such that

$$A \rho(x) \leq \tilde{\varrho}(x) \leq B \rho(x), \qquad x \in \Omega,$$

and, for all $\alpha \in \mathbb{N}_0^d$,

$$\left| \partial^\alpha \tilde{\varrho}(x) \right| \leq C_\alpha \rho^{1-|\alpha|}(x), \qquad x \in \Omega,$$

see Stein [Ste70, Thm. VI.2.2] (the construction given there is valid for arbitrary closed subsets of $\mathbb{R}^d$).

(iv) By using the previous item (iii) and replacing ρ by $\tilde{\varrho}$ in the norm of $\mathcal{K}_{p,a}^m(\Omega, M)$ one can prove the following: Let $b \in \mathbb{R}$. Then the mapping $T_b : u \mapsto \tilde{\varrho}^b u$ yields an isomorphism of $\mathcal{K}_{p,a}^m(\Omega, M)$ onto $\mathcal{K}_{p,a+b}^m(\Omega, M)$.

(v) Let $a \geq 0$. Then $\mathcal{K}_{p,a}^m(\Omega, M) \hookrightarrow L_p(\Omega)$.

(vi) A function $\varphi \in C^m(\Omega)$ is a pointwise multiplier for $\mathcal{K}_{p,a}^m(\Omega, M)$, i.e., $\varphi u \in \mathcal{K}_{p,a}^m(\Omega, M)$ for all $u \in \mathcal{K}_{p,a}^m(\Omega, M)$ and

$$\|\varphi u | \mathcal{K}_{p,a}^m(\Omega, M)\| \leq c_\varphi \|u | \mathcal{K}_{p,a}^m(\Omega, M)\|. \tag{2.4.6}$$

(vii) Let $1 \leq p < \infty$. We assume

$$\lim_{\delta \downarrow 0} |\{x \in \Omega : \operatorname{dist}(x, M) < \delta\}| = 0.$$

Then $C_*^\infty(\Omega, M) = \{u|_\Omega : u \in C_0^\infty(\mathbb{R}^d \setminus M)\}$ is a dense subset of $\mathcal{K}_{p,a}^m(\Omega, M)$.

Furthermore, we shall need a lifting property for Kondratiev spaces. For classical Sobolev spaces by definition it is clear that

$$u \in W_p^m \quad \Longrightarrow \quad D^\alpha u \in W_p^{m-|\alpha|}, \tag{2.4.7}$$

for $\alpha \in \mathbb{N}_0^d$ with $|\alpha| \leq m$. For a generalization to Besov and Triebel-Lizorkin spaces we refer to [RS96, p. 22, Prop. 2] in this context. In the following theorem we study the behaviour of $u \to D^\alpha u$ in Kondratiev spaces, which turns out to be very similar as for Sobolev spaces.

Theorem 2.4.4 (Lift Property) *Let $a \in \mathbb{R}$, $1 < p < \infty$, and $m \in \mathbb{N}_0$. Then for $u \in \mathcal{K}_{p,a}^m(\Omega, M)$ and $\alpha \in \mathbb{N}_0^d$ with $|\alpha| \leq m$, we have*

$$D^\alpha u \in \mathcal{K}_{p,a-|\alpha|}^{m-|\alpha|}(\Omega, M).$$

Proof The result follows immediately from the following observation

$$\left\| D^\alpha u | \mathcal{K}^{m-|\alpha|}_{p,a-|\alpha|}(\Omega, M) \right\|^p = \sum_{|\beta| \le m-|\alpha|} \int_\Omega \rho^{p(|\beta|-(a-|\alpha|))}(x) |D^\beta (D^\alpha u(x))|^p dx$$

$$= \sum_{|\beta|+|\alpha| \le m} \int_\Omega \rho^{p(|\beta|+|\alpha|-a)}(x) |D^{\alpha+\beta} u(x)|^p dx$$

$$\lesssim \sum_{|\gamma| \le m} \int_\Omega \rho^{p(|\gamma|-a)}(x) |D^\gamma u(x)|^p dx = \|u | \mathcal{K}^m_{p,a}(\Omega, M)\|^p.$$

$\square$

2.4.2 Kondratiev Spaces on Domains of Polyhedral Type

For our analysis we make use of several properties of Kondratiev spaces that have been proved in [DHSS18b]. Therefore, in our later considerations, we will mainly be interested in the case that Ω is a bounded domain of polyhedral type. The precise definition will be given below in Definition 2.4.5. Essentially, we will consider domains for which the analysis of the associated Kondratiev spaces can be reduced to the following four basic cases:

- Smooth cones;
- Specific non-smooth cones;
- Specific dihedral domains;
- Polyhedral cones.

Let $d \ge 2$. As usual, an infinite smooth cone with vertex at the origin is the set

$$K := \{x \in \mathbb{R}^d : 0 < |x| < \infty, \, x/|x| \in \Omega\},$$

where Ω is a subdomain (i.e., simply connected strict subset) of the unit sphere S^{d-1} with C^∞ boundary.

Case I *Kondratiev spaces on smooth cones.* Let K' be an infinite smooth cone contained in $\mathbb{R}^d$ with vertex at the origin which is rotationally invariant with respect to the axis $\{(0, \ldots, 0, x_d) : x_d \in \mathbb{R}\}$. Then we define the truncated cone K by $K := K' \cap B_1(0)$. In this case we choose $M := \{0\}$ and define

$$\|u | \mathcal{K}^m_{p,a}(K, M)\|^p = \|u | \mathcal{K}^m_{p,a}(K, \{0\})\|^p = \sum_{|\alpha| \le m} \int_K |\, |x|^{|\alpha|-a} \partial^\alpha u(x)|^p \, dx.$$

$$(2.4.8)$$

Fig. 2.13 Truncated smooth
cone K

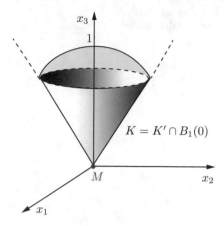

There is still one degree of freedom in the choice of the smooth cone, namely the
opening angle $\gamma \in (0, \pi)$ of the cone. Since this will be unimportant in what follows
we will not indicate this in the notation (Fig. 2.13).

Case II *Kondratiev spaces on specific non-smooth cones.* Let again K' denote a
rotationally symmetric cone as described in **Case I** with opening angle $\gamma \in (0, \pi)$.
Then we define the specific poyhedral cone P by $P = K' \cap I$, where I denotes the
unit cube

$$I := \{x \in \mathbb{R}^d \ : \ 0 < x_i < 1, \ i = 1, \ldots, d\}. \tag{2.4.9}$$

In this case, we choose $M = \Gamma := \{x \in \mathbb{R}^d, x = (0, \ldots, 0, x_d), \ 0 \le x_d \le 1\}$ and
see that

$$\|u|\mathcal{K}_{p,a}^m(P, \Gamma)\|^p = \sum_{|\alpha| \le m} \int_P |\rho^{|\alpha|-a}(x) \partial^\alpha u(x)|^p \, dx, \tag{2.4.10}$$

where $\rho(x)$ denotes the distance of x to Γ, i.e., $\rho(x) = |(x_1, \ldots, x_{d-1})|$. Again the
opening angle γ of the cone K' will be of no importance (Fig. 2.14).

Case III *Kondratiev spaces on specific dihedral domains.* Let $1 \le l < d$ and let I
be the unit cube defined in (2.4.9). For $x \in \mathbb{R}^d$ we write $x = (x', x'') \in \mathbb{R}^{d-l} \times \mathbb{R}^l$,
where $x' := (x_1, \ldots, x_{d-l})$ and $x'' := (x_{d-l+1}, \ldots, x_d)$. Hence $I = I' \times I''$ with
the obvious interpretation (Fig. 2.15).

Then we choose

$$M_l := \{x \in \overline{I} : x_1 = \ldots = x_{d-l} = 0, \ 0 \le x_i \le 1, \ i = d - l + 1, \ldots, d\} \tag{2.4.11}$$

Fig. 2.14 Specific
non-smooth cone P

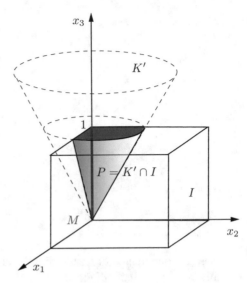

Fig. 2.15 Specific dihedral
domain I

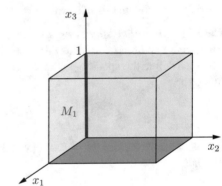

and define

$$\|u|\mathcal{K}_{p,a}^m(I, M_l)\|^p = \sum_{|\alpha| \le m} \int_I |\, |x'|^{|\alpha|-a} \partial^\alpha u(x)|^p \, dx. \qquad (2.4.12)$$

This time the set M_l is a subset of the singular set of I if, and only if, $l \le d - 2$.

Case IV *Kondratiev spaces on polyhedral cones.* Let

$$K'' := \{x \in \mathbb{R}^3 : 0 < |x| < \infty, \ x/|x| \in \Omega\},$$

be an infinite cone in $\mathbb{R}^3$. We suppose that the boundary $\partial K''$ consists of the vertex
$x = 0$, the edges (half lines) $M_1, \ldots, M_n$, and smooth faces $\Gamma_1, \ldots \Gamma_n$. Moreover,
$\Omega := K'' \cap S^2$ is a domain of polygonal type on the unit sphere with sides $\Gamma_k \cap S^2$.

Fig. 2.16 Polyhedral cone

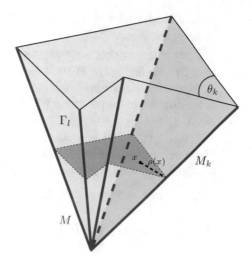

The angle at the edge M_k will be denoted by θ_k, $k = 1, \ldots, n$. Then we consider the truncated cone

$$K_0 := K'' \cap B_{r_0}(0),$$

for some real number $r_0 > 0$.

In this case, we choose $M := (M_1 \cup \ldots \cup M_n) \cap \overline{K_0}$ and define

$$\|u|\mathcal{K}^m_{p,a}(K_0, M)\|^p = \sum_{|\alpha| \leq m} \int_{K_0} |\rho^{|\alpha|-a}(x)\partial^\alpha u(x)|^p \, dx, \tag{2.4.13}$$

where $\rho(x)$ denotes the distance of x to M (Fig. 2.16).

Based on these four cases, we define domains of polyhedral type which will play a central role in our further considerations.

Definition 2.4.5 (Domains of Polyhedral Type) Let D be a domain in $\mathbb{R}^d$, $d \geq 2$, with singular set S. Then D is of *polyhedral type*, if there exists a finite covering $(U_i)_i$ of bounded open sets such that

$$\overline{D} \subset \left(\bigcup_{i \in \Lambda_1} U_i \right) \cup \left(\bigcup_{j \in \Lambda_2} U_j \right) \cup \left(\bigcup_{k \in \Lambda_3} U_k \right) \cup \left(\bigcup_{l \in \Lambda_4} U_l \right),$$

where

(i) $i \in \Lambda_1$ if U_i is a ball and $\overline{U_i} \cap S = \emptyset$.

(ii) $j \in \Lambda_2$ if there is a C^∞-diffeomorphism $\eta_j : \overline{U_j} \longrightarrow \eta_j(\overline{U_j}) \subset \mathbb{R}^d$ such that $\eta_j(U_j \cap D)$ is the smooth cone K as described in **Case I**. Moreover, we assume that for all $x \in U_j \cap D$ the distance to S is equivalent to the distance to the point $x^j := \eta_j^{-1}(0)$.

(iii) $k \in \Lambda_3$ if there exists a C^∞-diffeomorphism $\eta_k : \overline{U_k} \longrightarrow \eta_k(\overline{U_k}) \subset \mathbb{R}^d$ ($d \geq 3$) such that $\eta_k(U_k \cap D)$ is the non-smooth cone P as described in **Case II**. Moreover, we assume that for all $x \in U_k \cap D$ the distance to S is equivalent to the distance to the set $\Gamma^k := \eta_k^{-1}(\Gamma)$.

(iv) $l \in \Lambda_4$ if there exists a C^∞-diffeomorphism $\eta_l : \overline{U_l} \longrightarrow \eta_l(\overline{U_l}) \subset \mathbb{R}^d$ ($d \geq 3$) such that $\eta_l(U_l \cap D)$ is a specific dihedral domain as described in **Case III**. Moreover, we assume that for all $x \in U_l \cap D$ the distance to S is equivalent to the distance to the set $M^l := \eta_l^{-1}(M_n)$ for some $n \in \{1, \dots, d-1\}$.

Particularly in $d = 3$ we permit another type of subdomain: Here

$$\overline{D} \subset \left(\bigcup_{i \in \Lambda_1} U_i \right) \cup \left(\bigcup_{j \in \Lambda_2} U_j \right) \cup \left(\bigcup_{k \in \Lambda_3} U_k \right) \cup \left(\bigcup_{l \in \Lambda_4} U_l \right) \cup \left(\bigcup_{m \in \Lambda_5} U_m \right),$$

where

(v) $m \in \Lambda_5$ if there exists a C^∞-diffeomorphism $\eta_m : \overline{U_m} \longrightarrow \eta_m(\overline{U_m}) \subset \mathbb{R}^d$ such that $\eta_m(U_m \cap D)$ is a polyhedral cone as described in **Case IV**. Moreover, we assume that for all $x \in U_m \cap D$ the distance to S is equivalent to the distance to the set $M'_m := \eta_m^{-1}(M)$.

Remark 2.4.6

(i) In the literature many different types of polyhedral domains are considered. As was discussed in [DHSS18b], in our context only the **Cases I** and **III** are essential. Therefore, our definition (when $d = 3$) coincides with the one of Maz'ya and Rossmann [MR10, Def. 4.1.1], where a bounded domain $D \subset \mathbb{R}^3$ is defined to be of polyhedral type if the following holds:

    (a) The boundary ∂D consists of smooth (of class C^∞) open two-dimensional manifolds Γ_j (the faces of D), $j = 1, \dots, n$, smooth curves M_k (the edges), $k = 1, \dots, l$, and vertices $x^{(1)}, \dots, x^{(l')}$.

    (b) For every $\xi \in M_k$ there exists a neighbourhood U_ξ and a C^∞-diffeomorphism κ_ξ which maps $D \cap U_\xi$ onto $\mathcal{D}_\xi \cap B_1(0)$, where $\mathcal{D}_\xi$ is a dihedron, cf. Remark 2.1.7.

    (c) For every vertex $x^{(i)}$ there exists a neighbourhood U_i and a diffeomorphism κ_i mapping $D \cap U_i$ onto $K_i \cap B_1(0)$, where K_i is a polyhdral cone with edges and vertex at the origin (Fig. 2.17).

We will work with this characterization in Sect. 3.5 when dealing with operator pencils. Further variants of polyhedral domains can be found in Babuška, Guo [BG97], Bacuta, Mazzucato, Nistor, Zikatanov [BMNZ10] and Mazzucato, Nistor [MN10].

Fig. 2.17 Polyhedron

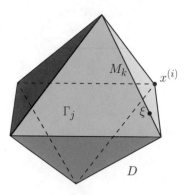

(ii) While the types of polyhedral domains coincide, in [MR10] more general weighted Sobolev spaces on those polyhedral domains are discussed. Our spaces $\mathcal{K}^m_{p,a}(D, S)$ coincide with the classes $V^{\tilde{l},p}_{\beta,\delta}(D)$ if $m = \tilde{l}$,

$$\beta = (\beta_1, \dots, \beta_{l'}) = (\tilde{l} - a, \dots, \tilde{l} - a) \qquad \text{and}$$

$$\delta = (\delta_1, \dots, \delta_l) = (\tilde{l} - a, \dots, \tilde{l} - a).$$

For the meaning of l and l' we refer to (a) above and [MR10, Def. 4.1.2].

(iii) As already mentioned in Remark 2.4.2, the set M according to Definition 2.4.1 will not be mentioned explicitly in case that it coincides with the singular set S. In the sequel, we will use the same convention if M is one of the specific sets introduced in **Case I—Case IV**, i.e., we simply write

$$\mathcal{K}^m_{p,a}(D) := \mathcal{K}^m_{p,a}(D, M).$$

Since for the specific domains in **Case I–Case IV** the set M does not coincide with the singular set S, this clearly causes some ambiguities. However, throughout the thesis this is not a serious problem since for the specific domains from **Case I–Case IV** Kondratiev spaces with respect to the whole singular set are not considered at all in what follows.

(iv) It is clear from the Definition 2.4.5 that a domain of polyhedral type is also a (general) polyhedral domain.

Below we give some examples of domains which are covered or excluded by Definition 2.4.5.

Examples 2.4.7 Numerically interesting domains of polyhedral type are e.g. L-shaped domains D_1 and 'the donut' D_2 in 2D. Moreover, in 3D we mention the Fichera corner D_3 and the icosahedron D_4 (Figs. 2.18 and 2.19).

Whereas the decomposition of the icosahedron into polyhedral cones and a smooth domain is obvious, we add a remark concerning D_3. An open neighbourhood of the Fichera corner can be obtained as the image of a diffeomorphic

Fig. 2.18 *L*-shaped domain

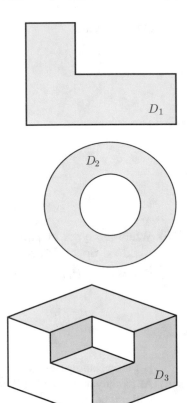

Fig. 2.19 Donut

Fig. 2.20 Fichera corner

map applied to a polyhedral cone (alternatively we could have defined polyhedral cones with an opening angle larger than π). Finally, domains with slits and cusps as illustrated below in examples D_5 and D_6, respectively, are excluded by Definition 2.4.5 and not of polyhedral type (Figs. 2.20 and 2.21).

There is one more interesting example of a polyhedral domain we wish to mention. Example D_7 is taken from [Dob10, Ex. 6.5] and consists of two cuboids lying on top of each other (Figs. 2.22 and 2.23). In particular, it is a domain of polyhedral type which is not Lipschitz (at the indicated point x it is not possible to turn the polyhedron such that the inner part of the domain lies behind the boundary).

2.4.3 Embeddings and Pointwise Multiplication

Throughout this subsection let $D \subset \mathbb{R}^d$ be a domain of polyhedral type as in Definition 2.4.5. For these type of domains we briefly collect some properties of Kondratiev spaces needed later on. In particular, we provide necessary and

Fig. 2.21 Icosahedron

Fig. 2.22 Not of polyhedral
type: Circle with a slit (D_5)
and Cusp (D_6)

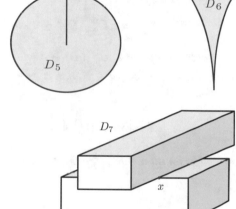

Fig. 2.23 Domain of
polyhedral type, which is not
Lipschitz

sufficient conditions for embeddings within the scale to hold and deal with pointwise
multiplication.

Embeddings of Kondratiev spaces have been discussed in [DHSS18b], but see
also Maz'ya and Rossmann [MR10] (Lemma 1.2.2, Lemma 1.2.3 (smooth cones),
Lemma 2.1.1 (dihedron), Lemma 3.1.3, Lemma 3.1.4 (cones with edges), Lemma
4.1.2 (domains of polyhedral type)).

The following embedding results with proofs may be found in [DHSS18b,
Sect. 3.1, 3.2].

Theorem 2.4.8 (Embeddings Within the Scale of Kondratiev Spaces) *Let* $1 \leq$
$p \leq q \leq \infty$, $a, a' \in \mathbb{R}$, *and* $m \in \mathbb{N}$.

(i) Let $q < \infty$ *or* $p = 1$ *and* $q = \infty$. *Then* $\mathcal{K}^m_{p,a}(D)$ *is embedded into* $\mathcal{K}^{m'}_{q,a'}(D)$ *if,*
and only if,

$$m - \frac{d}{p} \geq m' - \frac{d}{q} \qquad \text{and} \qquad a - \frac{d}{p} \geq a' - \frac{d}{q} \, .$$

(ii) *Let* $1 < p < \infty$ *and* $q = \infty$. *Then* $\mathcal{K}^m_{p,a}(D)$ *is embedded into* $\mathcal{K}^{m'}_{\infty,a'}(D)$ *if, and only if,*

$$m - \frac{d}{p} > m' \qquad and \qquad a - \frac{d}{p} \geq a' \,.$$

Concerning pointwise multiplication the following results follow from [DHSS18b, Cor. 31, 42; Thm. 44].

Theorem 2.4.9 (Pointwise Multiplication in Kondratiev Spaces)

(i) *Let* $m \in \mathbb{N}$, $a \geq \frac{d}{p}$, *and either* $1 < p < \infty$ *and* $m > \frac{d}{p}$ *or* $p = 1$ *and* $m \geq d$. *Then the Kondratiev space* $\mathcal{K}^m_{p,a}(D)$ *is an algebra with respect to pointwise multiplication, i.e., there exists a constant* c *such that*

$$\|uv|\mathcal{K}^m_{p,a}(D)\| \leq c\|u|\mathcal{K}^m_{p,a}(D)\| \cdot \|v|\mathcal{K}^m_{p,a}(D)\|$$

holds for all $u, v \in \mathcal{K}^m_{p,a}(D)$.

(ii) *Let* $\frac{d}{2} < p < \infty$, $m \in \mathbb{N}$, *and* $a \geq \frac{d}{p} - 1$. *Then there exists a constant* c *such that*

$$\|uv|\mathcal{K}^{m-1}_{p,a-1}(D)\| \leq c\|u|\mathcal{K}^{m+1}_{p,a+1}(D)\| \cdot \|v|\mathcal{K}^{m-1}_{p,a-1}(D)\|$$

holds for all $u \in \mathcal{K}^{m+1}_{p,a+1}(D)$ *and* $v \in \mathcal{K}^{m-1}_{p,a-1}(D)$.

2.4.4 Relations Between Kondratiev and Other Function Spaces

In this subsection we concentrate on embedding results showing the close relation between Kondratiev and Besov (or Triebel-Lizorkin) spaces. These will be our main tools when investigating the Besov regularity of solutions to the PDEs considered in Chaps. 4–6.

The following theorem is taken from [Han15, Sect. 5, Thm. 3].

Theorem 2.4.10 (Embeddings Between Kondratiev and Besov Spaces) *Let* D *be a bounded polyhedral domain in* $\mathbb{R}^d$. *Furthermore, let* $s, a \in \mathbb{R}$, $\gamma \in \mathbb{N}_0$, *and suppose* $\min(s, a) > \frac{\delta}{d}\gamma$, *where* δ *denotes the dimension of the singular set (i.e.* $\delta = 0$ *if there are only vertex singularities,* $\delta = 1$ *if there are edge and vertex singularities etc.). Then there exists some* $0 < \tau_0 \leq p$ *such that*

$$\mathcal{K}^\gamma_{p,a}(D) \cap B^s_{p,\infty}(D) \hookrightarrow B^\gamma_{\tau,\infty}(D) \hookrightarrow L_p(D), \qquad (2.4.14)$$

for all $\tau_* < \tau < \tau_0$, *where* $\frac{1}{\tau_*} = \frac{\gamma}{d} + \frac{1}{p}$.

Remark 2.4.11 It also holds that $u \in B_{\tau,\infty}^{\gamma}(D)$ for $\tau \leq \tau_*$ but these spaces are no longer embedded into $L_p(D)$. Moreover, because we are particularly interested in embeddings into the adaptivity scale $B_{\tau,\tau}^s(D)$ from (1.0.3) we later make use of the embedding

$$B_{\tau,\infty}^{\gamma}(D) \hookrightarrow B_{\tau,\tau}^{\gamma-\varepsilon}(D), \qquad \varepsilon > 0.$$

Since we want to study parabolic problems in Chap. 5, we show that Theorem 2.4.10 immediately generalizes to the function spaces defined in (2.4.4) as follows:

Theorem 2.4.12 (Embeddings Between Generalized Kondratiev and Besov Spaces) *Let D be some bounded polyhedral domain in $\mathbb{R}^d$ and assume $k \in \mathbb{N}_0$, $0 < q \leq \infty$. Furthermore, let $s, a \in \mathbb{R}$, $\gamma \in \mathbb{N}_0$, and suppose $\min(s, a) > \frac{\delta}{d}\gamma$, where δ denotes the dimension of the singular set. Then there exists some $0 < \tau_0 \leq p$ such that*

$$W_q^k([0, T], \mathcal{K}_{p,a}^{\gamma}(D)) \cap W_q^k([0, T], B_{p,\infty}^s(D)) \hookrightarrow W_q^k([0, T], B_{\tau,\infty}^{\gamma}(D)) \tag{2.4.15}$$

for all $\tau_* < \tau < \tau_0$, where $\frac{1}{\tau_*} = \frac{\gamma}{d} + \frac{1}{p}$.

Proof Put $X_1 := \mathcal{K}_{p,a}^{\gamma}(D)$, $X_2 := B_{p,\infty}^s(D)$, and $X = B_{\tau,\infty}^{\gamma}(D)$. Then Theorem 2.4.10 states that

$$X_1 \cap X_2 \hookrightarrow X,$$

i.e., for some $x \in X_1 \cap X_2$ we have $\|x|X\| \lesssim \|x|X_1 \cap X_2\| \sim \|x|X_1\| + \|x|X_2\|$. Using this we calculate for $I := [0, T]$ that

$$\|u|W_q^k([0, T], B_{\tau,\infty}^{\gamma}(D))\|$$

$$= \left\|u|W_q^k(I, X)\right\| = \left(\sum_{l=0}^{k} \|\partial_{t^l} u|L_q(I, X)\|^q\right)^{1/q}$$

$$\lesssim \left(\sum_{l=0}^{k} \int_I \|\partial_{t^l} u(t, \cdot)|X\|^q dt\right)^{1/q}$$

$$\lesssim \left(\sum_{l=0}^{k} \int_I \|\partial_{t^l} u(t, \cdot)|X_1 \cap X_2\|^q dt\right)^{1/q}$$

$$\sim \left(\sum_{l=0}^{k} \int_I \|\partial_{t^l} u(t, \cdot)|X_1\|^q dt\right)^{1/q} + \left(\sum_{l=0}^{k} \int_I \|\partial_{t^l} u(\cdot, t)|X_2\|^q dt\right)^{1/q}$$

$$= \left\| u | W_q^k (I, X_1) \right\| + \left\| u | W_q^k (I, X_2) \right\|$$

$$\sim \left\| u | W_q^k (I, X_1) \cap W_q^k (I, X_2) \right\|$$

$$= \left\| u | W_q^k ([0, T], \mathcal{K}_{p,a}^\gamma (D)) \cap W_q^k ([0, T], B_{p,\infty}^s (D)) \right\|,$$

which establishes (2.4.15).   □

Remark 2.4.13 For $k = 0$ the embedding (2.4.15) in Theorem 2.4.12 reads as

$$L_q([0, T], \mathcal{K}_{p,a}^\gamma (D)) \cap L_q([0, T], B_{p,\infty}^s (D)) \hookrightarrow L_q([0, T], B_{\tau,\infty}^\gamma (D)).$$
$$(2.4.16)$$

The special Lipschitz domains Ω from Definition 2.1.8 that we deal with in Chap. 5 are not bounded polyhedral domains as considered in Theorems 2.4.10 and 2.4.12. However, regarding embeddings of the Kondratiev spaces into the scale of Besov spaces, we modify the arguments from [Han15, Sect. 5, Thm. 3] and show that the above results can be generalized to this context as follows.

Theorem 2.4.14 (Embeddings Between Kondratiev and Besov Spaces) *Let $\Omega \subset \mathbb{R}^d$ be a special Lipschitz domain according to Definition 2.1.8. Then we have a continuous embedding*

$$\mathcal{K}_{p,a}^m (\Omega) \cap B_{p,p}^s (\Omega) \hookrightarrow B_{\tau,\tau}^r (\Omega), \qquad \frac{1}{\tau} = \frac{r}{d} + \frac{1}{p}, \qquad 1 < p < \infty, \qquad (2.4.17)$$

for all $0 \le r < \min(m, \frac{sd}{d-1})$ and $a > \frac{\delta}{d} r$, where $\delta = d - 2 = \dim(l_0)$.

Proof Since for $r = 0$ the result is clear, we assume in the sequel that $r > 0$ and $0 < \tau < p$. The proof is based on the wavelet characterization of Besov spaces presented in Sect. 2.3. Theorem 2.3.6 implies that it is enough to show

$$\left(\sum_{(I,\psi) \in \Lambda} |I|^{\left(\frac{1}{p} - \frac{1}{2}\right)\tau} |\langle \tilde{u}, \psi_I \rangle|^\tau \right)^{1/\tau} \le c \max\{ \|u | \mathcal{K}_{p,a}^m (\Omega)\|, \|u | B_{p,p}^s (\Omega)\| \}.$$

Step 1. We explain why the first term in (2.3.13) can be incorporated in the estimates that follow in Step 2. Since our domain Ω is Lipschitz, we can extend every $u \in B_{p,p}^s (\Omega)$ to some function $\tilde{u} = Eu \in B_{p,p}^s (\mathbb{R}^d)$. Then the first term reads as

$$\sum_{k \in \mathbb{Z}^d} \langle \tilde{u}, \phi(\cdot - k) \rangle \phi(\cdot - k).$$

Since ϕ shares the same smoothness and support properties as the wavelets ψ_I for $|I| = 1$ (note that below the vanishing moments of ψ_I only become relevant for $|I| < 1$), the coefficients $\langle \tilde{u}, \phi(\cdot - k)\rangle$ can be treated exactly like any of the coefficients $\langle \tilde{u}, \psi_I \rangle$ in Step 2.

Step 2. For our analysis we shall split the index set Λ as follows. For $j \in \mathbb{N}_0$ the refinement level j is denoted by

$$\Lambda_j := \{(I, \Psi) \in \Lambda : |I| = 2^{-jd}\}.$$

Furthermore, for $k \in \mathbb{N}_0$ put

$$\Lambda_{j,k} := \{(I, \psi) \in \Lambda_j : k2^{-j} \le \rho_I < (k+1)2^{-j}\},$$

where $\rho_I = \inf_{x \in Q(I)} \rho(x)$. In particular, we have $\Lambda_j = \bigcup_{k=0}^{\infty} \Lambda_{j,k}$ and $\Lambda = \bigcup_{j=0}^{\infty} \Lambda_j$.

We consider first the situation when $\rho_I > 0$ corresponding to $k \ge 1$ and therefore put $\Lambda_j^0 = \bigcup_{k \ge 1} \Lambda_{k,j}$. Moreover, we require $Q(I) \subset \Omega$. Recall Whitney's estimate regarding approximation with polynomials, cf. [DeV98, Sect. 6.1], which states that for every I there exists a polyomial P_I of degree less than m, such that

$$\|\tilde{u} - P_I | L_p(Q(I))\| \le c_0 |Q(I)|^{m/d} |\tilde{u}|_{W_p^m(Q(I))} \le c_1 |I|^{m/d} |\tilde{u}|_{W_p^m(Q(I))}$$

for some constant c_1 independent of I and u, where

$$|u|_{W_p^m(Q(I))} := \left(\int_{Q(I)} |\nabla^m u(x)|^p dx \right)^{1/p}.$$

Note that ψ_I satisfies moment conditions of order up to m, i.e., it is orthogonal to any polynomial of degree up to $m - 1$. Thus, using Hölder's inequality with $p > 1$ we estimate

$$|\langle \tilde{u}, \psi_I \rangle| = |\langle \tilde{u} - P_I, \psi_I \rangle| \le \|\tilde{u} - P_I | L_p(Q(I))\| \cdot \|\psi_I | L_{p'}(Q(I))\|$$

$$\le c_1 |I|^{m/d} |\tilde{u}|_{W_p^m(Q(I))} |I|^{\frac{1}{2} - \frac{1}{p}}$$

$$\le c_1 |I|^{\frac{m}{d} + \frac{1}{2} - \frac{1}{p}} \rho_I^{a-m} \left(\sum_{|\alpha|=m} \int_{Q(I)} |\rho(x)|^{m-a} \partial^\alpha \tilde{u}(x)|^p dx \right)^{1/p}$$

$$=: c_1 |I|^{\frac{m}{d} + \frac{1}{2} - \frac{1}{p}} \rho_I^{a-m} \mu_I. \tag{2.4.18}$$

Note that in the third step above we require $a < m$. On the refinement level j, using Hölder's inequality with $\frac{p}{\tau} > 1$, we find

$$
\sum_{(I,\psi)\in\Lambda_j^0} |I|^{\left(\frac{1}{p}-\frac{1}{2}\right)\tau} |\langle \tilde{u}, \psi_I\rangle|^\tau
$$

$$
\leq \sum_{(I,\psi)\in\Lambda_j^0} \left(|I|^{\frac{m}{d}} \rho_I^{a-m} \mu_I\right)^\tau
$$

$$
\leq c_1 \left(\sum_{(I,\psi)\in\Lambda_j^0} \left(|I|^{\frac{m}{d}\tau} \rho_I^{(a-m)\tau}\right)^{\frac{p}{p-\tau}}\right)^{\frac{p-\tau}{p}} \left(\sum_{(I,\psi)\in\Lambda_j^0} \mu_I^p\right)^{\tau/p}.
$$

For the second factor we observe that there is a controlled overlap between the cubes $Q(I)$, meaning each $x \in \Omega$ is contained in a finite number of cubes independent of x, such that we get

$$
\left(\sum_{(I,\psi)\in\Lambda_j^0} \mu_I^p\right)^{1/p} = \left(\sum_{(I,\psi)\in\Lambda_j^0} \sum_{|\alpha|=m} \int_{Q(I)} |\rho^{m-a}(x)\partial^\alpha \tilde{u}(x)|^p dx\right)^{1/p}
$$

$$
\leq c_2 \left(\sum_{|\alpha|=m} \int_\Omega |\rho^{m-a}(x)\partial^\alpha \tilde{u}(x)|^p dx\right)^{1/p} \leq c_2 \|u|\mathcal{K}_{p,a}^m(\Omega)\|.
$$

For the first factor by choice or ρ we always have $\rho_I \leq 1$, hence the index k is at most 2^j for the sets $\Lambda_{j,k}$ to be non-empty. The number of elements in $\Lambda_{j,k}$ is bounded by $k^{d-1-\delta}2^{j\delta}$. With this we find

$$
\left(\sum_{(I,\psi)\in\Lambda_j^0} \left(|I|^{\frac{m}{d}\tau} \rho_I^{(a-m)\tau}\right)^{\frac{p}{p-\tau}}\right)^{\frac{p-\tau}{p}}
$$

$$
\leq \left(\sum_{k=1}^{2^j} \sum_{(I,\psi)\in\Lambda_{j,k}} \left(2^{-jm\tau}(k2^{-j})^{(a-m)\tau}\right)^{\frac{p}{p-\tau}}\right)^{\frac{p-\tau}{p}}
$$

$$
\leq \left(\sum_{k=1}^{2^j} \sum_{(I,\psi)\in\Lambda_{j,k}} \left(2^{-ja\tau}k^{(a-m)\tau}\right)^{\frac{p}{p-\tau}}\right)^{\frac{p-\tau}{p}}
$$

$$\leq \left(c_3 2^{-ja\frac{p\tau}{p-\tau}} \sum_{k=1}^{2^j} k^{(a-m)\frac{p\tau}{p-\tau}} k^{d-1-\delta} 2^{j\delta} \right)^{\frac{p-\tau}{p}}$$

$$= c_4 2^{-ja\tau} 2^{j\delta\frac{p-\tau}{p}} \left(\sum_{k=1}^{2^j} k^{(a-m)\frac{p\tau}{p-\tau}+d-1-\delta} \right)^{\frac{p-\tau}{p}}.$$

Looking at the value of the exponent in the last sum we see that

$$(a-m)\frac{p\tau}{p-\tau} + d - 1 - \delta > -1 \quad \Longleftrightarrow \quad a - m + r\frac{d-\delta}{d} > 0,$$

which leads to

$$\left(\sum_{(I,\psi)\in\Lambda_j^0} \left(|I|^{\frac{m}{d}\tau} \rho_I^{(a-m)\tau} \right)^{\frac{p}{p-\tau}} \right)^{\frac{p-\tau}{p}}$$

$$\leq c_5 2^{-ja\tau} 2^{j\delta\frac{p-\tau}{p}} \begin{cases} 2^{j\left((a-m)\tau+(d-\delta)\frac{p-\tau}{p}\right)}, & a-m+r\frac{d-\delta}{d} > 0, \\ (j+1)^{\frac{p-\tau}{p}}, & a-m+r\frac{d-\delta}{d} = 0, \\ 1, & a-m+r\frac{d-\delta}{d} < 0. \end{cases}$$

$$(2.4.19)$$

The case $a > m$ can be treated in the same way as above by taking out $\tilde{\rho}_I^{a-m}$ with $\tilde{\rho}_I := \inf_{x\in Q(I)} \rho(x)$ instead of ρ_I^{a-m} in the integral appearing in (2.4.18). The values of ρ_I and $\tilde{\rho}_I$ are comparable in this situation, since we consider cubes which do not intersect with the boundary, i.e., we have $k \geq 1$. In particular, in (2.4.19) only the first case occurs if $a > m$.

Step 3. We now put $\Lambda^0 := \bigcup_{j\geq 0} \Lambda_j^0$. Summing the first line of the last estimate over all j, we obtain

$$\sum_{(I,\psi)\in\Lambda^0} |I|^{\left(\frac{1}{p}-\frac{1}{2}\right)\tau} |\langle \tilde{u}, \psi_I \rangle|^\tau$$

$$\leq c_6 \sum_{j=0}^{\infty} 2^{-j(m\tau-d\frac{p-\tau}{p})} \|u|\mathcal{K}_{p,a}^m(\Omega)\|^\tau \lesssim \|u|\mathcal{K}_{p,a}^m(\Omega)\|^\tau < \infty,$$

if the geometric series converges, which happens if

$$m\tau > d\frac{p-\tau}{p} \quad \Longleftrightarrow \quad m > d\frac{r}{d} \quad \Longleftrightarrow \quad m > r.$$

Similarly, in the second case we see that

$$\sum_{(I,\psi)\in\Lambda^0} |I|^{\left(\frac{1}{p}-\frac{1}{2}\right)\tau} |\langle \tilde{u}, \psi_I \rangle|^\tau$$

$$\leq c_7 \sum_{j=0}^{\infty} 2^{-j(a\tau-\delta\frac{p-\tau}{p})}(j+1)^{\frac{p-\tau}{p}} \|u|\mathcal{K}_{p,a}^m(\Omega)\|^\tau \lesssim \|u|\mathcal{K}_{p,a}^m(\Omega)\|^\tau < \infty,$$

where the series converges if

$$a\tau > \delta\frac{p-\tau}{p}, \quad \text{i.e.,} \quad a > \delta\frac{r}{d}, \quad \text{i.e.,} \quad m > r\frac{d-\delta}{d} + \frac{\delta}{d}r = r, \quad \text{i.e.,} \quad m > r,$$

which is the same condition as before. Finally, in the third case we find

$$\sum_{(I,\psi)\in\Lambda^0} |I|^{\left(\frac{1}{p}-\frac{1}{2}\right)\tau} |\langle \tilde{u}, \psi_I \rangle|^\tau$$

$$\leq c_8 \sum_{j=0}^{\infty} 2^{-j(a\tau-\delta\frac{p-\tau}{p})} \|u|\mathcal{K}_{p,a}^m(\Omega)\|^\tau \lesssim \|u|\mathcal{K}_{p,a}^m(\Omega)\|^\tau < \infty,$$

whenever

$$a\tau > \delta\frac{p-\tau}{p} \quad \Longleftrightarrow \quad a > \delta\frac{r}{d}$$

as in the second case above.

Step 4. We need to consider the sets $\Lambda_{j,0}$, i.e., the wavelets close to l_0. Here, we shall make use of the assumption $\tilde{u} \in B_{p,p}^s(\mathbb{R}^d)$. Since the number of elements in $\Lambda_{j,0}$ is bounded from above by $c_9 2^{j\delta}$ we estimate using Hölder's inequality with $\frac{p}{\tau} > 1$ and obtain

$$\sum_{(I,\psi)\in\Lambda_{j,0}} |I|^{\left(\frac{1}{p}-\frac{1}{2}\right)\tau} |\langle \tilde{u}, \psi_I \rangle|^\tau$$

$$\leq c_9^{\frac{p-\tau}{p}} 2^{j\delta\frac{p-\tau}{p}} 2^{-jd\left(\frac{1}{p}-\frac{1}{2}\right)\tau} \left(\sum_{(I,\psi)\in\Lambda_{j,0}} |\langle \tilde{u}, \psi_I \rangle|^p \right)^{\tau/p}$$

$$= c_9^{\frac{p-\tau}{p}} 2^{j\delta\frac{p-\tau}{p}} 2^{-js\tau} \left(\sum_{(I,\psi)\in\Lambda_{j,0}} 2^{j\left(s+\frac{d}{2}-\frac{d}{p}\right)p} |\langle \tilde{u}, \psi_I \rangle|^p \right)^{\tau/p}.$$

Summing up over j and once more using Hölder's inequality with $\frac{p}{\tau} > 1$ gives

$$\sum_{j=0}^{\infty} \sum_{(I,\psi) \in \Lambda_{j,0}} |I|^{\left(\frac{1}{p} - \frac{1}{2}\right)\tau} |\langle \tilde{u}, \psi_I \rangle|^{\tau}$$

$$\leq c_9^{\frac{p-\tau}{p}} \sum_{j=0}^{\infty} 2^{j\delta \frac{p-\tau}{p}} 2^{-js\tau} \left(\sum_{(I,\psi) \in \Lambda_{j,0}} 2^{j\left(s + \frac{d}{2} - \frac{d}{p}\right)p} |\langle \tilde{u}, \psi_I \rangle|^p \right)^{\tau/p}$$

$$\leq c_9^{\frac{p-\tau}{p}} \left(\sum_{j=0}^{\infty} 2^{j\delta} 2^{-js\tau \frac{p}{p-\tau}} \right)^{\frac{p-\tau}{p}} \cdot \left(\sum_{j=0}^{\infty} \sum_{(I,\psi) \in \Lambda_{j,0}} 2^{j\left(s + \frac{d}{2} - \frac{d}{p}\right)p} |\langle \tilde{u}, \psi_I \rangle| \right)^{\tau/p}$$

$$\lesssim \|\tilde{u}|B^s_{p,p}(\mathbb{R}^d)\|^{\tau} \lesssim \|u|B^s_{p,p}(\Omega)\|^{\tau},$$

under the condition

$$\delta < \frac{sp\tau}{p - \tau} \quad \Longleftrightarrow \quad \frac{s}{\delta} > \frac{1}{\tau} - \frac{1}{p} = \frac{r}{d} \quad \Longleftrightarrow \quad r < \frac{sd}{\delta}.$$

Step 5.   Finally, we need to consider those ψ_I whose support intersect $\partial\Omega$. In this case we can estimate similar as in Step 4 with δ replaced by $d - 1$. This results in the condition

$$\sum_{(I,\psi) \in \Lambda: \text{ supp } \psi_I \cap \partial\Omega \neq \emptyset} |I|^{\left(\frac{1}{p} - \frac{1}{2}\right)\tau} |\langle \tilde{u}, \psi_I \rangle|^{\tau} \lesssim \|\tilde{u}|B^s_{p,p}(\mathbb{R}^d)\|^{\tau} \lesssim \|u|B^s_{p,p}(\Omega)\|^{\tau}$$

if $r < \frac{sd}{d-1}$. Altogether, we have proved

$$\|u|B^r_{\tau,\tau}(\Omega)\| \leq \|\tilde{u}|B^r_{\tau,\tau}(\mathbb{R}^d)\| \lesssim \|u|B^s_{p,p}(\Omega)\| + \|u|\mathcal{K}^m_{p,a}(\Omega)\|,$$

with constants independent of u. $\square$

As an immediate consequence of Theorem 2.4.14 and the definition of corresponding function spaces on Ω_T, we have the following generalized embedding result.

Theorem 2.4.15 (Embeddings Between Generalized Kondratiev and Besov Spaces) *Let $\Omega \subset \mathbb{R}^d$ be a special Lipschitz domain according to Definition 2.1.8. Then for $1 < p < \infty$ and $0 < q \leq \infty$, we have*

$$C([0,T], \mathcal{K}^m_{p,a}(\Omega)) \cap C([0,T], B^s_{p,p}(\Omega)) \hookrightarrow C\left([0,T], B^r_{\tau,\tau}(\Omega)\right), \quad \frac{1}{\tau} = \frac{r}{d} + \frac{1}{p},$$

for all $0 \leq r < \min(m, \frac{sd}{d-1})$ and $a > \frac{\delta}{d}r$, where $\delta = d - 2 = \dim(l_0)$.

There is another embedding from Kondratiev into Triebel-Lizorkin spaces proved in [HS18, Thm. 4.9], which has the advantage compared to (2.4.14) that the intersection on the left-hand side can be avoided. Therefore, no knowledge in terms of fractional Soblev regularity (reflected in the term $B_{p,p}^s$ or $B_{p,\infty}^s$, respectively) of the solution is needed for the embedding to hold. This will be useful when studying the regularity of the nonlinear elliptic problem (4.1.1) in Chap. 4 and the fractional Sobolev regularity of Problem III in Chap. 5.

Theorem 2.4.16 (Embeddings Between Kondratiev and Triebel-Lizorkin Spaces) *Let* $D \subset \mathbb{R}^d$ *be some bounded Lipschitz domain of polyhedral type with singular set S of dimension δ and let* $1 < p < \infty, 0 < \tau < p, m_0 \in \mathbb{N}$, *and* $a > 0$. *Further assume*

$$m - a < (d - \delta)\left(\frac{1}{\tau} - \frac{1}{p}\right).$$

Then it holds

$$\mathcal{K}_{p,a}^m(D) \hookrightarrow F_{\tau,2}^m(D). \tag{2.4.20}$$

Remark 2.4.17 Theorem 2.4.16 in [HS18] is there stated for bounded Lipschitz domains with piecewise smooth boundary, which covers our bounded Lipschitz domains of polyhedral type.

In particular, the result in [HS18] extends to $m = 0$ since for $a > 0$ and $\tau < p$ we have

$$\mathcal{K}_{p,a}^0(D) \hookrightarrow L_p(D) = F_{p,2}^0(D) \hookrightarrow F_{\tau,2}^0(D),$$

where the identity follows from (2.3.4) and the second embedding from Proposition 2.3.4. Extensions of (2.4.20) to generalized functions spaces for time dependent problems are also possible, i.e., it holds

$$L_q([0, T], \mathcal{K}_{p,a}^m(D)) \hookrightarrow L_q([0, T], F_{\tau,2}^m(D)), \qquad 0 < q \leq \infty. \tag{2.4.21}$$

2.4.5 Fractional Kondratiev Spaces

We define families of fractional Kondratiev spaces by means of complex interpolation of the usual Kondratiev spaces. We refer to [Han17] and [Lot00] in this context, where these spaces are studied in detail. Concerning the basics in complex interpolation theory we furthermore refer to [BL76, Tri78].

Definition 2.4.18 (Fractional Kondratiev Spaces) Let Ω be either a bounded Lipschitz domain $\mathcal{O}$ with singular set $S = \partial\mathcal{O}$ or a bounded Lipschitz domain of polyhedral type D with the respective closed subset M of ∂D. Let $s \geq 0$,

$1 \leq p \leq \infty$, and $a \in \mathbb{R}$. For $s \in \mathbb{N}_0$, put $\mathfrak{K}_{p,a}^s(D) = \mathcal{K}_{p,a}^s(D)$. Otherwise, for $s \geq 0$ with $s \notin \mathbb{N}_0$, let $m = [s]$ denote its integer part and $\theta = \{s\} := s - [s]$. Then we define the fractional Kondratiev spaces via

$$\mathfrak{K}_{p,a}^s(\Omega) := \left[\mathcal{K}_{p,a}^m(\Omega), \mathcal{K}_{p,a}^{m+1}(\Omega) \right]_\theta.$$

Remark 2.4.19 From the results established in [Han17] and [Lot00], we can conclude a generalized interpolation result for arbitrary pairs of Kondratiev spaces within the full scale of parameters s and a. Let $1 < p < \infty$, $s_0, s_1 \geq 0$, $a_0, a_1 \in \mathbb{R}$, and $0 < \theta < 1$. Then

$$\mathfrak{K}_{p,a}^s(\Omega) = \left[\mathfrak{K}_{p,a_0}^{s_0}(\Omega), \mathfrak{K}_{p,a_1}^{s_1}(\Omega) \right]_\theta, \tag{2.4.22}$$

where $s = (1 - \theta)s_0 + \theta s_1$ and $a = (1 - \theta)a_0 + \theta a_1$. Thus, we can see that the Kondratiev spaces $\mathcal{K}_{p,a}^m$ are naturally included in the scale of fractional Kondratiev spaces, since (2.4.22) gives

$$\mathcal{K}_{p,a}^m(\Omega) = \left[\mathcal{K}_{p,a}^{m-1}(\Omega), \mathcal{K}_{p,a}^{m+1}(\Omega) \right]_{\frac{1}{2}}.$$

We collect some embedding properties of fractional Kondratiev spaces which will be important for our later considerations. In particular, for $\mathfrak{K}_{p,a}^s$ defined on general Lipschitz domains $\mathcal{O}$ we rewrite [Cio13, Thm 5.1] and obtain the following embedding of the spaces into the scale of Besov spaces.

Theorem 2.4.20 (Embeddings Between Generalized Fractional Kondratiev and Besov Spaces) *Let $\mathcal{O} \subset \mathbb{R}^d$ be a bounded Lipschitz domain. Fix $s \in (0, \infty)$, $p \in [2, \infty)$, and $a \in \mathbb{R}$. Then*

$$L_p([0, T], \mathfrak{K}_{p,a}^s(\mathcal{O})) \hookrightarrow L_p([0, T], B_{\tau,\tau}^\alpha(\mathcal{O})),$$

for all α and τ with

$$\frac{1}{\tau} = \frac{\alpha}{d} + \frac{1}{p} \quad and \quad 0 < \alpha < \min\left\{ s, a\frac{d}{d-1} \right\}.$$

Remark 2.4.21 In contrast to Theorems 2.4.10 and 2.4.12 the result in Theorem 2.4.20 is weaker, since here we have the restriction $\alpha < a\frac{d}{d-1}$. On the other hand, in the embedding above no knowledge about the Sobolev regularity (or regularity in the spaces $B_{p,\infty}^s$) is needed.

In terms of compact embeddings the following result is proven in [Han17].

Proposition 2.4.22 *Let D be a bounded Lipschitz domain of polyhedral type. Moreover, let $m \in \mathbb{N}$, $a \in \mathbb{R}$, and $1 < p < \infty$. Then the embedding*

$$\mathcal{K}_{p,a}^m(D) \hookrightarrow \mathfrak{R}_{p,a-\varepsilon}^{m-\varepsilon}(D) \tag{2.4.23}$$

is compact for arbitrary $0 < \varepsilon < 1$.

Remark 2.4.23 From (2.4.23) together with [CK95, Thm. 10] we obtain that the embedding

$$\mathfrak{R}_{p,a-\varepsilon'}^{m-\varepsilon'}(D) = \left[\mathcal{K}_{p,a-1}^{m-1}(D), \mathcal{K}_{p,a}^m(D) \right]_{1-\varepsilon'} \hookrightarrow \left[\mathfrak{R}_{p,a-1-\varepsilon}^{m-1-\varepsilon}(D), \mathfrak{R}_{p,a-\varepsilon}^{m-\varepsilon}(D) \right]_{1-\varepsilon'}$$

$$= \mathfrak{R}_{p,a-(\varepsilon+\varepsilon')}^{m-(\varepsilon+\varepsilon')}(D)$$

is also compact for arbitrary $0 < \varepsilon, \varepsilon' < 1$.

2.5   Smoothness Morrey Spaces and Their Generalizations

In this section we introduce generalized smoothness Morrey spaces which are based on a generalized version of the classical Morrey spaces, defined via a function parameter $\varphi(\cdot)$ that allows greater flexibility. The spaces were first introduced in [NNS16] and—as the name suggests—include the smoothness Morrey spaces studied in [YSY10] as well as many other scales of function spaces, cf. Remark 2.5.4. We study traces on C^k domains in this general framework in Chap. 9.

Generalized Morrey Spaces $\mathcal{M}_p^\varphi(\mathbb{R}^d)$ Morrey spaces were introduced by Morrey in [Mor38], when studying solutions of second order quasi-linear elliptic equations in the framework of Lebesgue spaces. They can be understood as a complement (generalization) of the Lebesgue spaces $L_p(\Omega)$. In particular, the classical Morrey space $\mathcal{M}_{u,p}(\mathbb{R}^d)$, $0 < p \le u < \infty$, is defined to be the set of all locally p-integrable functions $f \in L_p^{\mathrm{loc}}(\mathbb{R}^d)$ such that

$$\|f \mid \mathcal{M}_{u,p}(\mathbb{R}^d)\| := \sup_{Q \in \mathcal{Q}} |Q|^{\frac{1}{u} - \frac{1}{p}} \left(\int_Q |f(y)|^p \mathrm{d}y \right)^{\frac{1}{p}} < \infty. \tag{2.5.1}$$

Obviously, $\mathcal{M}_{p,p}(\mathbb{R}^d) = L_p(\mathbb{R}^d)$. As can be seen from the definition, Morrey spaces investigate the local behaviour of the L_p-norm, which makes them useful when describing the local behaviour of solutions of nonlinear partial differential equations, cf. [KY94, LeR07, LeR12, LeR13, LeR18, Maz03b, Maz03a]. Furthermore, applications in harmonic analysis and potential analysis can be found in the papers [AX04, AX11, AX12a, AX12b]. For more information we refer to the books [Ada15] and [SYY10].

In what follows we will mostly deal with generalized Morrey spaces according to the following definition.

Definition 2.5.1 Let $0 < p < \infty$ and $\varphi : (0, \infty) \to (0, \infty)$ be a function. Then $\mathcal{M}_p^\varphi(\mathbb{R}^d)$ is the set of all locally p-integrable functions $f \in L_p^{\mathrm{loc}}(\mathbb{R}^d)$ for which

$$\|f \mid \mathcal{M}_p^\varphi(\mathbb{R}^d)\| := \sup_{Q \in \mathcal{Q}} \varphi(\ell(Q)) \left(\frac{1}{|Q|} \int_Q |f(y)|^p \, dy \right)^{\frac{1}{p}} < \infty.$$

Remark 2.5.2 The above definition goes back to [Nak94]. When $\varphi(t) := t^{\frac{d}{u}}$ for $t > 0$ and $0 < p \leq u < \infty$ then $\mathcal{M}_p^\varphi(\mathbb{R}^d)$ coincides with $\mathcal{M}_{u,p}(\mathbb{R}^d)$, which in turn recovers the Lebesgue space $L_p(\mathbb{R}^d)$ when $u = p$. Another example of particular interest is the case of $\varphi(t) := (1 + t^d)^{\frac{1}{u}} \left(\log(e + t^{-d}) \right)^{-1}$, which arises naturally in the target space when studying embeddings of Sobolev-Morrey spaces in the critical case, cf. [SW13, Thm. 5.1].

Observe that in the quasi-norm (2.5.1), in the proper Morrey case of $p < u$, as $\ell(Q)$ increases the integral also increases while the remaining term $|Q|^{\frac{1}{u} - \frac{1}{p}} = \ell(Q)^{\frac{d}{u} - \frac{d}{p}}$ decreases. If we want to keep this feature in the generalized Morrey case it is natural to consider functions φ in the class $\mathcal{G}_p$, $0 < p < \infty$, where $\mathcal{G}_p$ is set of all nondecreasing functions $\varphi : (0, \infty) \to (0, \infty)$ such that $\varphi(t) t^{-d/p}$ is a nonincreasing function. This will be our restriction on the function parameter φ from now on when dealing with generalized smoothness Morrey spaces. In this context we also refer to [Nak00].

2.5.1 Spaces of Type $\mathcal{A}^s_{\mathcal{M}_{p,q}^\varphi}$

Now we introduce the generalized Besov-Morrey and Triebel-Lizorkin-Morrey spaces on $\mathbb{R}^d$ from [NNS16] and extend the concept also to C^k domains Ω and their boundary Γ. Moreover, we provide equivalent characterizations of these spaces in terms of atomic and quarkonial decompositions.

Definition 2.5.3 Let $0 < p < \infty$, $0 < q \leq \infty$, $s \in \mathbb{R}$, and $\varphi \in \mathcal{G}_p$. Let $\mu_0, \mu \in \mathcal{S}(\mathbb{R}^d)$ be nonnegative compactly supported functions satisfying

$$\mu_0(x) > 0 \quad \text{if} \quad x \in Q(2),$$

$$0 \notin \operatorname{supp} \mu \quad \text{and} \quad \mu(x) > 0 \quad \text{if} \quad x \in Q(2) \setminus Q(1).$$

For $j \in \mathbb{N}$, let $\mu_j(x) := \mu(2^{-j}x)$, where $x \in \mathbb{R}^d$.

(i) The *generalized Besov-Morrey space* $\mathcal{N}^s_{\mathcal{M}^\varphi_{p,q}}(\mathbb{R}^d)$ is defined to be the set of all $f \in \mathcal{S}'(\mathbb{R}^d)$ such that

$$\| f \mid \mathcal{N}^s_{\mathcal{M}^\varphi_{p,q}}(\mathbb{R}^d) \| := \left(\sum_{j=0}^{\infty} 2^{jsq} \| \mathcal{F}^{-1}(\mu_j \mathcal{F}f) \mid \mathcal{M}^\varphi_p(\mathbb{R}^d) \|^q \right)^{1/q} < \infty$$

(with the usual modification if $q = \infty$).

(ii) When $q < \infty$, assume that there exist $C, \varepsilon > 0$ such that

$$\frac{t^\varepsilon}{\varphi(t)} \leq C \frac{r^\varepsilon}{\varphi(r)} \quad \text{if} \quad t \geq r. \tag{2.5.2}$$

The *generalized Triebel-Lizorkin-Morrey space* $\mathcal{E}^s_{\mathcal{M}^\varphi_{p,q}}(\mathbb{R}^d)$ is defined to be the set of all $f \in \mathcal{S}'(\mathbb{R}^d)$ such that

$$\| f \mid \mathcal{E}^s_{\mathcal{M}^\varphi_{p,q}}(\mathbb{R}^d) \| := \left\| \left(\sum_{j=0}^{\infty} 2^{jsq} |\mathcal{F}^{-1}(\mu_j \mathcal{F}f)(\cdot)|^q \right)^{1/q} \mid \mathcal{M}^\varphi_p(\mathbb{R}^d) \right\| < \infty$$

(with the usual modification if $q = \infty$).

Remark 2.5.4

(i) We write $\mathcal{A}^s_{\mathcal{M}^\varphi_{p,q}}$ instead of $\mathcal{N}^s_{\mathcal{M}^\varphi_{p,q}}$ or $\mathcal{E}^s_{\mathcal{M}^\varphi_{p,q}}$, respectively, when both scales of spaces are meant simultaneously in some context, assuming always that there exist $C, \varepsilon > 0$ such that (2.5.2) holds, when $q < \infty$ and $\mathcal{A}^s_{\mathcal{M}^\varphi_{p,q}}$ denotes $\mathcal{E}^s_{\mathcal{M}^\varphi_{p,q}}$.

(ii) The above spaces were introduced in [NNS16]. There the authors proved that those spaces are independent of the choice of the functions μ_0 and μ considered in the definition, as different choices lead to equivalent quasi-norms, cf.[NNS16, Thm. 1.4]. In particular, the sequence $\{\mu_j\}_{j=0}^{\infty}$ constitutes a generalized version of a dyadic resolution of unity according to Remark 2.3.1.

(iii) When $\varphi(t) = t^{\frac{d}{u}}$ for $t > 0$ and $0 < p \leq u < \infty$, then

$$\mathcal{N}^s_{\mathcal{M}^\varphi_{p,q}}(\mathbb{R}^d) = \mathcal{N}^s_{u,p,q}(\mathbb{R}^d) \quad \text{and} \quad \mathcal{E}^s_{\mathcal{M}^\varphi_{p,q}}(\mathbb{R}^d) = \mathcal{E}^s_{u,p,q}(\mathbb{R}^d)$$

are the usual Besov-Morrey and Triebel-Lizorkin-Morrey spaces, which are studied in [YSY10] or in the recent survey papers by Sickel [Sic12, Sic13]. We remark that, in this particular case, the additional condition (2.5.2) on φ required in Definition 2.5.3(ii) for the generalized Triebel-Lizorkin-Morrey spaces is automatically fulfilled, as there always exist $0 < \varepsilon < d/u$. Of course, we can recover the classical Besov spaces $B^s_{p,q}(\mathbb{R}^d)$ and the classical Triebel-

Lizorkin spaces $F^s_{p,q}(\mathbb{R}^d)$ for any $0 < p < \infty$, $0 < q \le \infty$, and $s \in \mathbb{R}$, since

$$B^s_{p,q}(\mathbb{R}^d) = \mathcal{N}^s_{p,p,q}(\mathbb{R}^d) \quad \text{and} \quad F^s_{p,q}(\mathbb{R}^d) = \mathcal{E}^s_{p,p,q}(\mathbb{R}^d).$$

Furthermore, the Triebel-Lizorkin-type spaces $F^{s,\tau}_{p,q}(\mathbb{R}^d)$ defined in [YSY10] are also included in this scale when $0 \le \tau < 1/p$ as

$$F^{s,\tau}_{p,q}(\mathbb{R}^d) = \mathcal{E}^s_{u,p,q}(\mathbb{R}^d) \quad \text{with} \quad u = \frac{p}{1 - p\tau}, \tag{2.5.3}$$

for any $0 < p < \infty$, $0 < q \le \infty$, and $s \in \mathbb{R}$, cf. [YSY10, Cor. 3.3, p. 63]. Note that the corresponding Besov-type spaces $B^{s,\tau}_{p,q}(\mathbb{R}^d)$, see page 95 for the definition, are not covered by our approach. In particular, by [YSY10, Cor. 3.3, p. 64] we have

$$\mathcal{N}^s_{u,p,q}(\mathbb{R}^d) \hookrightarrow B^{s,\tau}_{p,q}(\mathbb{R}^d) \quad \text{with} \quad u = \frac{p}{1 - p\tau}, \tag{2.5.4}$$

and the embedding is proper if $\tau > 0$ and $q < \infty$. However, if $\tau = 0$ or $q = \infty$ then both spaces coincide. Besides the elementary embeddings

$$\mathcal{A}^{s+\varepsilon}_{\mathcal{M}^\varphi_p,q_1}(\mathbb{R}^d) \hookrightarrow \mathcal{A}^s_{\mathcal{M}^\varphi_p,q_2}(\mathbb{R}^d), \quad \varepsilon > 0,$$

and

$$\mathcal{A}^s_{\mathcal{M}^\varphi_p,q_1}(\mathbb{R}^d) \hookrightarrow \mathcal{A}^s_{\mathcal{M}^\varphi_p,q_2}(\mathbb{R}^d), \quad q_1 \le q_2,$$

cf. [NNS16, Prop. 3.3], we can also easily prove that

$$\mathcal{N}^s_{\mathcal{M}^\varphi_p,\min\{p,q\}}(\mathbb{R}^d) \hookrightarrow \mathcal{E}^s_{\mathcal{M}^\varphi_p,q}(\mathbb{R}^d) \hookrightarrow \mathcal{N}^s_{\mathcal{M}^\varphi_p,\infty}(\mathbb{R}^d).$$

In terms of coincidences for Triebel-Lizorkin-Morrey spaces it is known that

$$\mathcal{E}^0_{u,p,2}(\mathbb{R}^d) = \mathcal{M}_{u,p}(\mathbb{R}^d) \quad \text{for} \quad 1 < p \le u < \infty,$$

cf. [Maz03b, Prop. 4.1]. The following is the counterpart for the generalized version of the spaces.

Proposition 2.5.5 *Let* $1 < p < \infty$ *and* $\varphi \in \mathcal{G}_p$ *satisfy* (2.5.2). *If* φ *is strictly increasing, then*

$$\mathcal{E}^0_{\mathcal{M}^\varphi_p,2}(\mathbb{R}^d) = \mathcal{M}^\varphi_p(\mathbb{R}^d).$$

Proof This is a consequence of Theorem 3.12 and Proposition 3.18 of [YZY15] letting $\phi(x, r) := \varphi(r)^{-1} r^{\frac{d}{p}}$ for any $x \in \mathbb{R}^d$ and $r > 0$. $\square$

Atomic Decompositions An important tool in our later considerations is the characterization of the generalized Besov-Morrey and Triebel-Lizorkin-Morrey spaces by means of atomic decompositions. We follow [NNS16] and start by defining the appropriate sequence spaces and atoms.

Definition 2.5.6 Let $0 < p < \infty, 0 < q \leq \infty, s \in \mathbb{R}$, and $\varphi \in \mathcal{G}_p$.

(i) The generalized Besov-Morrey sequence space $\mathbf{n}^s_{\mathcal{M}^\varphi_{p,q}}(\mathbb{R}^d)$ is the set of all doubly indexed sequences $\lambda := \{\lambda_{j,m}\}_{j \in \mathbb{N}_0, m \in \mathbb{Z}^d} \subset \mathbb{C}$ for which the quasi-norm

$$\|\lambda \mid \mathbf{n}^s_{\mathcal{M}^\varphi_{p,q}}\| := \left(\sum_{j=0}^\infty 2^{jsq} \left\| \sum_{m \in \mathbb{Z}^d} \lambda_{j,m} \chi_{Q_{j,m}} \mid \mathcal{M}^\varphi_p(\mathbb{R}^d) \right\|^q \right)^{1/q}$$

is finite (with the usual modification if $q = \infty$).

(ii) Assume in addition (2.5.2) when $q < \infty$. The generalized Triebel-Lizorkin-Morrey sequence space $\mathbf{e}^s_{\mathcal{M}^\varphi_{p,q}}(\mathbb{R}^d)$ is the set of all doubly indexed sequences $\lambda := \{\lambda_{j,m}\}_{j \in \mathbb{N}_0, m \in \mathbb{Z}^d} \subset \mathbb{C}$ for which the quasi-norm

$$\|\lambda | \mathbf{e}^s_{\mathcal{M}^\varphi_{p,q}}\| := \left\| \left\{ \sum_{j=0}^\infty 2^{jsq} \left(\sum_{m \in \mathbb{Z}^d} |\lambda_{j,m}| \chi_{Q_{j,m}} \right)^q \right\}^{1/q} \mid \mathcal{M}^\varphi_p(\mathbb{R}^d) \right\|$$

is finite (with the usual modification if $q = \infty$).

Remark 2.5.7 We write $\mathbf{a}^s_{\mathcal{M}^\varphi_{p,q}}$ instead of $\mathbf{n}^s_{\mathcal{M}^\varphi_{p,q}}$ or $\mathbf{e}^s_{\mathcal{M}^\varphi_{p,q}}$, for convenience, when both scales are meant simultaneously, assuming always that there exist $C, \varepsilon > 0$ such that (2.5.2) holds, when $q < \infty$ and $\mathbf{a} = \mathbf{e}$.

Definition 2.5.8 Let $L \in \mathbb{N}_0 \cup \{-1\}, K \in \mathbb{N}_0$, and $\tilde{d} > 1$. A C^K function $a : \mathbb{R}^d \to \mathbb{C}$ is said to be a (K, L)-atom centered at $Q_{j,m}$, where $j \in \mathbb{N}_0$ and $m \in \mathbb{Z}^d$, if

$$2^{-j|\alpha|} |D^\alpha a(x)| \leq \chi_{\tilde{d} Q_{j,m}}(x) \tag{2.5.5}$$

for all $x \in \mathbb{R}^d$ and for all $\alpha \in \mathbb{N}_0^d$ with $|\alpha| \leq K$, and when for $j \in \mathbb{N}$ it holds

$$\int_{\mathbb{R}^d} x^\beta a(x) dx = 0, \tag{2.5.6}$$

for all $\beta \in \mathbb{N}_0^d$ with $|\beta| \leq L$ when $L \geq 0$. If $L = -1$ then no moment conditions according to (2.5.6) are required. In the sequel we write $a_{j,m}$ instead of a if the atom is located at $Q_{j,m}$, i.e., supp $a_{j,m} \subset \tilde{d} Q_{j,m}$.

The following coincides with [NNS16, Thm. 4.4, 4.5], cf. also [NNS16, Rem. 4.3].

Theorem 2.5.9 (Atomic Decomposition for $\mathcal{A}^s_{\mathcal{M}^\varphi_{p,q}}(\mathbb{R}^d)$) *Let $0 < p < \infty$, $0 < q \leq \infty$, $s \in \mathbb{R}$, and $\varphi \in \mathcal{G}_p$. Assume in addition that φ satisfies (2.5.2) when $q < \infty$ and $\mathcal{A} = \mathcal{E}$. Let also $\tilde{d} > 1$, $L \in \mathbb{N}_0 \cup \{-1\}$ and $K \in \mathbb{N}_0$ be such that*

$$K \geq [1+s]_+ \quad and \quad L \geq \begin{cases} \max(-1, [\sigma_p - s]), & if \ \mathcal{A} = \mathcal{N}, \\ \max(-1, [\sigma_{p,q} - s]), & if \ \mathcal{A} = \mathcal{E}. \end{cases}$$

(i) Let $f \in \mathcal{A}^s_{\mathcal{M}^\varphi_{p,q}}(\mathbb{R}^d)$. Then there exists a family $\{a_{j,m}\}_{j \in \mathbb{N}_0, m \in \mathbb{Z}^d}$ of (K,L)-atoms and a sequence $\lambda = \{\lambda_{j,m}\}_{j \in \mathbb{N}_0, m \in \mathbb{Z}^d} \in \mathbf{a}^s_{\mathcal{M}^\varphi_{p,q}}(\mathbb{R}^d)$ such that

$$f = \sum_{j=0}^{\infty} \sum_{m \in \mathbb{Z}^d} \lambda_{j,m} a_{j,m} \quad in \quad \mathcal{S}'(\mathbb{R}^d)$$

and

$$\|\lambda \mid \mathbf{a}^s_{\mathcal{M}^\varphi_{p,q}}(\mathbb{R}^d)\| \lesssim \|f \mid \mathcal{A}^s_{\mathcal{M}^\varphi_{p,q}}(\mathbb{R}^d)\|.$$

(ii) Let $\{a_{j,m}\}_{j \in \mathbb{N}_0, m \in \mathbb{Z}^d}$ be a family of (K,L)-atoms and $\lambda = \{\lambda_{j,m}\}_{j \in \mathbb{N}_0, m \in \mathbb{Z}^d} \in \mathbf{a}^s_{\mathcal{M}^\varphi_{p,q}}(\mathbb{R}^d)$. Then

$$f = \sum_{j=0}^{\infty} \sum_{m \in \mathbb{Z}^d} \lambda_{j,m} a_{j,m}$$

converges in $\mathcal{S}'(\mathbb{R}^d)$ and belongs to $\mathcal{A}^s_{\mathcal{M}^\varphi_{p,q}}(\mathbb{R}^d)$. Furthermore

$$\|f \mid \mathcal{A}^s_{\mathcal{M}^\varphi_{p,q}}(\mathbb{R}^d)\| \lesssim \|\lambda \mid \mathbf{a}^s_{\mathcal{M}^\varphi_{p,q}}(\mathbb{R}^d)\|.$$

Remark 2.5.10 Note that in comparison with the atomic decompositions for the Besov spaces $\mathbf{B}^s_{p,q}(\mathbb{R}^d)$ presented in Sect. 2.3.4 our smooth atoms in Definition 2.5.8 are now required to satisfy moment conditions (2.5.6) when $s \leq \sigma_p$ in Theorem 2.5.9 above. In particular, Theorem 2.5.9 provides an atomic decomposition for the spaces $B^s_{p,q}(\mathbb{R}^d)$ because the special choice $\varphi(t) = t^{\frac{d}{p}}, t > 0$, yields $\mathcal{N}^s_{p,p,q}(\mathbb{R}^d) = B^s_{p,q}(\mathbb{R}^d)$. Since the spaces $B^s_{p,q}(\mathbb{R}^d)$ differ from $\mathbf{B}^s_{p,q}(\mathbb{R}^d)$ if $s < \sigma_p$, cf. (2.3.17), one sees that the moment conditions are indispensable in this context. For a more detailed discussion on this subject we refer to [Sch09b].

Furthermore, we remark that in [NNS16] the atomic decompositions for the spaces $A^s_{\mathcal{M}^\varphi_{p,q}}(\mathbb{R}^d)$ were based on atoms $a_{j,m}$ supported in cubes $\tilde{Q}_{j,m}$ with lower left corner $2^{-j}m$ and side length 2^{-j}, whereas the atoms in Definition 2.5.8 (as well as in Definition 2.3.15) are supported in cubes centered at $2^{-j}m$ with side length 2^{-j+1}. These changes are immaterial for the atomic decompositions. We prefer this notation in this thesis in order to have a unified approach for atomic decompositions in good agreement with Definition 2.3.15.

Quarkonial Decompositions The consideration of special atoms, so-called quarks, and subatomic or quarkonial decompositions goes back to [Tri97]. For the quarkonial decomposition for the spaces $A^s_{\mathcal{M}^\varphi_{p,q}}(\mathbb{R}^d)$ we follow [NNS16].

Throughout this section the function $\theta \in \mathcal{S}(\mathbb{R}^d)$ is fixed so that it has compact support and $\{\theta(\cdot - m)\}_{m \in \mathbb{Z}^d}$ forms a partition of unity:

$$\sum_{m \in \mathbb{Z}^d} \theta(x - m) = 1 \quad \text{for} \quad x \in \mathbb{R}^d \tag{2.5.7}$$

and, for some $R > 0$,

$$\operatorname{supp} \theta \subset 2^R Q_{0,0}. \tag{2.5.8}$$

Definition 2.5.11 Let $\beta \in \mathbb{N}^n_0$, $v \in \mathbb{N}_0$, and $m \in \mathbb{Z}^d$. Then the function θ^β and the quark $(\beta qu)_{v,m}$ are defined by

$$\theta^\beta(x) := x^\beta \theta(x) \quad \text{and} \quad (\beta qu)_{v,m}(x) := \theta^\beta(2^v x - m) \quad \text{for} \quad x \in \mathbb{R}^d.$$

The following coincides with [NNS16, Thm. 4.18].

Theorem 2.5.12 (Quarkonial Decomposition for $A^s_{\mathcal{M}^\varphi_{p,q}}(\mathbb{R}^d)$) *Let $0 < p < \infty$, $0 < q \leq \infty$, $s \in \mathbb{R}$, and $\varphi \in \mathcal{G}_p$. Assume in addition that φ satisfies (2.5.2) when $q < \infty$ and $\mathcal{A} = \mathcal{E}$. Suppose further that*

$$s > \begin{cases} \sigma_p, & \text{if } \mathcal{A} = \mathcal{N}, \\ \sigma_{p,q}, & \text{if } \mathcal{A} = \mathcal{E}, \end{cases}$$

and let ρ be such that $\rho > R$, where R is a constant as in (2.5.8).

(i) Let $f \in A^s_{\mathcal{M}^\varphi_{p,q}}(\mathbb{R}^d)$. Then there exists a triply indexed complex sequence

$$\lambda := \{\lambda^\beta_{v,m}\}_{\beta \in \mathbb{N}^d_0, v \in \mathbb{N}_0, m \in \mathbb{Z}^d}$$

such that

$$f = \sum_{\beta \in \mathbb{N}_0^d} \sum_{v=0}^{\infty} \sum_{m \in \mathbb{Z}^n} \lambda_{v,m}^{\beta} \, (\beta q u)_{v,m}, \tag{2.5.9}$$

convergence being in $\mathcal{S}'(\mathbb{R}^d)$, and

$$\|\lambda \mid \mathbf{a}_{\mathcal{M}_{p,q}^{\varphi}}^{s}(\mathbb{R}^d)\|_{\rho} := \sup_{\beta \in \mathbb{N}_0^d} 2^{\rho |\beta|} \, \|\lambda^{\beta} \mid \mathbf{a}_{\mathcal{M}_{p,q}^{\varphi}}^{s}(\mathbb{R}^d)\| \lesssim \|f \mid \mathcal{A}_{\mathcal{M}_{p,q}^{\varphi}}^{s}(\mathbb{R}^d)\|.$$

The numbers $\lambda_{v,m}^{\beta}$ depend continuously and linearly on f.

(ii) If $\lambda := \{\lambda_{v,m}^{\beta}\}_{\beta \in \mathbb{N}_0^d, v \in \mathbb{N}_0, m \in \mathbb{Z}^d}$ satisfies $\|\lambda \mid \mathbf{a}_{\mathcal{M}_{p,q}^{\varphi}}^{s}(\mathbb{R}^d)\|_{\rho} < \infty$, then

$$f = \sum_{\beta \in \mathbb{N}_0^d} \sum_{v=0}^{\infty} \sum_{m \in \mathbb{Z}^n} \lambda_{v,m}^{\beta} \, (\beta q u)_{v,m} \tag{2.5.10}$$

converges in $\mathcal{S}'(\mathbb{R}^d)$ and belongs to $\mathcal{A}_{\mathcal{M}_{p,q}^{\varphi}}^{s}(\mathbb{R}^d)$. Furthermore

$$\|f \mid \mathcal{A}_{\mathcal{M}_{p,q}^{\varphi}}^{s}(\mathbb{R}^d)\| \lesssim \|\lambda \mid \mathbf{a}_{\mathcal{M}_{p,q}^{\varphi}}^{s}(\mathbb{R}^d)\|_{\rho}.$$

Spaces of Type $\mathcal{A}_{\mathcal{M}_{p,q}^{\varphi}}^{s}(\Omega)$, $\mathcal{A}_{\mathcal{M}_{p,q}^{\varphi}}^{s}(\Gamma)$ on Domains and Boundaries We define generalized smoothness Morrey spaces on domains in the usual way by restriction. Recall that $\mathcal{D}'(\Omega)$ is the collection of all complex-valued distributions on Ω. If $g \in \mathcal{S}'(\mathbb{R}^d)$ then the restriction of g to Ω is an element of $\mathcal{D}'(\Omega)$, which will be denoted by $g|_{\Omega}$.

Definition 2.5.13 Let $0 < p < \infty$, $0 < q \leq \infty$, $s \in \mathbb{R}$, and $\varphi \in \mathcal{G}_p$. Additionally assume that φ satisfies (2.5.2) when $q < \infty$ and $\mathcal{A} = \mathcal{E}$. The space $\mathcal{A}_{\mathcal{M}_{p,q}^{\varphi}}^{s}(\Omega)$ is defined as the restriction of the corresponding space $\mathcal{A}_{\mathcal{M}_{p,q}^{\varphi}}^{s}(\mathbb{R}^d)$ to Ω, quasi-normed by

$$\|f | \mathcal{A}_{\mathcal{M}_{p,q}^{\varphi}}^{s}(\Omega)\| := \inf \|g | \mathcal{A}_{\mathcal{M}_{p,q}^{\varphi}}^{s}(\mathbb{R}^d)\|,$$

where the infimum is taken over all $g \in \mathcal{A}_{\mathcal{M}_{p,q}^{\varphi}}^{s}(\mathbb{R}^d)$ with $g|_{\Omega} = f$ in the sense of $\mathcal{D}'(\Omega)$.

We now define the generalized smoothness Morrey spaces on the boundary Γ of a C^k domain Ω. Then $\mathcal{D}'(\Gamma)$ stands for the distributions on the compact C^k manifold Γ. The explanations on page 24, in particular Remark 2.1.5 (since our setting now is

$\mathcal{S}'(\mathbb{R}^d)$ rather than $L_p(\mathbb{R}^d)$), lead to the following precise definition for smoothness Morrey spaces $\mathcal{A}^s_{\mathcal{M}^\varphi_{p,q}}(\Gamma)$.

Definition 2.5.14 Let $d \geq 2$, and let Ω be a bounded C^k domain in $\mathbb{R}^d$ with boundary Γ, and φ_j, $\psi^{(j)}$, W_j be as given on page 24. Assume $0 < p < \infty$, $0 < q \leq \infty$, $s \in \mathbb{R}$, and $\varphi \in \mathcal{G}_p$. Additionally assume that φ satisfies (2.5.2) when $q < \infty$ and $\mathcal{A} = \mathcal{E}$. Then we introduce

$$\mathcal{A}^s_{\mathcal{M}^\varphi_{p,q}}(\Gamma) := \{f \in \mathcal{D}'(\Gamma) : g_j \in \mathcal{A}^s_{\mathcal{M}^\varphi_{p,q}}(W_j), \ j = 1, \ldots, N\},$$

equipped with the quasi-norm

$$\|f|\mathcal{A}^s_{\mathcal{M}^\varphi_{p,q}}(\Gamma)\| := \sum_{j=1}^{N} \|g_j|\mathcal{A}^s_{\mathcal{M}^\varphi_{p,q}}(W_j)\|.$$

Remark 2.5.15 The spaces $\mathcal{A}^s_{\mathcal{M}^\varphi_{p,q}}(\Gamma)$ turn out to be independent of the particular choice of the covering $\{K_j\}_{j=1}^N$, the resolution of unity $\{\varphi_j\}_{j=1}^N$, and the local diffeomorphisms $\{\psi^{(j)}\}_{j=1}^N$ (the proof is similar to the proof of [Tri83, Prop. 3.2.3(ii)], making use of Theorem 2.5.19 and Proposition 2.5.20 below).

Note that we could furthermore replace W_j in the definition of the norm above by $\mathbb{R}^{d-1}$ if we extend g_j outside W_j with zero, i.e.,

$$\|f|\mathcal{A}^s_{\mathcal{M}^\varphi_{p,q}}(\Gamma)\| \sim \sum_{j=1}^{N} \|g_j|\mathcal{A}^s_{\mathcal{M}^\varphi_{p,q}}(\mathbb{R}^{d-1})\|.$$

2.5.2   Properties

We collect several properties for the generalized smoothness Morrey spaces that will be useful for our investigations later on. In particular, for $\mathcal{A}^s_{\mathcal{M}^\varphi_{p,q}}(\mathbb{R}^d)$ we need results concerning Fourier multipliers, embeddings into the spaces of uniformly continuous functions, as well as diffeomorphisms and pointwise multipliers. The results can be found in [NNS16].

Additionally, for the spaces defined on domains $\mathcal{A}^s_{\mathcal{M}^\varphi_{p,q}}(\Omega)$ we now derive an equivalent characterization via corresponding spaces on hyperplanes. This property will be crucial in Chap. 9 in order to reduce the trace problem on Ω to hyperplanes $\mathbb{R}^d_+$.

Fourier Multipliers The following result is a direct consequence of Theorem 2.19 from [NNS16].

Theorem 2.5.16 *Let $0 < p < \infty$, $0 < q \leq \infty$, $s \in \mathbb{R}$, $\varphi \in \mathcal{G}_p$, and assume*

$$\nu > \frac{d}{\min(1, p, q)} + \frac{d}{2}.$$

Suppose that for each $j \in \mathbb{N}$ we are given a compact set K_j of $\mathbb{R}^d$ with diameter d_j, $H_j \in \mathcal{S}(\mathbb{R}^d)$ and $f_j \in \mathcal{M}_p^\varphi(\mathbb{R}^d) \cap \mathcal{S}'(\mathbb{R}^d)$ with supp $\mathcal{F}f_j \subset K_j$.

(i) The inequality

$$\|2^{js}\mathcal{F}^{-1}(H_j\mathcal{F}f_j) \mid \mathcal{M}_p^\varphi(\mathbb{R}^d)\| \lesssim \left(\sup_{k \in \mathbb{N}} \|H_k(d_k\cdot) \mid H_2^\nu(\mathbb{R}^d)\|\right)$$

$$\times \|2^{js}f_j \mid \mathcal{M}_p^\varphi(\mathbb{R}^d)\|$$

holds for all $j \in \mathbb{N}$.

(ii) Assume (2.5.2) in addition when $q < \infty$. If the collection of measurable functions $\{f_j\}_{j=1}^\infty$ satisfies

$$\left\|\left(\sum_{j=1}^\infty 2^{jsq}|f_j|^q\right)^{1/q} \mid \mathcal{M}_p^\varphi(\mathbb{R}^d)\right\| < \infty,$$

then we have

$$\left\|\left(\sum_{j=1}^\infty 2^{jsq}|\mathcal{F}^{-1}(H_j\mathcal{F}f_j)|^q\right)^{1/q} \mid \mathcal{M}_p^\varphi(\mathbb{R}^d)\right\|$$

$$\lesssim \left(\sup_{k \in \mathbb{N}} \|H_k(d_k\cdot) \mid H_2^\nu(\mathbb{R}^d)\|\right) \left\|\left(\sum_{j=1}^\infty 2^{jsq}|f_j|^q\right)^{1/q} \mid \mathcal{M}_p^\varphi(\mathbb{R}^d)\right\|.$$

Embeddings into $C(\overline{\mathbb{R}^d})$ The next result was proved in [NNS16, Lem. 3.4] and will be used in Theorem 9.2.1 for the construction of a suitable extension operator.

Proposition 2.5.17 *Let $0 < p < \infty$, $0 < q \leq \infty$, $s \in \mathbb{R}$, and $\varphi \in \mathcal{G}_p$. Assume in addition that φ satisfies (2.5.2) when $q < \infty$ and $\mathcal{A} = \mathcal{E}$. If $s > 0$ is such that*

$$\sum_{j=1}^\infty \frac{1}{2^{sj}\varphi(2^{-j})} < \infty, \tag{2.5.11}$$

then

$$\mathcal{A}_{\mathcal{M}_{p,q}^\varphi}^s(\mathbb{R}^d) \hookrightarrow B_{\infty,1}^0(\mathbb{R}^d) \hookrightarrow C(\overline{\mathbb{R}^d}).$$

Remark 2.5.18 Since $\varphi \in \mathcal{G}_p$, it is clear that (2.5.11) is satisfied when $s > \frac{d}{p}$. Moreover, (2.5.11) is also necessary in order to have $\mathcal{N}_{\mathcal{M}^{\varphi}_{p,q},\infty}(\mathbb{R}^d) \hookrightarrow B^0_{\infty,1}(\mathbb{R}^d)$, cf. [NNS16, Rem. 3.5].

Diffeomorphisms and Pointwise Multipliers The following theorem about diffeomorphisms and pointwise multiplication can be found in [NNS16, Thm. 5.4, 5.5].

Theorem 2.5.19 *Let* $0 < p < \infty$, $0 < q \le \infty$, $s \in \mathbb{R}$, *and* $\varphi \in \mathcal{G}_p$. *Additionally assume that* φ *satisfies* (2.5.2) *when* $q < \infty$ *and* $\mathcal{A} = \mathcal{E}$. *Moreover, let*

$$k > s > \begin{cases} \sigma_p, & \text{if } \mathcal{A} = \mathcal{N}, \\ \sigma_{p,q}, & \text{if } \mathcal{A} = \mathcal{E}. \end{cases}$$

(i) *Let* $g \in C^k(\mathbb{R}^d)$. *Then* $f \to gf$ *is a linear and bounded operator from* $\mathcal{A}^s_{\mathcal{M}^{\varphi}_{p,q}}(\mathbb{R}^d)$ *into itself, i.e., there exists a positive constant* $C(k)$ *such that*

$$\|gf|\mathcal{A}^s_{\mathcal{M}^{\varphi}_{p,q}}(\mathbb{R}^d)\| \le C(k)\|g|C^k(\mathbb{R}^d)\| \cdot \|f|\mathcal{A}^s_{\mathcal{M}^{\varphi}_{p,q}}(\mathbb{R}^d)\|.$$

(ii) *Let* ψ *be a* k-*diffeomorphism. Then* $f \to f \circ \psi$ *is a linear and bounded operator from* $\mathcal{A}^s_{\mathcal{M}^{\varphi}_{p,q}}(\mathbb{R}^d)$ *into itself. In particular, we have for some positive constant* $C(\psi)$,

$$\|f \circ \psi|\mathcal{A}^s_{\mathcal{M}^{\varphi}_{p,q}}(\mathbb{R}^d)\| \le C(\psi)\|f|\mathcal{A}^s_{\mathcal{M}^{\varphi}_{p,q}}(\mathbb{R}^d)\|.$$

An Equivalent Norm For later purposes we provide an equivalent norm for $\mathcal{A}^s_{\mathcal{M}^{\varphi}_{p,q}}(\Omega)$ via corresponding spaces defined on hyperplanes $\mathcal{A}^s_{\mathcal{M}^{\varphi}_{p,q}}(\mathbb{R}^d_+)$.

Proposition 2.5.20 *Let* $0 < p < \infty$, $0 < q \le \infty$, $s \in \mathbb{R}$, *and* $\varphi \in \mathcal{G}_p$. *Additionally assume that* φ *satisfies* (2.5.2) *when* $q < \infty$ *and* $\mathcal{A} = \mathcal{E}$. *Furthermore, let* $\Omega \subset \mathbb{R}^d$ *be a bounded* C^k *domain with*

$$k > s > \begin{cases} \sigma_p, & \text{if } \mathcal{A} = \mathcal{N}, \\ \sigma_{p,q}, & \text{if } \mathcal{A} = \mathcal{E}. \end{cases}$$

Then

$$\|\varphi_0 f|\mathcal{A}^s_{\mathcal{M}^{\varphi}_{p,q}}(\mathbb{R}^d)\| + \sum_{j=1}^{N} \|(\varphi_j f)(\psi^{(j)-1}(\cdot))|\mathcal{A}^s_{\mathcal{M}^{\varphi}_{p,q}}(\mathbb{R}^d_+)\| \qquad (2.5.12)$$

is an equivalent quasi-norm in $\mathcal{A}^s_{\mathcal{M}^{\varphi}_{p,q}}(\Omega)$, *where we extended* $\varphi_0 f$ *by zero outside*
K_0 *and* $(\varphi_j f)(\psi^{(j)^{-1}}(\cdot))$ *by zero from* $\psi^{(j)}(K_j \cap \Omega)$ *to* $\mathbb{R}^d_+$ *for* $j = 1, \ldots, N$.

Proof The proof is the same as for Proposition 2.3.36 relying now on Theorem 2.5.19.
□

2.5.3 Spaces of Type $B^{s,\tau}_{p,q}$

When studying traces in Chap. 9 we also deal with the Besov-type spaces $B^{s,\tau}_{p,q}(\mathbb{R}^d)$ from [YSY10], since they are closely related with the generalized smoothness Morrey spaces from above.

However, they are not covered by our generalized approach with a function parameter, cf. Remark 2.5.4(iii), in particular, formula (2.5.4), which is why we introduce these spaces separately now in this subsection.

In particular, in order to study traces for these spaces in Chap. 9 we will need an equivalent characterization via quarkonial decompositions. Since this does not follow automatically from the results in Theorem 2.5.12, we give a detailed proof in Theorem 2.5.23, which is interesting on its own. Moreover, when studying traces later on we need some properties in terms of diffeomorphisms and pointwise multipliers stated in Theorem 2.5.25.

Let $s, \tau \in \mathbb{R}$ and $0 < p, q \leq \infty$. The inhomogeneous Besov-type space $B^{s,\tau}_{p,q}(\mathbb{R}^d)$ is defined to be the set of all $f \in \mathcal{S}'(\mathbb{R}^d)$ such that

$$\| f \mid B^{s,\tau}_{p,q}(\mathbb{R}^d)\|$$

$$:= \sup_{Q \in \mathcal{Q}} \frac{1}{|Q|^\tau} \left(\sum_{j=\max(j_Q,0)}^{\infty} \left[\int_Q \left(2^{js} |\mathcal{F}^{-1}[\mu_j(\xi)\mathcal{F}f(\xi)](x)| \right)^p dx \right]^{q/p} \right)^{1/q}$$

$$< \infty, \tag{2.5.13}$$

where the functions μ_j are as in Definition 2.5.3. In this case it follows from [YSY10, Cor. 2.1] that the definition of $B^{s,\tau}_{p,q}(\mathbb{R}^d)$ is independent of the choice of μ_j.

Corresponding spaces $B^{s,\tau}_{p,q}(\Omega)$ on domains $\Omega \subset \mathbb{R}^d$ are defined via restriction as in Definition 2.5.13, whereas on the boundary $\Gamma = \partial\Omega$ we use Definition 2.5.14 and obtain the spaces via localization and pull-back onto $\mathbb{R}^{d-1}$ with the help of suitable diffeomorphisms (recall Definition 2.1.1).

Quarkonial Decompositions We establish the quarkonial decomposition for spaces $B^{s,\tau}_{p,q}(\mathbb{R}^d)$. Recall the definition of quarks given in Definition 2.5.11, with θ satisfying (2.5.7) and (2.5.8).

We start by defining the corresponding sequence spaces and provide an auxiliary lemma.

Definition 2.5.21 Let $0 < p \leq \infty$, $0 < q \leq \infty$, $s \in \mathbb{R}$, and $\tau \geq 0$. The Besov-type sequence space $b_{p,q}^{s,\tau}(\mathbb{R}^d)$ is the set of all doubly indexed sequences $\lambda := \{\lambda_{j,m}\}_{j \in \mathbb{N}_0, m \in \mathbb{Z}^d} \subset \mathbb{C}$ for which the quasi-norm

$$\|\lambda \mid b_{p,q}^{s,\tau}(\mathbb{R}^d)\| := \sup_{P \in \mathcal{Q}} \frac{1}{|P|^\tau} \left(\sum_{j=\max(j_P,0)} 2^{j(s-\frac{d}{p})q} \left(\sum_{m \in \mathbb{Z}^d : Q_{j,m} \subset P} |\lambda_{j,m}|^p \right)^{q/p} \right)^{1/q}$$

is finite (with the usual modification if $p = \infty$ or $q = \infty$).

Lemma 2.5.22 Let $0 < p \leq \infty$, $0 < q \leq \infty$, $s \in \mathbb{R}$, and $\tau \geq 0$. There exists a positive constant c such that

$$\|\lambda^l \mid b_{p,q}^{s,\tau}(\mathbb{R}^d)\| \leq c \langle l \rangle^{d\tau} \|\lambda \mid b_{p,q}^{s,\tau}(\mathbb{R}^d)\|$$

for all $\lambda = \{\lambda_{j,m}\}_{j \in \mathbb{N}_0, m \in \mathbb{Z}^d}$ and all $l \in \mathbb{Z}^d$, where $\lambda^l := \{\lambda_{j,m+l}\}_{j \in \mathbb{N}_0, m \in \mathbb{Z}^d}$.

Proof Let $P \in \mathcal{Q}$, $P = Q_{v,k}$ centered at $2^{-v}k$ with side length 2^{-v+1} for some $v \in \mathbb{Z}$ and $k \in \mathbb{Z}^d$. Denoting by $x_{j,m} = (x_{j,m}^1, \cdots, x_{j,m}^d)$ a point of the cube $Q_{j,m}$, if the cube $Q_{j,m}$ is contained in P, then

$$j \geq v \quad \text{and} \quad |x_{j,m}^i - 2^{-v}k_i| \leq 2^{-v} \quad \text{for all} \quad i \in \{1, \ldots, d\}.$$

Then, for a point in the cube $Q_{j,m+l}$ centered at $2^{-j}(m+l)$ and $x_{j,m+l} \in Q_{j,m+l}$, we have

$$|x_{j,m+l}^i - 2^{-v}k_i| = |x_{j,m+l}^i - x_{j,m}^i + x_{j,m}^i - 2^{-v}k_i| \leq |x_{j,m}^i - 2^{-v}k_i| + 2^{-j}|l_i|$$

$$\leq 2^{-v} + 2^{-v}|l_i| \leq \langle l \rangle 2^{-v},$$

and hence $Q_{j,m+l} \subset \langle l \rangle P$.

Let $r \in \mathbb{N}$ be such that $2^r \leq \langle l \rangle < 2^{r+1}$ and put $P^* := 2^{r+1}P$. Then we have

$$\frac{1}{|P|^\tau} \left(\sum_{j=\max(j_P,0)} 2^{j(s-\frac{d}{p})q} \left(\sum_{m \in \mathbb{Z}^d : Q_{j,m} \subset P} |\lambda_{j,m+l}|^p \right)^{q/p} \right)^{1/q}$$

$$\leq \frac{|P^*|^\tau}{|P|^\tau} \frac{1}{|P^*|^\tau} \left(\sum_{j=\max(j_P,0)} 2^{j(s-\frac{d}{p})q} \left(\sum_{m \in \mathbb{Z}^d : Q_{j,m+l} \subset P^*} |\lambda_{j,m+l}|^p \right)^{q/p} \right)^{1/q}$$

$$= 2^{(r+1)d\tau} \frac{1}{|P^*|^\tau} \left(\sum_{j=\max(j_{P^*},0)} 2^{j(s-\frac{d}{p})q} \left(\sum_{m \in \mathbb{Z}^d : Q_{j,m} \subset P^*} |\lambda_{j,m}|^p \right)^{q/p} \right)^{1/q}$$

$$\leq 2^{d\tau} \langle l \rangle^{d\tau} \|\lambda \mid b_{p,q}^{s,\tau}(\mathbb{R}^d)\|.$$

By taking the supremum over all $P \in Q$ we arrive at the desired inequality. □

With this we now obtain the following quarkonial decomposition for the Besov-type spaces.

Theorem 2.5.23 (Quarkonial Decomposition for $B_{p,q}^{s,\tau}(\mathbb{R}^d)$) *Let* $0 < p < \infty$, $0 < q \leq \infty$, $0 \leq \tau \leq \frac{1}{p}$, *and* $s > \sigma_p$. *Let* ρ *be such that* $\rho > R$, *where* R *is a constant as in* (2.5.8).

(i) *If* $f \in B_{p,q}^{s,\tau}(\mathbb{R}^d)$ *then there exists a triply indexed complex sequence*

$$\lambda := \{\lambda_{\nu,m}^{\beta}\}_{\beta \in \mathbb{N}_0^d, \nu \in \mathbb{N}_0, m \in \mathbb{Z}^d}$$

such that

$$f = \sum_{\beta \in \mathbb{N}_0^d} \sum_{\nu=0}^{\infty} \sum_{m \in \mathbb{Z}^n} \lambda_{\nu,m}^{\beta} (\beta q u)_{\nu,m},$$

convergence being in $\mathcal{S}'(\mathbb{R}^d)$, *and*

$$\|\lambda \mid b_{p,q}^{s,\tau}(\mathbb{R}^d)\|_\rho := \sup_{\beta \in \mathbb{N}_0^d} 2^{\rho|\beta|} \|\lambda^{\beta} \mid b_{p,q}^{s,\tau}(\mathbb{R}^d)\| \lesssim \|f \mid B_{p,q}^{s,\tau}(\mathbb{R}^d)\|.$$

The numbers $\lambda_{\nu,m}^{\beta}$ *depend continuously and linearly on* f.

(ii) *If* $\lambda := \{\lambda_{\nu,m}^{\beta}\}_{\beta \in \mathbb{N}_0^d, \nu \in \mathbb{N}_0, m \in \mathbb{Z}^d}$ *satisfies* $\|\lambda \mid b_{p,q}^{s,\tau}(\mathbb{R}^d)\|_\rho < \infty$, *then*

$$f = \sum_{\beta \in \mathbb{N}_0^d} \sum_{\nu=0}^{\infty} \sum_{m \in \mathbb{Z}^n} \lambda_{\nu,m}^{\beta} (\beta q u)_{\nu,m} \tag{2.5.14}$$

converges in $\mathcal{S}'(\mathbb{R}^d)$ *and belongs to* $B_{p,q}^{s,\tau}(\mathbb{R}^d)$. *Furthermore,*

$$\|f \mid B_{p,q}^{s,\tau}(\mathbb{R}^d)\| \lesssim \|\lambda \mid b_{p,q}^{s,\tau}(\mathbb{R}^d)\|_\rho.$$

Proof We start by proving (i) and follow the proof presented in [Tri97, Sect. 14.15] in the context of classical Besov spaces. Let $(\phi_j)_{j \in \mathbb{N}_0}$ be a smooth dyadic partition of unity as in (2.3.1), i.e.,

$$\phi_0(x) = 1 \quad \text{if} \quad |x| \leq 1 \quad \text{and} \quad \text{supp}\, \phi_0 \subset \{x \in \mathbb{R}^d : |x| \leq 2\}.$$

For $\nu \in \mathbb{N}$ we put

$$\phi_\nu(x) := \phi_0(2^{-\nu}x) - \phi_0(2^{-\nu+1}x), \quad x \in \mathbb{R}^d.$$

Then

$$\sum_{\nu=0}^{\infty} \phi_\nu(x) = 1, \quad x \in \mathbb{R}^d,$$

and, for any $f \in \mathcal{S}'(\mathbb{R}^d)$, it follows that

$$f = \sum_{\nu=0}^{\infty} \mathcal{F}^{-1}(\phi_\nu \mathcal{F} f) \quad \text{with convergence in} \quad \mathcal{S}'(\mathbb{R}^d).$$

Let $\kappa \in \mathcal{S}(\mathbb{R}^d)$ be such that $\kappa(x) = 1$ if $|x| \le 2$ and $\operatorname{supp} \kappa \subset Q(\pi)$. For $(\nu, k) \in \mathbb{N}_0 \times \mathbb{Z}^d$, let $\Lambda_{\nu,k} := c[\mathcal{F}^{-1}(\phi_\nu \mathcal{F} f)](2^{-\nu} k)$. Then we have, for any $x \in \mathbb{R}^d$,

$$[\mathcal{F}^{-1}(\phi_\nu \mathcal{F} f)](x) = \sum_{k \in \mathbb{Z}^d} \Lambda_{\nu,k} (\mathcal{F}^{-1} \kappa)(2^\nu x - k)$$

$$= \sum_{k \in \mathbb{Z}^d} \Lambda_{\nu,k} \sum_{m \in \mathbb{Z}^d} (\mathcal{F}^{-1} \kappa)(2^\nu x - k)\, \theta(2^{\nu+\rho} x - m),$$

where the last equality is due to $\sum_{m \in \mathbb{Z}^d} \theta(x - m) = 1$ for all $x \in \mathbb{R}^d$. Expanding $(\mathcal{F}^{-1} \kappa)(2^\nu \cdot -k)$ in a Taylor series at the point $2^{-(\nu+\rho)} m$, we obtain

$$(\mathcal{F}^{-1} \kappa)(2^\nu x - k) = \sum_{\beta \in \mathbb{N}_0^d} \frac{2^{\nu|\beta|}}{\beta!} [\partial^\beta (\mathcal{F}^{-1} \kappa)](2^{-\rho} m - k)\, (x - 2^{-(\nu+\rho)} m)^\beta$$

thus,

$$[\mathcal{F}^{-1}(\phi_\nu \mathcal{F} f)](x)$$

$$= \sum_{k \in \mathbb{Z}^d} \Lambda_{\nu,k} \sum_{m \in \mathbb{Z}^d} \sum_{\beta \in \mathbb{N}_0^d} \frac{2^{-\rho|\beta|}}{\beta!} [\partial^\beta (\mathcal{F}^{-1} \kappa)](2^{-\rho} m - k)\, \theta^\beta(2^{\nu+\rho} x - m),$$

and hence

$$f = \sum_{\nu=0}^{\infty} \sum_{m \in \mathbb{Z}^d} \sum_{\beta \in \mathbb{N}_0^d} \theta^\beta(2^{\nu+\rho} x - m) \sum_{k \in \mathbb{Z}^d} \Lambda_{\nu,k} \frac{2^{-\rho|\beta|}}{\beta!} [\partial^\beta (\mathcal{F}^{-1} \kappa)](2^{-\rho} m - k)$$

$$= \sum_{\nu=0}^{\infty} \sum_{m \in \mathbb{Z}^d} \sum_{\beta \in \mathbb{N}_0^d} \lambda_{\nu+\rho,m}^\beta (\beta q u)_{\nu+\rho,m}$$

with

$$\lambda^{\beta}_{\nu+\rho,m} := \frac{2^{-\rho|\beta|}}{\beta!} \sum_{k\in\mathbb{Z}^d} [\partial^{\beta}(\mathcal{F}^{-1}\kappa)](2^{-\rho}m - k)\,\Lambda_{\nu,k}.$$

As a consequence of the Paley-Wiener-Schwartz theorem and iterative application of Cauchy's representation formula one can prove that

$$|\partial^{\beta}(\mathcal{F}^{-1}\kappa)(x)| \le c(\eta)\,\beta!\,\langle x\rangle^{-\eta} \qquad \text{for any } \eta > 0,$$

where $c(\eta)$ is a positive constant independent of $x \in \mathbb{R}^d$ and of the multi-index $\beta \in \mathbb{N}_0^d$. Then, for $l \in \mathbb{Z}^d$ and l_0 a lattice point in $[0, 2^{\rho})^d$, we obtain

$$|\lambda^{\beta}_{\nu+\rho,2^{\rho}l+l_0}| \lesssim 2^{-\rho|\beta|} \sum_{k\in\mathbb{Z}^d} \langle l - k\rangle^{-\eta}\,|\Lambda_{\nu,k}| = 2^{-\rho|\beta|} \sum_{k\in\mathbb{Z}^d} \langle k\rangle^{-\eta}|\Lambda_{\nu,l+k}|.$$

For each $k \in \mathbb{Z}^d$ let $\Lambda^k := \{|\Lambda_{\nu,l+k}|\}_{\nu\in\mathbb{N}_0, l\in\mathbb{Z}^d}$. By $\mathbb{R}^d = 2^{\rho}\mathbb{Z}^d + [0, 2^{\rho})^d$ we see that with $r := \min(1, p, q)$ and by Lemma 2.5.22, we have

$$\|\lambda^{\beta} \mid b^{s,\tau}_{p,q}(\mathbb{R}^d)\| \lesssim 2^{-\rho|\beta|}\Big\| \sum_{k\in\mathbb{Z}^d} \langle k\rangle^{-\eta}\Lambda^k \mid b^{s,\tau}_{p,q}(\mathbb{R}^d)\Big\|$$

$$\lesssim 2^{-\rho|\beta|}\Big(\sum_{k\in\mathbb{Z}^d} \langle k\rangle^{-\eta r}\|\Lambda^k \mid b^{s,\tau}_{p,q}(\mathbb{R}^d)\|^r \Big)^{1/r}$$

$$\lesssim 2^{-\rho|\beta|}\Big(\sum_{k\in\mathbb{Z}^d} \langle k\rangle^{(d\tau-\eta)r}\|\Lambda \mid b^{s,\tau}_{p,q}(\mathbb{R}^d)\|^r \Big)^{1/r}.$$

Hence, choosing η large enough such that $(d\tau - \eta)r < -1$, it follows that

$$\|\lambda^{\beta} \mid b^{s,\tau}_{p,q}(\mathbb{R}^d)\|_{\rho} \lesssim \|\Lambda \mid b^{s,\tau}_{p,q}(\mathbb{R}^d)\|.$$

Finally we have to prove that $\|\Lambda \mid b^{s,\tau}_{p,q}(\mathbb{R}^d)\| \lesssim \|f \mid B^{s,\tau}_{p,q}(\mathbb{R}^d)\|$. Note that, for $\nu \in \mathbb{N}_0$ and $k \in \mathbb{Z}^d$ and any $y \in Q_{\nu,k}$, we have

$$|\Lambda_{\nu,k}| = |[\mathcal{F}^{-1}(\phi_{\nu}\mathcal{F}f)](2^{-\nu}k)| = (1 + 2^{\nu}|y - 2^{-\nu}k|)^a \frac{|[\mathcal{F}^{-1}(\phi_{\nu}\mathcal{F}f)](2^{-\nu}k)|}{(1 + 2^{\nu}|y - 2^{-\nu}k|)^a}$$

$$\lesssim (\phi_{\nu}^* f)_a(y),$$

where $a > \frac{d}{p}$ and $(\phi_{\nu}^* f)_a$ are the Peetre's maximal functions defined by

$$(\phi_{\nu}^* f)_a(x) := \sup_{y\in\mathbb{R}^d} \frac{|[\mathcal{F}^{-1}(\phi_{\nu}\mathcal{F}f)](y)|}{(1 + 2^{\nu}|x - y|)^a}, \qquad x \in \mathbb{R}^d.$$

Then

$$\|\Lambda \mid b_{p,q}^{s,\tau}(\mathbb{R}^d)\|$$

$$= \sup_{P \in \mathcal{Q}} \frac{1}{|P|^\tau} \left(\sum_{j=\max(j_P,0)}^{} 2^{jsq} \left(\sum_{m \in \mathbb{Z}^d : Q_{j,m} \subset P} \int_{Q_{j,m}} |\Lambda_{j,m}|^p \, dy \right)^{q/p} \right)^{1/q}$$

$$\leq \sup_{P \in \mathcal{Q}} \frac{1}{|P|^\tau} \left(\sum_{j=\max(j_P,0)}^{} 2^{jsq} \left(\sum_{m \in \mathbb{Z}^d : Q_{j,m} \subset P} \int_{Q_{j,m}} (\phi_j^* f)_a(y)^p \, dy \right)^{q/p} \right)^{1/q}$$

$$\leq \sup_{P \in \mathcal{Q}} \frac{1}{|P|^\tau} \left(\sum_{j=\max(j_P,0)}^{} 2^{jsq} \left(\int_P (\phi_j^* f)_a(y)^p \, dy \right)^{q/p} \right)^{1/q}$$

$$\lesssim \|f \mid B_{p,q}^{s,\tau}(\mathbb{R}^d)\|.$$

The last step is justified by the equivalent characterization of $B_{p,q}^{s,\tau}(\mathbb{R}^d)$, cf. [LSUYY12, Thm. 3.6] (in the homogeneous case) and [YSY10, Lem. 4.1].

Now we prove (ii). We decompose the representation (2.5.14) as

$$f = \sum_{\beta \in \mathbb{N}_0^d} f^\beta$$

with

$$f^\beta := \sum_{\nu=0}^{\infty} \sum_{m \in \mathbb{Z}^n} \lambda_{\nu,m}^\beta \, (\beta q u)_{\nu,m}.$$

Note that for all $\nu \in \mathbb{N}_0$, $m \in \mathbb{Z}^d$, and $\beta \in \mathbb{N}_0^d$ it holds

$$\operatorname{supp} (\beta q u)_{\nu,m} \subset 2^R Q_{\nu,m}$$

and

$$|\partial^\alpha (\beta q u)_{\nu,m}(x)| \lesssim 2^{|\alpha|\nu + (R+\varepsilon)|\beta|}, \quad x \in \mathbb{R}^d,$$

for any $\varepsilon > 0$. Applying [YSY10, Thm. 3.3] we can conclude that $f^\beta \in B_{p,q}^{s,\tau}(\mathbb{R}^d)$ and

$$\|f^\beta \mid B_{p,q}^{s,\tau}(\mathbb{R}^d)\| \leq c_1 \, 2^{(R+\varepsilon)|\beta|} \|\lambda^\beta \mid b_{p,q}^{s,\tau}(\mathbb{R}^d)\|,$$

where $c_1 > 0$ is independent of β. So, with $0 < \varepsilon < \rho - R$,

$$\|f^\beta \mid B_{p,q}^{s,\tau}(\mathbb{R}^d)\| \leq c_1 \, 2^{(R+\varepsilon-\rho)|\beta|} \sup_{\beta \in \mathbb{N}_0^d} 2^{\rho|\beta|} \|\lambda^\beta \mid b_{p,q}^{s,\tau}(\mathbb{R}^d)\|,$$

and applying the r-triangle inequality, where $r := \min(1, p, q)$, we get

$$\Big\| \sum_{\beta \in \mathbb{N}_0^d} f^\beta \mid B_{p,q}^{s,\tau}(\mathbb{R}^d)\Big\| \leq \Big(\sum_{\beta \in \mathbb{N}_0^d} \|f^\beta \mid B_{p,q}^{s,\tau}(\mathbb{R}^d)\|^r \Big)^{1/r}$$

$$\leq c_1 \Big(\sum_{\beta \in \mathbb{N}_0^d} 2^{(R+\varepsilon-\rho)|\beta|r} \Big)^{1/r} \sup_{\beta \in \mathbb{N}_0^d} 2^{\rho|\beta|} \|\lambda^\beta \mid b_{p,q}^{s,\tau}(\mathbb{R}^d)\|$$

$$\leq c_2 \|\lambda \mid b_{p,q}^{s,\tau}(\mathbb{R}^d)\|_\rho.$$

$\square$

Remark 2.5.24 The restriction $0 \leq \tau \leq \frac{1}{p}$ in Theorem 2.5.23 can be replaced by $0 \leq \tau < \tau_{s,p}$ with $\tau_{s,p}$ defined as in [YSY10, formula (1.6)], which follows from the atomic decomposition theorem for the Besov-type spaces, cf. [YSY10, Thm. 3.3], we rely on in the proof. For simplicity we restrict ourselves to $\tau \leq \frac{1}{p}$ here, since in [YY13] the remarkable result was proven that

$$B_{p,q}^{s,\tau}(\mathbb{R}^d) = B_{\infty,\infty}^{s+(\tau-\frac{1}{p})}(\mathbb{R}^d)$$

whenever $\tau > \frac{1}{p}$ or $\tau = \frac{1}{p}$ and $q = \infty$. Hence, concerning traces, only $\tau \in [0, 1/p]$ is of interest.

Properties In terms of diffeomorphisms and pointwise multipliers the following result was proved in [YSY10, Thm. 6.1].

Theorem 2.5.25 *Let $s \in \mathbb{R}$, $0 < p, q \leq \infty$, and $0 \leq \tau \leq \frac{1}{p}$. Moreover, assume that $k \in \mathbb{N}$ is sufficiently large.*

(i) *Let $g \in C^k(\mathbb{R}^d)$. Then $f \to gf$ is a linear and bounded operator from $B_{p,q}^{s,\tau}(\mathbb{R}^d)$ into itself, i.e., there exists a positive constant $C(k)$ such that*

$$\|gf \mid B_{p,q}^{s,\tau}(\mathbb{R}^d)\| \leq C(k) \|g \mid C^k(\mathbb{R}^d)\| \cdot \|f \mid B_{p,q}^{s,\tau}(\mathbb{R}^d)\|.$$

(ii) *Let ψ be a k-diffeomorphism. Then $f \to f \circ \psi$ is a linear and bounded operator from $B_{p,q}^{s,\tau}(\mathbb{R}^d)$ into itself. In particular, we have for some positive constant $C(\psi)$,*

$$\|f \circ \psi \mid B_{p,q}^{s,\tau}(\mathbb{R}^d)\| \leq C(\psi) \|f \mid B_{p,q}^{s,\tau}(\mathbb{R}^d)\|.$$

2.6 Function Spaces on Manifolds

In this section we introduce function spaces on manifolds with particular focus on Sobolev, Besov, and Triebel-Lizorkin spaces. On order to come from the euclidean setting to manifolds we will rely on local charts and diffeomorphisms which enable us to define spaces on manifolds M with bounded geometry via corresponding spaces defined on $\mathbb{R}^d$. Usually, one uses geodesic coordinates and the exponential map in this context as was e.g. done in [Tri92]. However, we want to be more flexible later on when studying traces on submanifolds N by using Fermi coordinates adapted to N.

Therefore, in the beginning of this section we recall in detail the concepts of bounded geometry, atlas of the manifold, geodesic coordinates and define our spaces in a more general way by relaxing the assumptions on the coordinates. In particular, the triple consisting of the cover of M, local coordinates and a partition of unity subordinate to the cover is called *trivialization*. We study under which restrictions on the trivializations we obtain spaces which are equivalent to those defined via geodesic coordinates. We speak about *admissible trivializations* in this context. We start our considerations for fractional Sobolev spaces, but the results are generalized afterwards to Besov- and Triebel-Lizorkin spaces as well.

Let (M^d, g) be an d-dimensional complete manifold with Riemannian metric g (we often simply write M if the dimension of the manifold is clear). We denote the volume element on M with respect to the metric g by dvol_g. For $1 < p < \infty$ the L_p-norm of a compactly supported smooth function $v \in \mathcal{D}(M)$ is given by

$$\|v|L_p(M)\| = \left(\int_M |v|^p \mathrm{dvol}_g \right)^{\frac{1}{p}}.$$

The set $L_p(M)$ is then the completion of $\mathcal{D}(M)$ with respect to the L_p-norm. The space of distributions on M is denoted by $\mathcal{D}'(M)$.

A *cover* $(U_\alpha)_{\alpha \in I}$ of M is a collection of open subsets $U_\alpha \subset M$ where α runs over an index set I. The cover is called *locally finite* if each U_α is intersected by at most finitely many U_β. The cover is called *uniformly locally finite* if there exists a constant $L > 0$ such that each U_α is intersected by at most L sets U_β.

A *chart* on U_α is given by local coordinates—a diffeomorphism $\kappa_\alpha : x = (x^1, \ldots, x^d) \in V_\alpha \subset \mathbb{R}^d \to \kappa_\alpha(x) \in U_\alpha$. We will always assume our charts (the diffeomorphisms κ_α) to be smooth. A collection $\mathcal{A} = (U_\alpha, \kappa_\alpha)_{\alpha \in I}$ is called an *atlas* of M.

Moreover, a collection of smooth functions $(h_\alpha)_{\alpha \in I}$ on M with

$$\mathrm{supp}\ h_\alpha \subset U_\alpha, \qquad 0 \le h_\alpha \le 1 \qquad \text{and} \qquad \sum_\alpha h_\alpha = 1 \quad \text{on} \quad M$$

is called a *partition of unity subordinated to the cover* $(U_\alpha)_{\alpha \in I}$. The triple $\mathcal{T} := (U_\alpha, \kappa_\alpha, h_\alpha)_{\alpha \in I}$ is called a *trivialization* of the manifold M (Fig. 2.24).

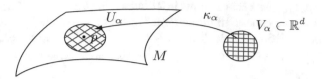

Fig. 2.24 Local charts κ_α

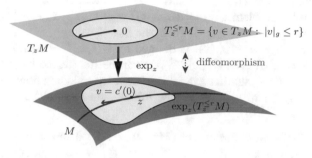

Fig. 2.25 Exponential map $\exp_z$

Using the standard Euclidean coordinates $x = (x^1, \ldots, x^d)$ on $V_\alpha \subset \mathbb{R}^d$, we introduce an orthonormal frame $(e_i^\alpha)_{1 \le i \le d}$ on TU_α by $e_i^\alpha := (\kappa_\alpha)_*(\partial_i)$. In case we talk about a fixed chart we will often leave out the superscript α. Then, in those local coordinates the metric g is expressed via the matrix coefficients $g_{ij} (= g_{ij}^\alpha)$: $V_\alpha \to \mathbb{R}$ defined by $g_{ij} \circ \kappa_\alpha^{-1} = g(e_i, e_j)$ and the corresponding Christoffel symbols $\Gamma_{ij}^k (= {}^\alpha \Gamma_{ij}^k) : V_\alpha \to \mathbb{R}$ are defined by $\nabla_{e_i}^M e_j = (\Gamma_{ij}^k \circ \kappa_\alpha^{-1}) e_k$, where ∇^M denotes the Levi-Civita connection of (M, g). In local coordinates,

$$\Gamma_{ij}^k = \frac{1}{2} g^{kl} (\partial_j g_{il} + \partial_i g_{jl} - \partial_l g_{ij}) \tag{2.6.1}$$

where g^{ij} is the inverse matrix of g_{ij}. If $\alpha, \beta \in I$ with $U_\alpha \cap U_\beta \ne \varnothing$, we define the transition function $\mu_{\alpha\beta} = \kappa_\beta^{-1} \circ \kappa_\alpha : \kappa_\alpha^{-1}(U_\alpha \cap U_\beta) \to \kappa_\beta^{-1}(U_\alpha \cap U_\beta)$. Then,

$$g_{ij}^\alpha(x) = \partial_i \mu_{\alpha\beta}^k(x) \partial_j \mu_{\alpha\beta}^l(x) g_{kl}^\beta(\mu_{\alpha\beta}(x)). \tag{2.6.2}$$

Example 2.6.1 (Geodesic Normal Coordinates) Let (M^d, g) be a complete Riemannian manifold. Fix $z \in M$ and let $r > 0$. For $v \in T_z^{\le r} M := \{w \in T_z M \mid g_z(w, w) = |w|_g^2 \le r^2\}$, we denote by $c_v : [-1, 1] \to M$ the unique geodesic with $c_v(0) = z$ and $c_v'(0) = v$. Then, the exponential map $\exp_z^M : T_z^{\le r} M \to M$ is defined by $\exp_z^M(v) := c_v(1)$ (Fig. 2.25). The largest radius r for which the exponential map at z is a diffeomorphism is called injectivity radius at z. The injectivity radius of M is then the infimum over all injectivity radii at all points. We will always choose $r > 0$ smaller than the injectivity radius of M.

Let $S = \{p_\alpha\}_{\alpha \in I}$ be a set of points in M such that $(U_\alpha^{\text{geo}} := B_r(p_\alpha))_{\alpha \in I}$ covers M. For each p_α we choose an orthonormal frame of $T_{p_\alpha} M$ and call the resulting identification $\lambda_\alpha : \mathbb{R}^d \to T_{p_\alpha} M$. Then, $\mathcal{A}^{\text{geo}} = (U_\alpha^{\text{geo}}, \kappa_\alpha^{\text{geo}} = \exp_{p_\alpha}^M \circ \lambda_\alpha : V_\alpha^{\text{geo}} := B_r^d \to U_\alpha^{\text{geo}})_{\alpha \in I}$ is an atlas of M—called geodesic atlas. Note that λ_α^{-1} equals the tangent map $(d\kappa_\alpha^{\text{geo}})^{-1}$ at p_α.

Flows Let $x'(t) = F(t, x(t))$ be a system of *ordinary differential equations* (=*ODEs*) with $t \in \mathbb{R}$, $x(t) \in \mathbb{R}^d$ and $F \in C^\infty(\mathbb{R} \times \mathbb{R}^d, \mathbb{R}^d)$. Let the solution of the initial value problem $x'(t) = F(t, x(t))$ with $x(0) = x_0 \in \mathbb{R}^d$ be denoted by $x_{x_0}(t)$ and exist for $0 \le t \le t_0(x_0)$. Then, the flow $\Phi : \text{dom} \subset \mathbb{R} \times \mathbb{R}^d \to \mathbb{R}^d$ with $\text{dom} \subset \{(t, x) \mid 0 \le t \le t_0(x)\}$ is defined by $\Phi(t, x_0) = x_{x_0}(t)$. Higher order ODEs $x^{(n)}(t) = F(t, x(t), \ldots, x^{(n-1)}(t))$ can be transferred back to first order systems by introducing auxiliary variables. The corresponding flow then obviously depends not only on $x_0 = x(0)$ but the initial values $x(0), x'(0), \ldots, x^{(n-1)}(0)$: $\Phi(t, x(0), \ldots, x^{(n-1)}(0))$.

Example 2.6.2 (Geodesic Flow) Let (M^d, g) be a Riemannian manifold. Let $z \in M$, $v \in T_z M$. Let $\kappa : V \subset \mathbb{R}^d \to U \subset M$ be a chart around z. The corresponding coordinates on V are denoted by $x = (x^1, \ldots, x^d)$. We consider the geodesic equation in coordinates: $(x^k)'' = -\Gamma_{ij}^k (x^i)'(x^j)'$ with initial values $x(0) = \kappa^{-1}(z) \in \mathbb{R}^d$ and $x'(0) = \kappa^*(v)(= d\kappa^{-1}(v))$. Here Γ_{ij}^k are the Christoffel symbols with respect to the coordinates given by κ. Let $x(t)$ be the unique solution and $\Phi(t, x(0), x'(0))$ denote the corresponding flow. Then, $c_v(t) = \kappa(x(t))$ is the geodesic described in Example 2.6.1 and $\exp_z^M(v) = \kappa \circ \Phi(1, \kappa^{-1}(z), \kappa^*(v))$.

The following lemma may be found in [Sch01, Lem. 3.4, Cor. 3.5] and provides estimates for the derivatives of a flow.

Lemma 2.6.3 *Let $x'(t) = F(t, x(t))$ be a system of ODEs as described above. Suppose that $\Phi(t, x)$ is the flow of this equation. Then there is a universal expression* $\text{Expr}_{\mathfrak{a}}$ *only depending on the multi-index $\mathfrak{a}$ such that*

$$|D_x^{\mathfrak{a}} \Phi(t, x_0)| \le \text{Expr}_{\mathfrak{a}} \left(\sup_{0 \le \tau \le t} \left\{ \left| D_x^{\mathfrak{a}'} F(\tau, \Phi(\tau, x_0)) \right| \right\} \, \middle| \, \mathfrak{a}' \le \mathfrak{a}, \, t \right)$$

for all $t \ge 0$ where $\Phi(t, x_0)$ is defined. Moreover, a corresponding statement also holds for ODEs of order n.

2.6.1 Sobolev Spaces on Manifolds

From now on let M always be an d-dimensional manifold with Riemannian metric g. We want to define Sobolev spaces on manifolds. For this we require the manifold

M to be of bounded geometry. The definition is given below and follows [Shu92, Def. A.1.1].

Definition 2.6.4 (Manifold with Bounded Geometry) A Riemannian manifold (M^d, g) is of bounded geometry if the following two conditions are satisfied:

 (i) The injectivity radius r_M of (M, g) is positive.
(ii) Every covariant derivative of the Riemann curvature tensor R^M of M is bounded, i.e., for all $k \in \mathbb{N}_0$ there is a constant $C_k > 0$ such that

$$|(\nabla^M)^k R^M|_g \leq C_k.$$

Remark 2.6.5

 (i) Note that Definition 2.6.4(i) implies that M is complete, cf. [Eic07, Prop. 1.2a]. The figure below illustrates an example of a manifold which has injectivity radius $r_M = 0$ and is therefore excluded by the above definition (Fig. 2.26).
(ii) Property (ii) of Definition 2.6.4 can be replaced by the following equivalent property which will be more convenient later on, cf. [Shu92, Def. A.1.1 and below]: Consider a geodesic atlas $\mathcal{A}^{\text{geo}} = (U_\alpha^{\text{geo}}, \kappa_\alpha^{\text{geo}})_{\alpha \in I}$ as in Example 2.6.1.
    For all $k \in \mathbb{N}$ there are constants C_k such that for all $\alpha, \beta \in I$ with $U_\alpha^{\text{geo}} \cap U_\beta^{\text{geo}} \neq \varnothing$ we have for the corresponding *transition functions* $\mu_{\alpha\beta} := (\kappa_\beta^{\text{geo}})^{-1} \circ \kappa_\alpha^{\text{geo}}$ that

$$|D^a \mu_{\alpha\beta}| \leq C_k,$$

for all $a \in \mathbb{N}_0^d$ with $|a| \leq k$ and all charts (Fig. 2.27).

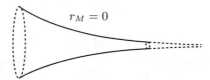

Fig. 2.26 Manifold with injectivity radius 0

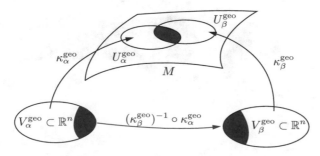

Fig. 2.27 Transition functions

(iii) Consider a geodesic atlas $\mathcal{A}^{\text{geo}}$ as above. Let g_{ij} denote the metric in these coordinates and g^{ij} its inverse. Then, property (ii) of Definition 2.6.4 can be replaced by the following equivalent property, cf. [Eic91, Thm. A and below]: For all $k \in \mathbb{N}_0$ there is a constant C_k such that

$$|D^{\mathfrak{a}} g_{ij}| \le C_k, \quad |D^{\mathfrak{a}} g^{ij}| \le C_k, \qquad \text{for all} \quad \mathfrak{a} \in \mathbb{N}_0^d \quad \text{with} \quad |\mathfrak{a}| \le k.$$
$$(2.6.3)$$

Example 2.6.6 (Geodesic Trivialization) Let (M, g) be of bounded geometry (this includes the case of closed manifolds). Then, there exists a geodesic atlas, see Example 2.6.1, that is uniformly locally finite:

Let S be a maximal set of points $\{p_\alpha\}_{\alpha \in I} \subset M$ such that the metric balls $B_{\frac{r}{2}}(p_\alpha)$ are pairwise disjoint. Then, the balls $\{B_r(p_\alpha)\}_{\alpha \in I}$ cover M, and we obtain a (uniformly locally finite) geodesic atlas $\mathcal{A}^{\text{geo}} = (U_\alpha^{\text{geo}} := B_r(p_\alpha), \kappa_\alpha^{\text{geo}})_{\alpha \in I})$. For an argument concerning the uniform local finiteness of the cover we refer to Remark 10.1.6(ii).

Moreover, there is a partition of unity h_α^{geo} subordinated to $(U_\alpha^{\text{geo}})_{\alpha \in I}$ such that for all $k \in \mathbb{N}_0$ there is a constant $C_k > 0$ such that $|D^{\mathfrak{a}}(h_\alpha^{\text{geo}} \circ \kappa_\alpha^{\text{geo}})| \le C_k$ for all multi-indices $\mathfrak{a}$ with $|\mathfrak{a}| \le k$, cf. [Tri92, Prop. 7.2.1] and the references therein (Fig. 2.28). The resulting trivialization is denoted by $\mathcal{T}^{\text{geo}} = (U_\alpha^{\text{geo}}, \kappa_\alpha^{\text{geo}}, h_\alpha^{\text{geo}})_{\alpha \in I}$ and referred to as *geodesic trivialization*.

Sobolev Spaces Using the Geodesic Trivialization Following [Tri92, Ch. 7], we give a definition for Sobolev spaces on the manifold M of bounded geometry using local descriptions in terms of geodesic normal coordinates and norms of corresponding spaces defined on $\mathbb{R}^d$.

Definition 2.6.7 Let (M^d, g) be a Riemannian manifold of bounded geometry with geodesic trivialization $\mathcal{T}^{\text{geo}} = (U_\alpha^{\text{geo}}, \kappa_\alpha^{\text{geo}}, h_\alpha^{\text{geo}})_{\alpha \in I}$ as above. Furthermore, let $s \in$

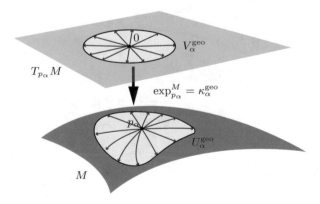

Fig. 2.28 Geodesic atlas

$\mathbb{R}$ and $1 < p < \infty$. Then the space $H_p^s(M)$ contains all distributions $f \in \mathcal{D}'(M)$ such that

$$\left(\sum_{\alpha \in I} \| (h_\alpha^{\mathrm{geo}} f) \circ \kappa_\alpha^{\mathrm{geo}} | H_p^s(\mathbb{R}^d) \|^p \right)^{\frac{1}{p}} \tag{2.6.4}$$

is finite. Note that although $\kappa_\alpha^{\mathrm{geo}}$ is only defined on $V_\alpha^{\mathrm{geo}} \subset \mathbb{R}^d$, $(h_\alpha^{\mathrm{geo}} f) \circ \kappa_\alpha^{\mathrm{geo}}$ is viewed as a function on $\mathbb{R}^d$ extended by zero, since $\mathrm{supp}\,(h_\alpha^{\mathrm{geo}} f) \subset U_\alpha^{\mathrm{geo}}$.

Remark 2.6.8

(i) The spaces $H_p^s(M)$ generalize in a natural way the classical Sobolev spaces $W_p^k(M), k \in \mathbb{N}_0, 1 < p < \infty$, on Riemannian manifolds M: Let

$$\| f | W_p^k(M) \| := \sum_{l=0}^{k} \| \nabla^l f | L_p(M) \|,$$

then $W_p^k(M)$ is the completion of $\mathcal{D}(M)$ in the $W_p^k(M)$-norm, cf. [Aub76], [Aub82]. As in the Euclidean case, on manifolds M of bounded geometry one has the coincidence

$$W_p^k(M) = H_p^k(M), \qquad k \in \mathbb{N}_0, \quad 1 < p < \infty, \tag{2.6.5}$$

cf. [Tri92, Sect. 7.4.5].

(ii) Alternatively, the fractional Sobolev spaces $H_p^s(M)$ on manifolds with bounded geometry can be characterized with the help of the Laplace-Beltrami operator, cf. [Tri92, Sect. 7.2.2, Thm. 7.4.5]. This approach was originally used by [Str83] and later on slightly modified in [Tri92, Sect. 7.4.5] in the following way: Let $1 < p < \infty$ and $\rho > 0$. Let $s > 0$, then $H_p^s(M)$ is the collection of all $f \in L_p(M)$ such that $f = (\rho\,\mathrm{Id} - \Delta)^{-s/2} h$ for some $h \in L_p(M)$, with the norm $\| f | H_p^s(M) \| = \| h | L_p(M) \|$. Let $s < 0$, then $H_p^s(M)$ is the collection of all $f \in \mathcal{D}'(M)$ having the form $f = (\rho\,\mathrm{Id} - \Delta)^l h$ with $h \in H_p^{2l+s}(M)$, where $l \in \mathbb{N}$ such that $2l + s > 0$, and $\| f | H_p^s(M) \| = \| h | H_p^{2l+s}(M) \|$. Let $s = 0$, then $H_p^0(M) = L_p(M)$.

In particular, the spaces $H_p^s(M)$ with $s < 0$ are independent of the number l appearing in their definition in the sense of equivalent norms, cf. [Str83, Def. 4.1]. The additional parameter $\rho > 0$ used by Triebel ensures that (2.6.5) holds in this context as well. In particular, for $2 \leq p < \infty$ one can choose $\rho = 1$, cf. [Tri92, Rem. 1.4.5/1, p. 301].

Technically, it is possible to extend Definition 2.6.7 to the limiting cases when $p = 1$ and $p = \infty$. However, already in the classical situation when $M = \mathbb{R}^d$ the outcome is not satisfactory: the resulting spaces $H_p^s(\mathbb{R}^d)$ have not enough Fourier multipliers, cf. [Tri92, p. 6, p. 13], and there is no hope for a coincidence in the sense of (2.6.5). Therefore, we restrict ourselves to $1 < p < \infty$ (as we already did when defining the fractional Sobolev spaces $H_p^s(\mathbb{R}^d)$ in Sect. 2.2.1), but emphasize that the boundary cases are included when dealing with Besov and Triebel-Lizorkin spaces on manifolds in Sect. 2.6.2.

Sobolev Spaces Using Other Trivializations For many applications the norm given in (2.6.4) is very useful. In particular, it enables us to transfer many results known on $\mathbb{R}^d$ to manifolds M of bounded geometry. The choice of geodesic coordinates, however, often turns out to be far too restrictive if one needs to adapt the underlying coordinates to a certain problem, e.g. to submanifolds N of M in order to study traces. Therefore, in order to replace the geodesic trivializations in (2.6.4) we now want to look for other *'good'* trivializations which will result in equivalent norms (and hence yield the same spaces).

Definition 2.6.9 Let (M^d, g) be a Riemannian manifold together with a uniformly locally finite trivialization $\mathcal{T} = (U_\alpha, \kappa_\alpha, h_\alpha)_{\alpha \in I}$. Furthermore, let $s \in \mathbb{R}$ and $1 < p < \infty$. Then the space $H_p^{s,\mathcal{T}}(M)$ contains all distributions $f \in \mathcal{D}'(M)$ such that

$$\|f|H_p^{s,\mathcal{T}}(M)\| := \left(\sum_{\alpha \in I} \|(h_\alpha f) \circ \kappa_\alpha | H_p^s(\mathbb{R}^d)\|^p \right)^{\frac{1}{p}} \tag{2.6.6}$$

is finite. Here again $(h_\alpha f) \circ \kappa_\alpha$ is viewed as function on $\mathbb{R}^d$, cf. (2.6.4) and below.

In general, the spaces $H_p^{s,\mathcal{T}}(M)$ do depend on the underlying trivialization $\mathcal{T}$. In what follows we investigate under which conditions on $\mathcal{T}$ the norm in (2.6.6) is equivalent to the $H_p^s(M)$-norm. For that we will use the following terminology.

Definition 2.6.10 (Admissible Trivialization) Let (M^d, g) be a Riemannian manifold of bounded geometry. Moreover, let a uniformly locally finite trivialization $\mathcal{T} = (U_\alpha, \kappa_\alpha, h_\alpha)_{\alpha \in I}$ be given. We say that $\mathcal{T}$ is *admissible* if the following conditions are fulfilled:

(B1) $\mathcal{A} = (U_\alpha, \kappa_\alpha)_{\alpha \in I}$ is compatible with geodesic coordinates, i.e., for $\mathcal{A}^{\text{geo}} = (U_\beta^{\text{geo}}, \kappa_\beta^{\text{geo}})_{\beta \in J}$ being a geodesic atlas of M as in Example 2.6.1 there are constants $C_k > 0$ for $k \in \mathbb{N}_0$ such that for all $\alpha \in I$ and $\beta \in J$ with $U_\alpha \cap U_\beta^{\text{geo}} \neq \emptyset$ and all $\mathfrak{a} \in \mathbb{N}_0^d$ with $|\mathfrak{a}| \leq k$,

$$|D^{\mathfrak{a}}(\mu_{\alpha\beta} = (\kappa_\alpha)^{-1} \circ \kappa_\beta^{\text{geo}})| \leq C_k \qquad \text{and}$$

$$|D^{\mathfrak{a}}(\mu_{\beta\alpha} = (\kappa_\beta^{\text{geo}})^{-1} \circ \kappa_\alpha)| \leq C_k.$$

(B2) For all $k \in \mathbb{N}$ there exist $c_k > 0$ such that for all $\alpha \in I$ and all multi-indices $\mathfrak{a}$ with $|\mathfrak{a}| \leq k$

$$|D^{\mathfrak{a}}(h_\alpha \circ \kappa_\alpha)| \leq c_k.$$

Remark 2.6.11

(i) If (B1) is true for some geodesic atlas, it is true for any refined geodesic atlas. This follows immediately from Remark 2.6.5(ii).

(ii) Condition (B1) implies in particular the compatibility of the charts in $\mathcal{T}$ among themselves, i.e., for all $k \in \mathbb{N}_0$ there are constants $C_k > 0$ such that for all multi-indices $\mathfrak{a}$ with $|\mathfrak{a}| \leq k$ and all $\alpha, \beta \in I$ with $U_\alpha \cap U_\beta \neq \varnothing$ we have $|D^{\mathfrak{a}}(\kappa_\alpha^{-1} \circ \kappa_\beta)| \leq C_k$. This is seen immediately when choosing $z \in U_\alpha \cap U_\beta$, considering the exponential map κ_z^{geo} around z, and applying the chain rule to $D^{\mathfrak{a}}(\kappa_\alpha^{-1} \circ \kappa_\beta) = D^{\mathfrak{a}}((\kappa_\alpha^{-1} \circ \kappa_z^{\text{geo}}) \circ ((\kappa_z^{\text{geo}})^{-1} \circ \kappa_\beta))$. The same works for charts belonging to different admissible trivializations.

Theorem 2.6.12 *Let (M, g) be a Riemannian manifold of bounded geometry, and let $\mathcal{T} = (U_\alpha, \kappa_\alpha, h_\alpha)_{\alpha \in I}$ be an admissible trivialization of M. Furthermore, let $s \in \mathbb{R}$ and $1 < p < \infty$. Then,*

$$H_p^{s,\mathcal{T}}(M) = H_p^s(M),$$

i.e., for admissible trivializations of M the resulting Sobolev spaces $H_p^{s,\mathcal{T}}(M)$ do not depend on $\mathcal{T}$.

Proof The proof is based on pointwise multiplier assertions and diffeomorphism properties of the spaces $H_p^s(\mathbb{R}^d)$, see Lemma 2.2.2. Let $\mathcal{T} = (U_\alpha, \kappa_\alpha, h_\alpha)_{\alpha \in I}$ be an admissible trivialization. Let a geodesic trivialization $\mathcal{T}^{\text{geo}} = (U_\beta^{\text{geo}}, \kappa_\beta^{\text{geo}}, h_\beta^{\text{geo}})_{\beta \in J}$ of M, see Example 2.6.6, be given. If $\alpha \in I$ is given, the index set $A(\alpha)$ collects all $\beta \in J$ for which $U_\alpha \cap U_\beta^{\text{geo}} \neq \varnothing$. The cardinality of $A(\alpha)$ can be estimated from above by a constant independent of α since the covers are uniformly locally finite.

We assume $f \in H_p^s(M)$. By Lemma 2.2.2 and Definition 2.6.10 we have for all $\alpha \in I$,

$$\|(h_\alpha f) \circ \kappa_\alpha | H_p^s(\mathbb{R}^d)\| = \left\| \sum_{\beta \in A(\alpha)} (h_\alpha h_\beta^{\text{geo}} f) \circ \kappa_\alpha | H_p^s(\mathbb{R}^d) \right\|$$

$$\leq \sum_{\beta \in A(\alpha)} \left\| (h_\alpha h_\beta^{\text{geo}} f) \circ \kappa_\alpha | H_p^s(\mathbb{R}^d) \right\|$$

$$= \sum_{\beta \in A(\alpha)} \left\| (h_\alpha h_\beta^{\text{geo}} f) \circ (\kappa_\beta^{\text{geo}} \circ (\kappa_\beta^{\text{geo}})^{-1}) \circ \kappa_\alpha | H_p^s(\mathbb{R}^d) \right\|$$

$$\lesssim \sum_{\beta \in A(\alpha)} \left\| (h_\alpha h_\beta^{\text{geo}} f) \circ \kappa_\beta^{\text{geo}} | H_p^s(\mathbb{R}^d) \right\|$$

$$\lesssim \sum_{\beta \in A(\alpha)} \left\| (h_\beta^{\text{geo}} f) \circ \kappa_\beta^{\text{geo}} | H_p^s(\mathbb{R}^d) \right\|.$$

In particular, the involved constant can be chosen independently of α. Then

$$
\|f\,|\,H_p^{s,\mathcal{T}}(M)\| = \left(\sum_{\alpha \in I} \|(h_\alpha f) \circ \kappa_\alpha \,|\, H_p^s(\mathbb{R}^d)\|^p \right)^{1/p}
$$

$$
\lesssim \left(\sum_{\alpha \in I, \beta \in A(\alpha)} \|(h_\beta^{\mathrm{geo}} f) \circ \kappa_\beta^{\mathrm{geo}} \,|\, H_p^s(\mathbb{R}^d)\|^p \right)^{1/p} \lesssim \|f\,|\,H_p^s(M)\|
$$

where the last estimate follows from $\sum_{\alpha \in I, \beta \in A(\alpha)} = \sum_{\beta \in J, \alpha \in A(\beta)}$ and the fact that the covers are uniformly locally finite. The reverse inequality is obtained in the same way. Thus, we see that $H_p^{s,\mathcal{T}}(M) = H_p^s(M)$. $\square$

In view of Remark 2.6.5(iii), we would like to have a similar result for trivializations satisfying condition (B1).

Lemma 2.6.13 *Let (M, g) be a Riemannian manifold with positive injectivity radius, and let $\mathcal{T} = (U_\alpha, \kappa_\alpha, h_\alpha)_{\alpha \in I}$ be a uniformly locally finite trivialization. Let g_{ij} be the coefficient matrix of g and g^{ij} its inverse with respect to the coordinates κ_α. Then (M, g) is of bounded geometry and $\mathcal{T}$ fulfills (B1) if, and only if, the following is fulfilled:*

For all $k \in \mathbb{N}_0$ there is a constant $C_k > 0$ such that for all multi-indices $\mathfrak{a}$ with $|\mathfrak{a}| \leq k$,

$$
|D^\mathfrak{a} g_{ij}| \leq C_k \quad and \quad |D^\mathfrak{a} g^{ij}| \leq C_k \tag{2.6.7}
$$

holds in all charts κ_α.

Proof Let (2.6.7) be fulfilled. Then (M, g) is of bounded geometry since R^M in local coordinates is given by a polynomial in g_{ij}, g^{ij} and its derivatives. Moreover, condition (B1) follows from [Sch01, Lem. 3.8]—we shortly sketch the argument here: Let Γ_{ij}^k denote the Christoffel symbols with respect to coordinates κ_α for $\alpha \in I$. By (2.6.1) and (2.6.7), there are constants $C_k > 0$ for $k \in \mathbb{N}_0$ such that $|D^\mathfrak{a} \Gamma_{ij}^k| \leq C_k$ for all $\alpha \in I$ and all $\mathfrak{a} \in \mathbb{N}_0^d$ with $|\mathfrak{a}| \leq k$. Moreover, let $\mathcal{A}^{\mathrm{geo}} = (U_\beta^{\mathrm{geo}} = B_r(p_\beta^{\mathrm{geo}}), \kappa_\beta^{\mathrm{geo}})_{\beta \in J}$ be a geodesic atlas of M, where $r > 0$ is fixed and smaller than the injectivity radius. We get that $(\kappa_\alpha)^{-1} \circ \kappa_\beta^{\mathrm{geo}}(x) = \Phi(1, \kappa_\alpha^{-1}(p_\beta), \kappa_\alpha^*(\lambda_\beta(x)))$ where Φ is the geodesic flow. Then, together with Lemma 2.6.3 it follows that $(\kappa_\alpha)^{-1} \circ \kappa_\beta^{\mathrm{geo}}$ and all its derivatives are uniformly bounded independent on α and β. Moreover, note that $(\kappa_\beta^{\mathrm{geo}})^{-1} \circ \kappa_\alpha : \kappa_\alpha^{-1}(U_\alpha \cap U_\beta^{\mathrm{geo}}) \subset B_r^d \to (\kappa_\beta^{\mathrm{geo}})^{-1}(U_\alpha \cap U_\beta^{\mathrm{geo}}) \subset B_r^d$ is bounded by r. Hence, together with the chain rule applied to $((\kappa_\beta^{\mathrm{geo}})^{-1} \circ \kappa_\alpha) \circ ((\kappa_\alpha)^{-1} \circ \kappa_\beta^{\mathrm{geo}}) = \mathrm{Id}$ condition (B1) follows for all (α, β).

Conversely, let (M, g) be of bounded geometry, and let condition (B1) be fulfilled. Then, by Remark 2.6.5(iii) and the transformation formula (2.6.2) for $\alpha \in I$ and $\beta \in J$, condition (2.6.7) follows. □

2.6.2 Besov and Triebel-Lizorkin Spaces on Manifolds

After having studied fractional Sobolev spaces on manifolds, we now want to turn our attention to Besov and Triebel-Lizorkin spaces. On $\mathbb{R}^d$ one usually gives priority to Besov spaces, and they are mostly considered to be the simpler spaces compared to Triebel-Lizorkin spaces. However, the situation is different on manifolds, since B-spaces lack the so-called *localization principle*, cf. [Tri92, Thm. 2.4.7(i)], which we relied on when defining the Sobolev spaces on manifolds in Definition 2.6.9.

Therefore, we now proceed as follows. First, we define F-spaces on M by generalizing Definition 2.6.9, i.e., replacing the $H_p^s(\mathbb{R}^d)$-norm with the corresponding norm for F-spaces. Afterwards, Besov spaces on M are introduced via real interpolation of Triebel-Lizorkin spaces.

Definition 2.6.14 Let (M^d, g) be a Riemannian manifold with an admissible trivialization $\mathcal{T} = (U_\alpha, \kappa_\alpha, h_\alpha)_{\alpha \in I}$ and let $s \in \mathbb{R}$.

(i) Let either $0 < p < \infty, 0 < q \leq \infty$ or $p = q = \infty$. Then the space $F_{p,q}^{s,\mathcal{T}}(M)$ contains all distributions $f \in \mathcal{D}'(M)$ such that

$$\|f|F_{p,q}^{s,\mathcal{T}}(M)\| := \left(\sum_{\alpha \in I} \|(h_\alpha f) \circ \kappa_\alpha | F_{p,q}^s(\mathbb{R}^d)\|^p \right)^{\frac{1}{p}} \tag{2.6.8}$$

is finite (with the usual modification if $p = \infty$).

(ii) Let $0 < p, q \leq \infty$, and let $-\infty < s_0 < s < s_1 < \infty$. Then

$$B_{p,q}^{s,\mathcal{T}}(M) := \left(F_{p,p}^{s_0,\mathcal{T}}(M), F_{p,p}^{s_1,\mathcal{T}}(M) \right)_{\Theta,q} \tag{2.6.9}$$

with $s = (1 - \Theta)s_0 + \Theta s_1$.

Remark 2.6.15

(i) Restricting ourselves to geodesic trivializations $\mathcal{T}^{\text{geo}}$, the spaces from Definition 2.6.14 coincide with the spaces $F_{p,q}^s(M)$ and $B_{p,q}^s(M)$, introduced in [Tri92, Def. 7.2.2, 7.3.1]. The space $B_{p,q}^{s,\mathcal{T}}(M)$ is independent of the chosen numbers $s_0, s_1 \in \mathbb{R}$ and, furthermore, for $s \in \mathbb{R}$ and $0 < p \leq \infty$ we have the coincidence

$$B_{p,p}^{s,\mathcal{T}}(M) = F_{p,p}^{s,\mathcal{T}}(M). \tag{2.6.10}$$

This follows from [Tri92, Thm. 7.3.1], since the arguments presented there are based on interpolation and completely oblivious of the chosen trivialization $\mathcal{T}$. In particular, (2.6.10) yields that for $f \in B^s_{p,p}(M)$ a quasi-norm is given by

$$\| f | B^{s,\mathcal{T}}_{p,p}(M) \| = \left(\sum_{\alpha \in I} \| (h_\alpha f) \circ \kappa_\alpha | B^s_{p,p}(\mathbb{R}^d) \|^p \right)^{\frac{1}{p}}. \tag{2.6.11}$$

(ii) Note that similar to the situation on $\mathbb{R}^d$ we alternatively could have defined Besov spaces $B^s_{p,p}(M)$ on manifolds via real interpolation of fractional Sobolev spaces $H^s_p(M)$, i.e.,

$$B^s_{p,p}(M) := \left(H^{s_0}_p(M), H^{s_1}_p(M) \right)_{\Theta,p},$$

where $s_0, s_1 \in \mathbb{R}$, $1 < p < \infty$, $0 < \Theta < 1$, and $s = \Theta s_0 + (1 - \Theta) s_1$.

Now we can transfer Theorem 2.6.12 to F- and B-spaces.

Theorem 2.6.16 *Let (M^d, g) be a Riemannian manifold with an admissible trivialization $\mathcal{T} = (U_\alpha, \kappa_\alpha, h_\alpha)_{\alpha \in I}$. Furthermore, let $s \in \mathbb{R}$ and let $0 < p, q \leq \infty$ ($0 < p < \infty$ or $p = q = \infty$ for F-spaces). Then*

$$F^{s,\mathcal{T}}_{p,q}(M) = F^s_{p,q}(M) \qquad and \qquad B^{s,\mathcal{T}}_{p,q}(M) = B^s_{p,q}(M).$$

Proof For F-spaces the proof is the same as the one of Theorem 2.6.12. The claim for B-spaces then follows from Definition 2.6.14(ii)   □

Remark 2.6.17 By Theorem 2.6.16, we can omit the dependency on the trivializations $\mathcal{T}$ (as long as they are admissible) from our notations in (2.6.8) and (2.6.9), since resulting norms are equivalent and yield the same spaces.

Part I
Besov and Fractional Sobolev Regularity of PDEs

Chapter 3
Theory and Background Material for PDEs

This chapter surveys the principal theoretical issues concerning the solving of partial differential equations (PDEs). In the first section we recall the notions of elliptic, parabolic, and hyperbolic PDEs, since we are going to study the regularity of solutions for all these types of equations in Chaps. 4–6. In the subsequent sections we study weak formulations of our PDEs. In a weak formulation an equation is no longer required to hold absolutely and has instead weak solutions only with respect to certain 'test functions'. We present the Lax-Milgram Theorem theorem which guarantees the existence and uniqueness of such a weak solution. In particular, it turns out that for elliptic problems with zero Dirichlet boundary conditions, the Sobolev spaces $\mathring{H}^m(\Omega)$ (and appropriate generalizations for parabolic and hyperbolic problems) are 'good' spaces among which we can look for weak solutions to a given PDE. Furthermore, we provide sufficient conditions on the coefficients of the involved differential operators and the functions on the right hand side of our equations, which guarantee the existence and uniqueness of a weak solution.

Moreover, for elliptic problems we also present regularity results of the weak solution in the fractional Sobolev scale $H^s(\Omega)$. For parabolic problems this will be investigated later on (and compared with the results presented here) in Chap. 5.

Finally, we discuss the concept of operator pencils generated by (elliptic) boundary value problems, since singularities of solutions on non-smooth domains can be described in terms of spectral properties of certain pencils. We use these pencils in Chap. 5 when proving some of the regularity results in Kondratiev spaces.

3.1 Definitions and Classifications

In this section we review basic definitions and classifications related with partial differential equations and give some examples.

© The Author(s), under exclusive license to Springer Nature Switzerland AG 2021
C. Schneider, *Beyond Sobolev and Besov*, Lecture Notes in Mathematics 2291,
https://doi.org/10.1007/978-3-030-75139-5_3

A *partial differential equation (PDE)* is an equation involving an unknown function of two or more variables and certain of its partial derivatives. PDEs can be classified according to the highest order of derivatives of the unknown function which appear in the equation. We give a precise definition below.

Definition 3.1.1 (Partial Differential Equation of Order m**)** Let $\Omega \subset \mathbb{R}^d$ with $d \geq 2$ be a domain and $m \in \mathbb{N}$. An expression of the form

$$F(D^m u(x), D^{m-1} u(x), \ldots, Du(x), u(x), x) = 0, \qquad x \in \Omega, \tag{3.1.1}$$

is called a *m-th order partial differential equation*, where

$$F : \mathbb{R}^{d^m} \times \mathbb{R}^{d^{m-1}} \times \cdots \times \mathbb{R}^d \times \mathbb{R} \times \Omega \to \mathbb{R}$$

is given and $u : \Omega \to \mathbb{R}$ is the unknown function.

We solve the PDE if we find all u verifying (3.1.1). In particular, u is called a *classical solution* if $u \in C^m(\Omega)$ and u satisfies (3.1.1).

PDEs can furthermore be distinguished by whether the equation is linear w.r.t. all or at least the highest order derivatives. The following definition collects the usual notation in this regard.

Definition 3.1.2 (Linear, Semilinear, Nonlinear PDEs)

 (i) The PDE (3.1.1) is called *linear* if it is linear w.r.t. the unknown function u and all its derivatives and the corresponding coefficients only depend on the variable x, i.e., it has the form

$$\sum_{|\alpha| \leq m} a_\alpha(x) D^\alpha u = f(x)$$

for given functions a_α and f. This linear PDE is *homogeneous* if $f \equiv 0$.
 (ii) The PDE (3.1.1) is called *semilinear* if it is linear w.r.t. the highest derivatives $D^m u$ of the unknown function u and the corresponding coefficients only depend on the variable x, i.e., it has the form

$$\sum_{|\alpha| = m} a_\alpha(x) D^\alpha u + F_0(D^{m-1} u, \ldots, Du, u, x) = 0$$

with functions F_0 and a_α.
 (iii) The PDE (3.1.1) is called *nonlinear* if it depends nonlinearly upon the highest order derivatives.

Remark 3.1.3 From the definition it is clear that we have the following inclusions

$$\{\text{linear PDEs}\} \subsetneq \{\text{semilinear PDEs}\} \subsetneq \{\text{nonlinear PDEs}\}.$$

Moreover, we can distinguish different types of PDEs. We start with second order linear PDEs and the classifications of elliptic, parabolic or hyperbolic equations. Afterwards the concepts are generalized to higher order PDEs.

Definition 3.1.4 (Elliptic, Hyperbolic, Parabolic PDEs of Second Order) Let $\Omega \subset \mathbb{R}^d$ be a domain. A second order linear PDE can be written in the form

$$-\sum_{i,j=1}^{d} a_{ij}(x)\frac{\partial^2 u}{\partial x_i \partial x_j} + \sum_{i=1}^{d} a_i(x)\frac{\partial u}{\partial x_i} + a(x)u = f(x). \tag{3.1.2}$$

We can always assume that the matrix $A(x) = \left(a_{ij}(x)\right)_{i,j=1}^{d}$ is symmetric and therefore has real eigenvalues.

(i) (3.1.2) is *elliptic* in $x \in \Omega$ if, and only if, the matrix $A(x)$ has eigenvalues with the same sign.

(ii) (3.1.2) is *hyperbolic* in $x \in \Omega$ if, and only if, the matrix $A(x)$ has $d-1$ eigenvalues with the same sign and one eigenvalue with the opposite sign.

(iii) (3.1.2) is *parabolic* in $x \in \Omega$ if, and only if, for the matrix $A(x)$ one eigenvalue vanishes, $d-1$ eigenvalues have the same sign, and rank $(A(x), \tilde{a}(x)) = d$, where $\tilde{a}(x) = (a_1(x), \ldots, a_d(x))^T$.

(iv) (3.1.2) is called elliptic (parabolic or hyperbolic, respectively) in $\Omega \subset \mathbb{R}^d$ if (3.1.2) is elliptic (parabolic or hyperbolic, respectively) for all $x \in \Omega$.

Remark 3.1.5

(i) One also writes $Lu = f$ with differential operator

$$L = -\sum_{i,j=1}^{d} a_{ij}\frac{\partial^2}{\partial x_i \partial x_j} + \sum_{i=1}^{d} a_i\frac{\partial}{\partial x_i} + a. \tag{3.1.3}$$

In this case we say that the operator L is given in *non-divergence form*. By Definition 3.1.4 the operator L is elliptic in $x \in \Omega$ if w.l.o.g. $A(x)$ has positive eigenvalues, which is equivalent to saying that $A(x)$ is *positive definite*, i.e.,

$$\sum_{i,j=1}^{d} a_{ij}(x)\xi_i\xi_j > 0 \qquad \text{for all} \quad \xi \in \mathbb{R}^d \setminus \{0\}.$$

Moreover, L is elliptic in Ω if

$$\sum_{i,j=1}^{d} a_{ij}(x)\xi_i\xi_j > c(x)|\xi|^2 \qquad \text{where} \quad c(x) > 0, \quad \xi \in \mathbb{R}^d, \quad x \in \Omega,$$

with $c(x) = \min \left\{ \sum_{i,j=1}^{d} a_{ij}(x)\xi_i\xi_j : |\xi| = 1 \right\}$ being the smallest eigenvalue

of $A(x)$.

(ii) Moreover, the *main part of* L is the operator

$$L_0 = - \sum_{i,j=1}^{d} a_{ij} \frac{\partial^2}{\partial x_i \partial x_j},$$

which contains only the highest derivatives of L. We see from Definition 3.1.4 that hyperbolicity and ellipticity only depend on the main part of L.

(iii) If the highest order coefficients a_{ij}, $i, j = 1, \ldots, d$, are sufficiently smooth (C^1 functions), then the operator from (3.1.3) given in non-divergence form can be written as

$$L = - \sum_{i,j=1}^{d} \frac{\partial}{\partial x_j} \left(a_{ij} \frac{\partial}{\partial x_i} \right) + \sum_{i=1}^{d} \tilde{a}_i \frac{\partial}{\partial x_i} + a$$

for $\tilde{a}_i := a_i + \sum_{j=1}^{d} \frac{\partial}{\partial x_j} a_{ij}$. This is called *divergence form*. Both representations of L have their advantages. In general, the divergence form is more appropriate when dealing with the variational formulation of a PDE based upon integration by parts.

(iv) The connection between the different types of PDEs is easily understood if we relate elliptic equations w.r.t. the variables $x_1, \ldots, x_d$ with parabolic and hyperbolic equations w.r.t. the variables $x_1, \ldots, x_d, t$. In this case for an elliptic operator L w.r.t. the variables $x_1, \ldots, x_d$, which w.l.o.g. has positive eigenvalues, we see that

$$\frac{\partial u}{\partial t} + Lu = f$$

is parabolic, whereas

$$\frac{\partial^2 u}{\partial t^2} + Lu = f$$

is hyperbolic. The parameter t is often related with the time.

We recall the standard examples for elliptic, parabolic, and hyperbolic PDEs.

Examples 3.1.6

(i) The Poisson equation

$$-\Delta u = f$$

with $L = -\Delta = -\sum_{i=1}^{d} \frac{\partial^2}{\partial x_i^2}$ is elliptic since

$$A(x) = \begin{pmatrix} -1 & 0 & \cdots & 0 \\ 0 & -1 & \cdots & 0 \\ & & \ddots & \\ 0 & 0 & \cdots & -1 \end{pmatrix}.$$

(ii) The wave equation

$$\frac{\partial^2 u}{\partial t^2} - \Delta u = 0$$

is hyperbolic since

$$A(x) = \begin{pmatrix} 1 & 0 & \cdots & 0 \\ 0 & -1 & \cdots & 0 \\ & & \ddots & \\ 0 & 0 & \cdots & -1 \end{pmatrix}.$$

(iii) The heat equation

$$\frac{\partial}{\partial t} u - \Delta u = 0$$

is parabolic since

$$A(x) = \begin{pmatrix} 0 & 0 & \cdots & 0 \\ 0 & -1 & \cdots & 0 \\ & & \ddots & \\ 0 & 0 & \cdots & -1 \end{pmatrix} \quad \text{and} \quad (A(x), \tilde{a}(x)) = \begin{pmatrix} 0 & 0 & \cdots & 0 & 1 \\ 0 & -1 & \cdots & 0 & 0 \\ & & \ddots & \\ 0 & 0 & \cdots & -1 & 0 \end{pmatrix}.$$

So far we have explained the different types only for second order PDEs. We now generalize the concept of ellipticity to higher order differential equations.

Definition 3.1.7 (General Elliptic Operator) Let $\Omega \subset \mathbb{R}^d$ be a domain and L be a linear differential operator of order $2m$, i.e.,

$$L = \sum_{|\alpha| \leq 2m} a_\alpha(x) D^\alpha, \qquad x \in \Omega. \tag{3.1.4}$$

(i) L is *elliptic* in $x \in \Omega$ if

$$\sum_{|\alpha|=2m} a_\alpha(x)\xi^\alpha \neq 0 \qquad \text{for all} \quad \xi \in \mathbb{R}^d \setminus \{0\}. \tag{3.1.5}$$

(ii) L is *uniformly elliptic* in Ω if

$$(-1)^m \sum_{|\alpha|=2m} a_\alpha(x)\xi^\alpha \geq c(x)|\xi|^{2m} \qquad \text{for all} \quad \xi \in \mathbb{R}^d, \quad x \in \Omega, \tag{3.1.6}$$

and $\inf\{c(x): x \in \Omega\} > 0$.

Remark 3.1.8

(i) If the coefficients a_α are smooth enough we can write the operator L in (3.1.4) as

$$L = \sum_{|\alpha|,|\beta|=0}^{m} (-1)^{|\beta|} D^\beta \left(a_{\alpha\beta}(x) D^\alpha\right). \tag{3.1.7}$$

The main part of L in this case is given by

$$L_0 = (-1)^m \sum_{|\alpha|=|\beta|=m} D^\beta \left(a_{\alpha\beta}(x) D^\alpha\right).$$

According to the definition L is uniformly elliptic if we replace (3.1.6) by

$$\sum_{|\alpha|=|\beta|=m} a_{\alpha\beta}(x)\xi^{\alpha+\beta} \geq c(x)|\xi|^{2m} \qquad \text{for all} \quad \xi \in \mathbb{R}^d, \quad x \in \Omega. \tag{3.1.8}$$

(ii) Elliptic operators of odd order are not common. Let $L = \sum_{|\alpha| \leq k} a_\alpha D^\alpha$ be elliptic at $x_0 \in \mathbb{R}^d$ with $d \geq 2$ and assume $a_\alpha(x_0)$ to be real for $|\alpha| = k$. Then k must be even. This can be seen as follows: Since the set $\mathbb{R}^d \setminus \{0\}$ is connected we deduce from (3.1.5) that for $P(x_0, \xi) := \sum_{|\alpha|=k} a_\alpha(x_0)\xi^\alpha$ we either have $P(x_0, \xi) > 0$ or $P(x_0, \xi) < 0$ for all $\xi \in \mathbb{R}^d \setminus \{0\}$. The fact that $P(x_0, -\xi) = (-1)^k P(x_0, \xi)$ forces k to be even.

(iii) We say that the PDE $Lu = f$ is given in *divergence form* if the differential operator L is given by (3.1.7), and is in *non-divergence form* provided L is given by (3.1.4).

(iv) Definition 3.1.7 together with Remark 3.1.5(iv) suggests a generalization of parabolic and hyperbolic PDEs to higher order equations: Let L be defined as in (3.1.4) and assume L to be elliptic. Then the equation

$$\frac{\partial u}{\partial t} + Lu = f \qquad \text{is } parabolic,$$

whereas

$$\frac{\partial^2 u}{\partial t^2} + Lu = f \qquad \text{is } hyperbolic.$$

3.2 Elliptic Boundary Value Problems

In this section we focus on elliptic problems. There may be a unique solution to a given PDE if we impose boundary values to the problem under consideration. We start by discussing the general boundary value setting below, although in our later considerations we focus on zero Dirichlet boundary conditions. In order to show the existence and uniqueness of a solution to our problem, we construct a general scheme for setting up elliptic boundary value problems in terms of bilinear forms. The reason for this is as follows: In the modern, functional-analytic approach to studying PDEs, one does not attempt to solve a given PDE directly, but merely uses weak formulations of the equations and corresponding bilinear forms. With this point of view an equation is no longer required to hold absolutely and has instead weak solutions only with respect to certain 'test functions'. We present the Lax-Milgram Theorem, which under certain assumptions on the bilinear from yields existence and uniqueness of a weak solution to an elliptic boundary value problem (with zero Dirichlet boundary conditions). Finally, we discuss the smoothness of this weak solution in the scale of fractional Sobolev spaces. It turns out that the shape of the underlying domain is crucial in this context.

The general elliptic boundary value problem we consider is stated below.

Definition 3.2.1 (Elliptic Boundary Value Problem) Let L be elliptic of order $2m$. Then

$$\left\{ \begin{array}{rl} Lu = f & \text{in } \Omega, \\ B_j u = \displaystyle\sum_{|\alpha| \le m_j} b_{j_\alpha} D^\alpha u = g_j & \text{on } \partial\Omega, \quad 0 \le m_j < 2m, \quad j = 1, \dots, m, \end{array} \right\}$$

is called *general elliptic boundary value problem of order* $2m$.

Remark 3.2.2

(i) The operators B_j cannot be arbitrarily chosen. They have to be independent and constitute a so-called 'normal system', cf. [LM72, p. 113] and [Wlo82, p. 214]. In particular, the orders m_j have to differ pairwise.

(ii) We have *Dirichlet boundary conditions* if

$$B_j = \left(\frac{\partial}{\partial \nu}\right)^{j-1}, \quad j = 1, \ldots, m,$$

where ν denotes the exterior normal to $\partial\Omega$ (for the normal to exist we have to assume that the boundary of Ω is sufficiently smooth). In particular, the *homogeneous Dirichlet problem* asks for functions u satisfying $Lu = f$ in Ω with boundary conditions

$$\left(\frac{\partial}{\partial \nu}\right)^{j-1} u = 0 \quad \text{on} \quad \partial\Omega, \quad j = 1, \ldots, m.$$

Throughout this thesis we shall mainly deal with zero Dirichlet boundary conditions. However, it is planned to study more general boundary value problems in the future.

(iii) If $m = 1$ then

$$B = \sum_{i=1}^{d} b_i(x)\frac{\partial}{\partial x_i} + b_0(x) = b^T \cdot \nabla + b_0, \qquad x \in \partial\Omega. \qquad (3.2.1)$$

Then we have

Dirichlet boundary conditions	if $b = 0$, $b_0 \neq 0$,
Neumann boundary conditions	if $b = \nu$, $b_0 = 0$,
Mixed boundary conditions	if $\langle b, \nu \rangle \neq 0$, $b_0 \neq 0$.

3.2.1 Bilinear Forms and Lax-Milgram Theorem

The aim of this subsection is to provide a general scheme for setting up elliptic boundary value problems in terms of bilinear forms. The Lax-Milgram Theorem, which is presented below, can be seen as a theoretical cornerstone in order to establish the existence and uniqueness for solutions of equations in the weak formulation. We do not provide the most general formulation but use a version which is valid for so-called V-elliptic bilinear forms set on a Hilbert space.

Definition 3.2.3 (*V*-Elliptic Bilinear Form) Let V be a Hilbert space. A bilinear form $B(\cdot, \cdot) : V \times V \to \mathbb{R}$ is called *V-elliptic* if it is continuous, i.e., for some constants $M, c > 0$ we have

$$|B(x, y)| \leq M \|x|V\| \cdot \|y|V\| \qquad \text{for all} \quad x, y \in V, \tag{3.2.2}$$

and

$$B(x, x) \geq c \|x|V\|^2 \qquad \text{for all} \quad x \in V. \tag{3.2.3}$$

Example 3.2.4 The scalar product on $L_2(\Omega)$ is $L_2(\Omega)$-elliptic with constants $c = M = 1$.

Remark 3.2.5 We can assign a unique operator $A \in \mathcal{L}(V, V')$ to every continous bilinear form B via

$$B(x, y) = \langle Ax, y \rangle_{V' \times V} \qquad \text{for all} \quad x, y \in V.$$

Moreover, if $B(\cdot, \cdot)$ satisfies (3.2.3) then the operator A is invertible with $A^{-1} \in \mathcal{L}(V', V)$.

V-elliptic bilinear forms have the following remarkable property.

Theorem 3.2.6 (Lax-Milgram Theorem) *Let $B(\cdot, \cdot)$ be V-elliptic. Then for any $f \in V'$ there exists a unique solution $x \in V$ to the equation*

$$B(x, y) = \langle f, y \rangle \qquad \textit{for all} \quad y \in V.$$

Moreover, $x = A^{-1} f$ with

$$\|x|V\| \leq c^{-1} \|f|V'\|,$$

where c is the constant from (3.2.3).

According to the Lax-Milgram Theorem, the task is always to identify the "good" Hilbert space V of functions among which we look for solutions to a given PDE, and to check the validity of the assumptions on B. As we shall see in the next subsection the bilinear form emerges naturally when testing the PDE against some test functions (in fact, one looks for weak solutions).

The notation of V-ellipticity seems to indicate that elliptic boundary value problems correspond to V-elliptic bilinear forms. In general this is not the case. In fact it will be the coercive bilinear forms which are assigned to elliptic boundary value problems. One could say that coercivity encodes some kind of Sobolev embedding. We briefly recall the concept.

Let U be a Hilbert space, V be a reflexive Banach space, and assume the embedding $V \subset U$ is continous and dense. Then by Riesz's Representation theorem we can identify U with its dual U', i.e., $U = U'$ and since $V \subset U$ is continuous and dense we also have that U' is embedded continuously and densely in V', i.e., $U' \hookrightarrow V'$. The triple

$$V \subset U = U' \subset V' \qquad (V \subset U \text{ continuously and densely embedded})$$

is called a *Gelfand triple*.

Example 3.2.7 A typical example for a Gelfand triple is

$$\mathring{H}^m(\Omega) \subset L_2(\Omega) \subset H^{-m}(\Omega).$$

Remark 3.2.8 (Density in Gelfand Triples) Since in a Gelfand triple $V \hookrightarrow U = U' \hookrightarrow V'$ the spaces are continuously and densely embedded, we deduce that a dense subset S in V is also dense in V': Let $w \in V'$, then since $U = U' \hookrightarrow V'$ is dense there exists a sequence $\{u_n\}_{n \in \mathbb{N}} \in U$ and $N_\varepsilon \in \mathbb{N}$ such that

$$\|w - u_n|V'\| \leq \frac{\varepsilon}{2} \qquad \text{for all} \quad n \geq N_\varepsilon.$$

Moreover, since the embedding $V \hookrightarrow U$ is also dense, for $u_n \in U$ we find a sequence $\{v_{n_k}\}_{n_k \in \mathbb{N}} \in V$ and $\tilde{N}_\varepsilon \in \mathbb{N}$ such that

$$\|u_n - v_{n_k}|V'\| \leq c\|u_n - v_{n_k}|U\| \leq \frac{\varepsilon}{2} \qquad \text{for all} \quad n_k \geq \tilde{N}_\varepsilon.$$

From the above observations we obtain that V is dense in V' since

$$\|w - v_{n_k}|V'\| \leq \|w - u_n|V'\| + \|u_n - v_{n_k}|V'\| \leq \varepsilon \qquad \text{for all} \quad n_k \geq \max\{\tilde{N}_\varepsilon, N_\varepsilon\}.$$

Now approximating $v_{n_k} \in V$ by a sequence $\{\varphi_{n_k,j}\}_{j \in \mathbb{N}}$ from the dense subset S we calculate

$$\|w - \varphi_{n_k,j}|V'\| \leq \|w - v_{n_k}|V'\| + \|v_{n_k} - \varphi_{n_k,j}|V'\| \leq \varepsilon + \|v_{n_k} - \varphi_{n_k,j}|V\| \leq 2\varepsilon$$

for index $n_{k,j} \in \mathbb{N}$ chosen large enough. This shows that S is also dense in V'.

Definition 3.2.9 (V-Coercivity) Let $V \subset U \subset V'$ be a Gelfand-triple. A bilinear form $B(\cdot, \cdot)$ is called V-*coercive* if it is continuous and there exist constants $\mu_0 > 0$ and $\lambda_0 \geq 0$ such that

$$B(x, x) \geq \mu_0 \|x|V\|^2 - \lambda_0 \|x|U\|^2 \qquad \text{for all} \quad x \in V. \tag{3.2.4}$$

Remark 3.2.10 We see that V-coercivity differs from V-ellipticity by the second term $\lambda_0 \|x|U\|^2$ with $\lambda_0 \geq 0$. Because of the embedding $V \subset U$ the U-Norm is weaker compared to the V-norm.

If we put $\tilde{B}(x, y) := B(x, y) + \lambda_0 \langle x, y \rangle_U$ it follows that the V-coercivity of B implies V-ellipticity of $\tilde{B}$:

$$\tilde{B}(x, x) = B(x, x) + \lambda_0 \|x\|_U^2 \geq \mu_0 \|x|V\|^2 - \lambda_0 \|x|U\|^2 + \lambda_0 \|x|U\|^2 = \mu_0 \|x|V\|^2.$$

3.2.2 Existence of Weak Solutions

We now deal with elliptic problems of order $2m$ with homogeneous Dirichlet boundary conditions according to Definition 3.2.1. We give a weak formulation with associated bilinear form and deduce that $\mathring{H}^m(\Omega)$ is an appropriate Hilbert space such that the Lax-Milgram Theorem can be applied. This in turn yields the existence and uniqueness of a weak solution $u \in \mathring{H}^m(\Omega)$ for our problem. Moreover, we provide necessary conditions on the coefficients of the elliptic operator L to guarantee the existence and uniqueness of a weak solution. We briefly discuss the Poisson equation as an important example.

Problem 3.2.11 (Elliptic Problem) Given a function f on Ω find a function u satisfying

$$\left\{ \begin{array}{l} Lu = f \text{ in } \Omega, \\ \left. \dfrac{\partial^{k-1} u}{\partial \nu^{k-1}} \right|_{\partial \Omega} = 0, \quad k = 1, \ldots, m. \end{array} \right\} \tag{3.2.5}$$

Here Ω is an open bounded subset of $\mathbb{R}^d$, $u : \Omega \to \mathbb{R}$ is the unknown, $f : \Omega \to \mathbb{R}$ is given, and $L = \sum_{|\alpha|,|\beta|=0}^m (-1)^{|\beta|} D^\beta \left(a_{\alpha\beta}(x) D^\alpha \right)$ is a uniformly elliptic differential operator of order $2m$. Henceforth, we assume that the coefficients satisfy the symmetry condition $a_{\alpha\beta} = a_{\beta\alpha}$ for $|\alpha|, |\beta| \leq m$.

Our overall plan is to define and construct an appropriate weak solution of Problem 3.2.11 and later on investigate the smoothness properties of u. Assuming for the moment that $\partial\Omega$ is smooth and u is a classical solution of Problem 3.2.11, i.e., $u \in C^{2m}(\Omega)$, multiplying the PDE $Lu = f$ by a smooth test function $v \in C_0^\infty(\Omega)$ and integrating by parts over Ω yields

$$\sum_{|\alpha|,|\beta|=0}^m (-1)^{|\beta|} \int_\Omega D^\beta \left(a_{\alpha\beta}(x) D^\alpha u \right)(x) v(x) dx$$

$$= \sum_{|\alpha|,|\beta|=0}^m \int_\Omega a_{\alpha\beta} \left(D^\alpha u \right)(x) (D^\beta v)(x) dx = \int_\Omega f(x) v(x) dx, \qquad v \in C_0^\infty(\Omega).$$

$$\tag{3.2.6}$$

There are no boundary terms since $v = 0$ on $\partial\Omega$. Since $C_0^\infty(\Omega)$ is dense in $\mathring{H}^m(\Omega)$ we may take $v \in \mathring{H}^m(\Omega)$ in (3.2.6) and the resulting identity makes sense if only $u \in \mathring{H}^m(\Omega)$ (note that we choose the space $\mathring{H}^m(\Omega)$ to incorporate the zero boundary condition from (3.2.5)): In particular the boundary condition only makes sense if $\partial\Omega$ is smooth enough (otherwise the derivatives in normal direction do not exist). Since $u = 0$ on $\partial\Omega$ implies that the derivatives in tangential direction vanish, we deduce that not only the normal derivatives of order $k \le m - 1$ vanish but all derivatives of order $\le m - 1$, i.e.,

$$D^\alpha u = 0 \quad \text{on} \quad \partial\Omega \quad \text{for} \quad |\alpha| \le m - 1. \tag{3.2.7}$$

It follows that $u \in \mathring{H}^m(\Omega)$.

The bilinear form $B(\cdot, \cdot)$ associated with the elliptic operator L is given by

$$B(u, v) := \sum_{|\alpha|, |\beta|=0}^m \int_\Omega a_{\alpha\beta} \left(D^\alpha u\right)(x)(D^\beta v)(x)dx$$

for $u, v \in \mathring{H}^m(\Omega)$. Using (3.2.6), the *weak formulation* of Problem 3.2.11 now reads as

$$B(u, v) = \int_\Omega f(x)v(x)dx = \langle f, v \rangle \qquad \text{for all} \quad v \in \mathring{H}^m(\Omega),$$

which makes sense for any function $f \in H^{-m}(\Omega)$.

Definition 3.2.12 (Weak Solution) Given $f \in H^{-m}(\Omega)$ we say that $u \in \mathring{H}^m(\Omega)$ is a *weak solution* of Problem 3.2.11 if, and only if,

$$B(u, v) = \langle f, v \rangle$$

holds for all $v \in \mathring{H}^m(\Omega)$.

Remark 3.2.13 For a bounded domain Ω it can be shown that a classical solution $u \in C^{2m}(\Omega) \cap C^m(\overline\Omega)$ is a weak solution. For unbounded domains this is not always the case.

Concerning existence and uniqueness of the weak solution according to Definition 3.2.12 by the Lax-Milgram Theorem 3.2.6 we obtain the following.

Theorem 3.2.14 (Existence and Uniqueness of Weak Solutions) *Let $B(\cdot, \cdot)$ be a $\mathring{H}^m(\Omega)$-elliptic bilinear form. Then Problem 3.2.11 has a unique weak solution $u \in \mathring{H}^m(\Omega)$ of the form $u = L^{-1}f$, where L is defined via $B(u, v) = \langle Lu, v \rangle$.*

We see that the Lax-Milgram Theorem 3.2.6 guarantees unique solvability of Problem 3.2.11 if $B(\cdot, \cdot)$ is $\mathring{H}^m(\Omega)$-elliptic. The following theorem tells us under

under which conditions on the operator L and its coefficients this is the case, cf. [Hac92, Thm. 7.2.7].

Theorem 3.2.15 *Let Ω be a bounded domain and let L be the uniformly elliptic operator in Problem 3.2.11. Assume that the coefficients of the principal part of the operator L are constants, i.e., $a_{\alpha\beta} = $ const. for $|\alpha| = |\beta| = m$. Furthermore, assume that $a_{\alpha\beta} = 0$ for $0 < |\alpha| + |\beta| < 2m$, $a_{00} \geq 0$ for $\alpha = \beta = 0$. Then $a(\cdot, \cdot)$ is $\overset{\circ}{H}{}^m(\Omega)$-elliptic.*

Remark 3.2.16 If $m = 1$ then it suffices to assume that L coincides with the principal part L_0 and $a_{\alpha\beta} \in L_\infty(\Omega)$ for $|\alpha| = |\beta| = 1$ to ensure the corresponding bilinear form is $\overset{\circ}{H}{}^1(\Omega)$-elliptic, cf. [Hac92, Thm. 7.2.3]. The condition 'Ω bounded' may be dropped if for $\alpha = \beta = 0$ one assumes $a_{00} \geq \eta > 0$ (instead of $a_{00} = 0$).

Example 3.2.17 Consider the Poisson equation

$$\left\{ \begin{aligned} -\Delta u &= f \quad \text{in} \quad \Omega, \\ u\big|_{\partial\Omega} &= 0, \end{aligned} \right\}$$

where $\Omega \subset \mathbb{R}^d$ is a bounded Lipschitz domain. In particular, we see that the operator

$$L = -\Delta = -\sum_{j=1}^{d} \frac{\partial^2}{\partial x_j^2}$$

satisfies the requirements from Theorem 3.2.15. Therefore, the bilinear form

$$B(u, v) = \int_\Omega \nabla u \cdot \nabla v \, dx$$

is $\overset{\circ}{H}{}^1(\Omega)$-elliptic. Hence, by Theorem 3.2.14 there exists exactly one solution $u \in \overset{\circ}{H}{}^1(\Omega)$.

Without the above restrictions on the coefficients, in general the uniform ellipticity of the differential operator L acting on $\overset{\circ}{H}{}^m(\Omega)$ does not guarantee $\overset{\circ}{H}{}^m(\Omega)$-ellipticity but merely $\overset{\circ}{H}{}^m(\Omega)$-coercivity, cf. Definition 3.2.9. A proof of the following theorem may be found in [Wlo82, p. 282].

Theorem 3.2.18 (Gårding's Inequality) *Consider the Gelfand triple $\overset{\circ}{H}{}^m(\Omega) \subset L_2(\Omega) \subset H^{-m}(\Omega)$. Let L be uniformly elliptic according to (3.1.8) and $a_{\alpha\beta} \in L_\infty(\Omega)$. Moreover, assume that the coefficients $a_{\alpha\beta}$ are uniformly continuous for $|\alpha| = |\beta| = m$ in Ω. Then $B(\cdot, \cdot)$ is $\overset{\circ}{H}{}^m(\Omega)$-coercive, i.e.,*

$$B(x, x) \geq \mu_0 \|x|\overset{\circ}{H}{}^m(\Omega)\|^2 - \lambda_0 \|x|L_2(\Omega)\|^2.$$

There is also a converse: if $a_{\alpha\beta} \in C(\Omega)$ for $|\alpha| = |\beta| = m$ and $a_{\alpha\beta} \in L_\infty(\Omega)$ otherwise, then the $\overset{\circ}{H}{}^m(\Omega)$-coercivity implies uniform ellipticity (3.1.8).

3.2.3 Fractional Sobolev Regularity on Smooth, Convex and Lipschitz Domains

In this subsection we recall what is known in terms of regularity results in fractional Sobolev spaces H^s for elliptic boundary value problems. We will generalize these results to parabolic problems in Chap. 5 and compare our findings with what is stated below.

In particular, it turns out that the shape of the underlying domain domain is crucial in this context. For smooth domains we have a shift by $2m$ within the scale H^s of fractional Sobolev spaces, i.e., for right hand side $f \in H^{-m+s}$ the weak solution u of Problem 3.2.11 belongs to H^{m+s}. On general Lipschitz domains the situation is completely different: Here the smoothness s might be less or equal to $3/2$—even for smooth right hand side f. The precise results are stated below.

For smooth domains the following is shown in [Hac92, Thm. 9.1.16].

Theorem 3.2.19 (H^s**-Regularity on Smooth Domains**) *Let* Ω *be a bounded* C^{m+t} *domain for some* $t \in \mathbb{N}$. *Moreover, let the bilinear form*

$$B(u, v) = \sum_{|\alpha|, |\beta| \leq m} \int_\Omega a_{\alpha\beta} D^\alpha u D^\beta v \, dx$$

be $\mathring{H}^m(\Omega)$*-coercive. For the coefficients we assume that*

$$D^\gamma a_{\alpha\beta} \in L_\infty(\Omega) \quad \text{for all} \quad \alpha, \beta, \gamma \quad \text{with} \quad |\gamma| \leq \max(0, t + |\beta| - m).$$

Then each weak solution $u \in \mathring{H}^m(\Omega)$ *of Problem 3.2.11 with* $f \in H^{-m+s}(\Omega)$, $s \leq t$, $s + \frac{1}{2} \notin \{1, 2, \ldots, m\}$ *belongs to* $H^{m+s}(\Omega) \cap \mathring{H}^m(\Omega)$ *and satisfies the estimate*

$$\|u|H^{m+s}(\Omega)\| \lesssim \|f|H^{-m+s}(\Omega)\| + \|u|H^m(\Omega)\|.$$

By Theorem 3.2.18 the condition of $\mathring{H}^m(\Omega)$-coercivity can be replaced by that of uniform ellipticity.

If we weaken the condition on the smoothness of the domain in Theorem 3.2.19 and assume that Ω is Lipschitz instead, we obtain the following regularity result, cf. [Hac92, Thm. 9.1.21].

Theorem 3.2.20 (H^s**-Regularity on Lipschitz Domains**) *Let* $\Omega \subset \mathbb{R}^d$ *be a bounded Lipschitz domain. Let the bilinear form*

$$B(u, v) = \sum_{|\alpha|, |\beta| \leq m} \int_\Omega a_{\alpha\beta} D^\alpha u D^\beta v \, dx$$

be $\mathring{H}^m(\Omega)$-coercive. Furthermore, let $0 < s < t \leq \frac{1}{2}$ and assume that the coefficients $a_{\alpha\beta} \in L_\infty(\Omega)$ belong to $C^t(\overline{\Omega})$ if $|\beta| = m$. Then the weak solution $u \in \mathring{H}^m(\Omega)$ of Problem 3.2.11 with $f \in H^{-m+s}(\Omega)$ belongs to $H^{m+s}(\Omega)$ and satisfies

$$\|u|H^{m+s}(\Omega)\| \lesssim \|f|H^{-m+s}(\Omega)\| + \|u|H^m(\Omega)\|.$$

For the particular case of the Poisson equation the famous $H^{3/2}$-Theorem by Jerison and Kenig [JK95] provides a stronger result and shows that the regularity results from Theorem 3.2.20 are sharp.

Theorem 3.2.21 ($H^{3/2}$**-Theorem**) *Let $\Omega \subset \mathbb{R}^d$ be a bounded Lipschitz domain and consider the Poisson equation*

$$\Delta u = f \quad on \quad \Omega, \qquad u\big|_{\partial\Omega} = 0.$$

If $f \in L_2(\Omega)$ then the weak solution u satisfies $u \in H^{3/2}(\Omega)$. Moreover, for any $s > \frac{3}{2}$ there is a Lipschitz domain Ω and $f \in C^\infty(\overline{\Omega})$ such that the solution u to the inhomogeneous Dirichlet probem does not belong to $H^s(\Omega)$.

We give an example to illustrate our results below. First note that the Poisson problem with homogeneous Dirichlet boundary conditions

$$- \Delta u = f \quad on \quad \Omega, \qquad u\big|_{\partial\Omega} = 0 \tag{3.2.8}$$

can be reduced by the following general strategy to the Laplace equation with inhomogeneous Dirichlet boundary conditions

$$- \Delta u = 0 \quad on \quad \Omega, \qquad u\big|_{\partial\Omega} = g. \tag{3.2.9}$$

Suppose that f is in some space $X^s(\Omega)$ which can be a smoothness space like $H^s(\Omega)$ or $L_2(\Omega)$ (in case $s = 0$). We extend f to a compactly supported function $\tilde{f}$ on all of $\mathbb{R}^d$, which is in the space $X^s(\mathbb{R}^d)$. We solve (3.2.8) with f replaced by $\tilde{f}$ and with Ω replaced by a C^∞-domain $\tilde{\Omega}$ satisfying $\overline{\Omega} \subset \tilde{\Omega}$. For suitable X^s the solution $\tilde{u}$ will be in $X^{s+2}(\tilde{\Omega})$. We can write the solution u to (3.2.8) as

$$u = \tilde{u} - v \quad on \quad \Omega,$$

where v is the solution to the Dirichlet problem

$$-\Delta v = 0 \quad on \quad \Omega, \qquad v\big|_{\partial\Omega} = \tilde{u}\big|_{\partial\Omega} =: g.$$

Using a trace theorem to infer smoothness of g on $\partial\Omega$ a regularity theorem can be deduced for u from regularity theorems for v.

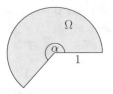

Fig. 3.1 Segment of circle with opening angle α

Fig. 3.2 Solutions u for $\alpha = 1, 2, \ldots, 6$

Example 3.2.22 Consider the following problem

$$\Delta u = 0 \quad \text{on} \quad \Omega, \qquad u|_{\partial\Omega} = \sin\left(\frac{\pi}{\alpha}\varphi\right),$$

where $\Omega \subset \mathbb{R}^2$ denotes the segment of the unit circle with opening angle α (Fig. 3.1), i.e.,

$$\Omega = \{(r\cos\varphi, r\sin\varphi) : 0 < r < 1, \ 0 < \varphi < \alpha\}.$$

The solution is given by

$$u = r^{\frac{\pi}{\alpha}} \sin\left(\frac{\pi}{\alpha}\varphi\right), \qquad r \in [0, 1], \quad \varphi \in [0, \alpha],$$

which can be verified by using the Laplace operator in polar coordinates, which gives

$$\Delta u = \frac{\partial^2 u}{\partial r^2} + \frac{1}{r}\frac{\partial u}{\partial r} + \frac{1}{r^2}\frac{\partial^2 u}{\partial^2 \varphi}$$

$$= \left[\frac{\pi}{\alpha}\left(\frac{\pi}{\alpha} - 1\right) + \frac{\pi}{\alpha} - \left(\frac{\pi}{\alpha}\right)^2\right] r^{\frac{\pi}{\alpha}-2} \sin\left(\frac{\pi}{\alpha}\varphi\right) = 0$$

Solutions for different opening angles α are illustrated below.

We see that even though the boundary conditions (reflected in the sinus function) are smooth, the solution not necessarily is, if the domain has corners. Already from Fig. 3.2 it is clear that the differentiability of the solution in 0 depends on the opening angle α. Precise computation yields

$$\partial_r u = \frac{\pi}{\alpha} r^{\frac{\pi}{\alpha}-1} \sin\left(\frac{\pi}{\alpha}\varphi\right)$$

and only for $\alpha \leq \pi$ exists the directional derivative $(\partial_\nu u)(0)$ if $\nu = (\cos(\alpha/2), \sin(\alpha/2))^T$. Moreover, concerning the regularity of the solution in the fractional Sobolev scale H^s we obtain the following: For solutions $u(r, \varphi) \sim r^\lambda$ the singularities of the derivatives behave like $D^\alpha u(r, \varphi) \sim r^{\lambda-|\alpha|}$, $|\alpha| = k \in \mathbb{N}_0$. Via interpolation this argument extends to non-integer values s. Therefore, we compute for $u \sim r^\lambda$,

$$\|u|H^s(\Omega)\|^2 \sim \int_0^c r^{2(\lambda-s)} r\, dr \sim \int_0^c r^{2\lambda-2s+1} dr \sim \left[r^{2\lambda-2s+2} \right]_0^c$$

i.e., u belongs to $H^s(\Omega)$ if, and only if, $2\lambda - 2s + 2 > 0$, i.e., $s < \lambda + 1$. For our concrete example we have $\lambda = \frac{\pi}{\alpha}$, thus,

$$u \in H^s(\Omega) \quad \text{for all} \quad s < \frac{\pi}{\alpha} + 1.$$

The worst case is obtained for $\alpha = 2\pi$ (circle with a slit), where $s < \frac{3}{2}$.

We conclude this subsection with the observation that for $m = 1$ it is known that convex domains (which are Lipschitz domains) permit stronger regularity results compared to what is stated in Theorem 3.2.20. For the following result we refer to [Hac92, Thm. 9.1.22].

Theorem 3.2.23 (H^s-Regularity on Convex Domains) *Let $\Omega \subset \mathbb{R}^d$ be bounded and convex. Let the bilinear form*

$$B(u, v) = \sum_{|\alpha|, |\beta| \leq m} \int_\Omega a_{\alpha\beta} D^\alpha u D^\beta v\, dx$$

be $\overset{\circ}{H}^1(\Omega)$-coercive and assume that the coefficients of the principal part are Lipschitz continuous, i.e.,

$$a_{\alpha\beta} \in \mathrm{Lip}(\Omega) \quad \text{for all} \quad |\alpha| = |\beta| = 1.$$

For the remaining coefficients we assume that

$$D^\gamma a_{\alpha\beta} \in L_\infty(\Omega) \quad \text{for all} \quad \alpha, \beta, \gamma \quad \text{with} \quad \gamma \leq |\beta|, \ |\alpha| + |\beta| < 1.$$

Then every weak solution $u \in \overset{\circ}{H}^1(\Omega)$ of Problem 3.2.11 with $f \in L_2(\Omega)$ belongs to $H^2(\Omega) \cap \overset{\circ}{H}^1(\Omega)$ and satisfies the estimate

$$\|u|H^2(\Omega)\| \leq c\|f|L_2(\Omega)\| + \|u|H^1(\Omega)\|,$$

where the appearing constant c only depends on the diameter of Ω.

Remark 3.2.24

(i) A generalization of Theorem 3.2.23 in the form of H^{m+1}-regularity for the biharmonic differential equation with $m = 2$ is known for convex polygons.

(ii) For practical applications one often considers polygonal domains Ω. Since polygons are Lipschitz domains, the Dirichlet problem according to Theorem 3.2.20, is H^{m+s}-regular with $0 \leq s < \frac{1}{2}$. If the polygon is convex (i.e, all inner angles are $\leq \pi$) then as in Theorem 3.2.23 one has H^2-regularity if $m = 1$. One obtains results between $H^{3/2}$ and H^2 if the maximal inner angle of the polygon lies between π and 2π (non-convex polygon).

3.3   Parabolic Boundary Value Problems

In this section we deal with parabolic problems of general order. In particular, we give a weak formulation and provide necessary conditions on the coefficients and the functions on the right hand side to guarantee the existence and uniqueness of a weak solution.

3.3.1   *Existence of Weak Solutions*

Let Ω be a bounded domain of $\mathbb{R}^d$ and set $\Omega_T := (0, T] \times \Omega$ for some fixed time $T > 0$.

Problem 3.3.1 (Parabolic Problem)   Consider the *parabolic initial-boundary value problem*

$$
\left\{
\begin{array}{rcl}
\frac{\partial}{\partial t}u + (-1)^m Lu & = & f \ \text{ in } \Omega_T, \\[2mm]
\left. \frac{\partial^{k-1} u}{\partial \nu^{k-1}} \right|_{[0,T] \times \partial\Omega} & = & 0, \ \ k = 1, \ldots, m, \\[2mm]
u(0, \cdot) & = & g \ \text{ in } \Omega.
\end{array}
\right\}
\tag{3.3.1}
$$

Here $f : \Omega_T \to \mathbb{R}$ and $g : \Omega \to \mathbb{R}$ are given functions, ν denotes the exterior normal to $\partial\Omega$ and $u : \overline{\Omega}_T \to \mathbb{R}$ is the unknown $u = u(t, x)$. The operator L is given in divergence form

$$
L = L(t, x, D_x) = \sum_{|\alpha|, |\beta| = 0}^{m} D_x^\alpha (a_{\alpha\beta}(t, x) D_x^\beta),
$$

for given coefficients $a_{\alpha\beta}$. Moreover, we assume $a_{\alpha\beta} = a_{\beta\alpha}$ for all $|\alpha|, |\beta| \le m$ and that for each fixed time $t \in [0, T]$ the operator L is uniformly elliptic according to (3.1.8) in the spatial variable x, i.e., there exists some constant $c > 0$ such that

$$\sum_{|\alpha|=|\beta|=m} a_{\alpha\beta}(t, x)\xi^{\alpha+\beta} \ge c|\xi|^{2m} \qquad \text{for all} \quad \xi \in \mathbb{R}^d, \quad (t, x) \in \Omega_T.$$

Similar as in Sect. 3.2.2 we now consider the *time-dependent bilinear form*

$$B(t, u, v) = \int_\Omega \sum_{|\alpha|,|\beta|=0}^m a_{\alpha\beta}(t, x)(D_x^\beta u)(D_x^\alpha v)dx \qquad (3.3.2)$$

for $u, v \in C_0^\infty(D)$. Then we have

$$(-1)^m (Lu, v)_{L_2(\Omega)} = B(t, u, v),$$

for all $u, v \in C_0^\infty(D)$ and a.e. $t \in [0, T]$ and the *weak formulation* of (3.3.1) reads as

$$\langle \partial_t u, v \rangle + B(t, u, v) = \langle f, v \rangle. \qquad (3.3.3)$$

Definition 3.3.2 (Weak Solution) Given $f \in L_2([0, T], H^{-m}(\Omega))$, a function u satisfying

$$u \in L_2([0, T], H^m(\Omega)), \quad \partial_t u \in L_2([0, T], H^{-m}(\Omega))$$

is called a *weak solution* of Problem 3.3.1 if, and only if, $u(0, \cdot) = g$ in Ω and the equality

$$\langle \partial_t u(t, \cdot), v \rangle + B(t, u, v) = \langle f(t, \cdot), v \rangle,$$

holds for all $v \in \overset{\circ}{H}{}^m(\Omega)$ and a.e. $t \in [0, T]$.

Remark 3.3.3 If $f \in L_2(\Omega_T)$ we have $\langle f, v \rangle = (f, v)$.

We need some assumptions on the functions f, g and the coefficients $a_{\alpha\beta}$ in order to guarantee the existence of a weak solution. The following Theorem is proven in [Eva10, Sect. 7.1, Thm. 3,4] and provides existence and uniqueness of a weak solution for second order parabolic problems.

Theorem 3.3.4 (Existence and Uniqueness of Weak Solutions) *Let Ω be a bounded domain, $m = 1$, and assume that*

$$a_{\alpha\beta} \in L_\infty(\Omega_T), \qquad f \in L_2(\Omega_T), \qquad g \in L_2(\Omega).$$

Then Problem 3.3.1 has a unique weak solution u.

This is the counterpart of Theorems 3.2.14 and 3.2.15 for parabolic problems of second order. In Theorem 5.2.2 of Sect. 5.2.1 we will show that the results in Theorem 3.3.4 can be generalized to higher order parabolic problems under suitable assumptions on the coefficients $a_{\alpha\beta}$ and the right hand side f (choosing $g = 0$), when the underlying domain is bounded and of polyhedral type. Moreover, in Theorem 5.2.3 we present some regularity estimates for the weak solution in Sobolev spaces with higher order time derivatives.

Remark 3.3.5 (Assumptions on the Time-Dependent Bilinear Form) In general, if we assume uniform ellipticity of L and $a_{\alpha\beta} \in L_\infty(\Omega_T)$ as above, the corresponding bilinear form $B(t, \cdot, \cdot)$ of our parabolic problem satisfies Gårding's inequality, cf. Theorem 3.2.18. To be precise, there exist constants $\mu > 0$, $\lambda_0 \geq 0$ such that

$$B(t, u, u) \geq \mu \|u|H^m(D)\|^2 - \lambda_0 \|u|L_2(D)\|^2 \qquad (3.3.4)$$

holds for all $u \in \mathring{H}^m(D)$ and a.e. $t \in [0, T]$. Note that in the parabolic case the constant λ_0 can be chosen to be 0, since by a substitution $v = e^{-\lambda_0 t} u$ the operator L can be transformed to $\tilde{L} = L + (-1)^m \lambda_0$ with time-dependent bilinear form $\tilde{B}(t, \cdot, \cdot)$ satisfying (3.3.4) with constant $\lambda_0 = 0$: Consider the equation

$$\partial_t v + (-1)^m \tilde{L}(x, t, D_x) v = e^{-\lambda_0 t} f,$$

thus, the weak formulation is given by

$$\langle \partial_t v, w \rangle + \tilde{B}(t, v, w) = \left\langle e^{-\lambda_0 t} f, w \right\rangle, \qquad (3.3.5)$$

where $\tilde{B}(t, v, w) = B(t, v, w) + \lambda_0 (v, w)$ and $w \in C_0^\infty(D)$. Putting $v = e^{-\lambda_0 t} u$ in (3.3.5) yields

$$\left\langle \frac{\partial e^{-\lambda_0 t} u}{\partial t}, w \right\rangle + B(t, e^{-\lambda_0 t} u, w) + \lambda_0(e^{-\lambda_0 t} u, w)$$

$$= \left\langle \frac{\partial u}{\partial t} e^{-\lambda_0 t} - \lambda_0 e^{-\lambda_0 t} u, w \right\rangle + e^{-\lambda_0 t} B(t, u, w) + e^{-\lambda_0 t} \lambda_0(u, w)$$

$$= e^{-\lambda_0 t} \left(\langle \partial_t u, w \rangle + B(t, u, w) \right) = e^{-\lambda_0 t} \langle f, w \rangle = \left\langle e^{-\lambda_0 t} f, w \right\rangle.$$

From the above calculations we see that u is a solution of (3.3.3) if, and only if, v is a solution of (3.3.5). Moreover, $\tilde{B}(t, v, w)$ is elliptic since

$$\tilde{B}(t, v, v) = B(t, v, v) + \lambda_0(v, v)$$

$$\geq \mu \|v|H^m(D)\|^2 - \lambda_0 \|v|L_2(D)\|^2 + \lambda_0 \|v|L_2(D)\|^2$$

$$= \mu \|v|H^m(D)\|^2.$$

Hence, when dealing with parabolic problems it will be reasonable to suppose that $B(t, \cdot, \cdot)$ satisfies

$$B(t, u, u) \geq \mu \|u|H^m(D)\|^2 \tag{3.3.6}$$

for all $u \in \mathring{H}^m(D)$ and a.e. $t \in [0, T]$.

3.4 Hyperbolic Boundary Value Problems

We now turn our attention to hyperbolic problems of second order and present a result concerning necessary conditions for the existence and uniqueness of a weak solution.

As in the previous section we write $\Omega_T := (0, T] \times \Omega$, where $T > 0$ and $\Omega \subset \mathbb{R}^d$ is a bounded domain.

Problem 3.4.1 (Second Order Hyperbolic Problem) We consider the following *hyperbolic initial-boundary value problem*

$$\left\{ \begin{array}{c} \frac{\partial^2}{\partial t^2} u + Lu = f \quad \text{in} \quad \Omega_T, \\ u\big|_{[0,T] \times \partial\Omega} = 0, \\ u(0, \cdot) = g, \quad \frac{\partial}{\partial t} u(0, \cdot) = h \quad \text{in} \quad \Omega. \end{array} \right\} \tag{3.4.1}$$

Here $f : \Omega_T \to \mathbb{R}$, $g, h : \Omega \to \mathbb{R}$ are given functions and $u : \overline{\Omega}_T \to \mathbb{R}$ is the unknown $u = u(t, x)$. Moreover, L denotes a linear differential operator of second order on Ω_T in divergence form

$$Lu = L(t, x, D_x)u = -\sum_{i,j=1}^{d} \frac{\partial}{\partial x_j} \left(a_{ij}(t, x) \frac{\partial u}{\partial x_i} \right) + \sum_{i=1}^{d} b_i(t, x) \frac{\partial u}{\partial x_i} + c(t, x)u,$$

for given coefficients a_{ij}, b_i, and c. Moreover, we assume $a_{ij} = a_{ji}$, where $i, j = 1, \ldots, d$, and that for each fixed time $t \in [0, T]$ the operator L is uniformly elliptic in the spatial variable x according to (3.1.8), i.e., there exists a constant $c > 0$ such that

$$\sum_{i,j=1}^{d} a_{ij}(t, x)\xi_i\xi_j \geq c|\xi|^2 \qquad \text{for all} \quad \xi \in \mathbb{R}^d, \quad (t, x) \in \Omega_T.$$

As in Sect. 3.3 we consider the time-dependent bilinear form

$$B(t, u, v) = \int_\Omega \left(\sum_{i,j=1}^d a_{ij}(t, x) \frac{\partial u}{\partial x_i} \frac{\partial v}{\partial x_j} + \sum_{i=1}^d b_i(t, x) \frac{\partial u}{\partial x_i} v + c(t, x) u v \right) dx,$$

for $u, v \in \overset{\circ}{H}{}^1(\Omega)$ and $t \in [0, T]$.

Definition 3.4.2 (Weak Solution) Given $f \in L_2([0, T], H^{-1}(\Omega))$ a function u satisfying

$$u \in L_2([0, T], \overset{\circ}{H}{}^1(\Omega)), \quad \partial_t u \in L_2([0, T], L_2(\Omega)), \quad \partial_{t^2} u \in L_2([0, T], H^{-1}(\Omega)),$$

is called a *weak solution* of Problem 3.4.1, if, and only if, $u(\cdot, 0) = g$, $\partial_t u(\cdot, 0) = h$, and the equality

$$\langle \partial_{t^2} u(t, \cdot), v \rangle + B(t, u(t), v) = \langle f(\cdot, t), v \rangle \tag{3.4.2}$$

holds for all $v \in \overset{\circ}{H}{}^1(\Omega)$ and a.e. $t \in [0, T]$.

Remark 3.4.3 If $f \in L_2(\Omega_T)$ we have $\langle f, v \rangle = (f, v)$ in (3.4.2).

We need some assumptions on the functions g, h and the coefficients a_{ij}, b_i, and c in order to guarantee the existence of a weak solution. The following Theorem is proven in [Eva10, Sect. 7.2, Thm. 3,4].

Theorem 3.4.4 (Existence and Uniqueness of Weak Solution) *Let Ω be a bounded domain and assume that*

$$a_{ij}, b_i, c \in C^1(\overline{\Omega}_T), \qquad f \in L_2(\Omega_T), \qquad g \in \overset{\circ}{H}{}^1(\Omega), \quad h \in L_2(\Omega).$$
$$\tag{3.4.3}$$

Then Problem 3.4.1 has a unique weak solution u.

3.5   Operator Pencils

In order to correctly state the global regularity results in Kondratiev spaces for Problems I and II in Chap. 5, we need to work with operator pencils generated by the corresponding elliptic problems in the polyhedral type domain $D \subset \mathbb{R}^3$. The reason for this is as follows: When establishing regularity results for parabolic PDEs we heavily rely on regularity results for related elliptic problems. However, in the elliptic theory for non-smooth domains most of the results are conditional since singularities of the solutions are described in terms of spectral properties of certain operator pencils (i.e., operators polynomially depending on a complex parameter λ) of boundary value problems on spherical domains. Therefore, results on the regularity of solutions are usually given under a priori conditions on the eigenvalues and (generalized) eigenvectors of these pencils.

In order to give an idea what kind of singularities we are dealing with and how they are related to the spectral theory of operator pencils, we consider the solution of an elliptic boundary value problem $Lu = f$ in a smooth cone $K \subset \mathbb{R}^d$. The solution, under certain conditions, behaves asymptotically near the vertex as

$$u(x) = |x|^{\lambda_0} \sum_{k=0}^{s} \frac{1}{k!} (\log |x|)^k U_{s-k}(x/|x|), \tag{3.5.1}$$

where λ_0 is an eigenvalue of an operator pencil—denoted by $\mathfrak{A}(\lambda)$—of the corresponding boundary value problem on the domain $\Omega = K \cap S^{d-1}$, which the cone cuts out on the unit sphere. Moreover, U_0 denotes an eigenvector and $U_1, \ldots, U_s$ generalized eigenvectors of the pencil corresponding to λ_0. Thus, one has been naturally led to the study of spectral properties of polynomial operator pencils.

Similar expressions with power-logarithmic terms as in (3.5.1) can also be used to characterize singularities near edges and vertices of polyhedral type domains.

The above mentioned operator pencil $\mathfrak{A}(\lambda)$ is obtained by constructing a so-called model problem from the original equation $Lu = f$ (depending on the parameter λ), which only involves the main part L_0 of the given differential operator L with coefficients frozen at a certain point: If one sets $u(x) = |x|^{\lambda} U(\omega)$, where $\omega = x/|x|$ in the corresponding elliptic equation $Lu = f$, one obtains a boundary value problem for the function U on the subdomain Ω of the unit sphere with the complex parameter λ. This boundary value problem is uniquely solvable for all λ except for a denumerable set of eigenvalues of the pencil $\mathfrak{A}(\lambda)$. In particular, it is known that the '*energy line*' $\mathrm{Re}\,\lambda = m - d/2$, is free of eigenvalues of the pencils and, hence, we are concerned with the widest strip in the λ-plane, free of eigenvalues and containing the line $\mathrm{Re}\,\lambda = m - d/2$. From this we can afterwards deduce that our original problem has a (unique) solution in weighted Sobolev spaces subject to some restrictions on the weight parameters depending on the (distribution of) eigenvalues of the operator pencils, cf. Assumption 5.2.5 and the discussion afterwards.

After having outlined the general ideas, we now briefly recall the basic facts needed in the sequel. For further information on this subject we refer to [KMR01] and [MR10, Sect. 2.3, 3.2., 4.1]. Consider Problem 3.2.11 on a domain $D \subset \mathbb{R}^3$ of polyhedral type according to Definition 2.4.5, i.e.,

$$\left\{ \begin{array}{l} Lu = f \ \text{ in } \ D, \\ \left. \frac{\partial^{k-1} u}{\partial \nu^{k-1}} \right|_{\partial D} = 0, \quad k = 1, \ldots, m. \end{array} \right\} \tag{3.5.2}$$

When introducing the operator pencils we shall use the equivalent characterization of the domain D from Remark 2.4.6(i) and assume that the boundary ∂D consists of smooth faces Γ_j, $j = 1, \ldots, n$, smooth curves M_k, $k = 1, \ldots, l$ (the edges), and vertices $x^{(1)}, \ldots, x^{(l')}$. The singular set S of D then is given by the boundary points $M_1 \cup \ldots \cup M_l \cup \{x^{(1)}, \ldots, x^{(l')}\}$. We do not exclude the cases $l = 0$ (corner domain) and $l' = 0$ (edge domain). In the last case, the set S consists only of smooth non-intersecting edges. Figure 3.3 gives examples of polyhedral domains without edges or corners, respectively.

The elliptic boundary value problem (3.5.2) on D generates two types of operator pencils for the edges M_k and for the vertices $x^{(i)}$ of the domain, respectively.

(1) Operator Pencil $A_\xi(\lambda)$ for Edge Points

The pencils $A_\xi(\lambda)$ for edge points $\xi \in M_k$ are defined as follows: By the characterization of polyhedral type domains, cf. Remark 2.4.6(i), there exists a neighborhood U_ξ of ξ and a diffeomorphism κ_ξ mapping $D \cap U_\xi$ onto $\mathcal{D}_\xi \cap B_1(0)$, where $\mathcal{D}_\xi$ is a dihedron.

Let $\Gamma_{k\pm}$ be the faces adjacent to M_k. Then by $\mathcal{D}_\xi$ we denote the dihedron which is bounded by the half-planes $\mathring{\Gamma}_{k\pm}$ tangent to $\Gamma_{k\pm}$ at ξ and the edge $M_\xi = \mathring{\Gamma}_{k_+} \cap \mathring{\Gamma}_k$ (Fig. 3.4). Furthermore, let r, φ be polar coordinates in the plane perpendicular to M_ξ such that

$$\mathring{\Gamma}_{k\pm} = \left\{ x \in \mathbb{R}^3 : r > 0, \ \varphi = \pm\frac{\theta_\xi}{2} \right\}.$$

Fig. 3.3 Corner domain D_c ($l = 0$) and edge domain D_e ($l' = 0$)

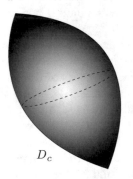

D_e

D_c

Fig. 3.4 Dihedron $\mathcal{D}_\xi$

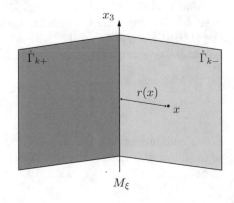

We define the *operator pencil* $A_\xi(\lambda)$ as follows:

$$A_\xi(\lambda)U(\varphi) = r^{2m-\lambda}L_0(0, D_x)u, \tag{3.5.3}$$

where $u(x) := r^\lambda U(\varphi)$, $\lambda \in \mathbb{C}$, U is a function on $I_\xi := \left(\frac{-\theta_\xi}{2}, \frac{\theta_\xi}{2}\right)$, and

$$L_0(\xi, D_x) = \sum_{|\alpha|=|\beta|=m} D_x^\alpha (a_{\alpha\beta}(\xi)D_x^\beta)$$

denotes the main part of the differential operator $L(x, D_x)$ with coefficients frozen at ξ. This way we obtain in (3.5.3) a boundary value problem for the function U on the 1-dimensional subdomain I_ξ with the complex parameter λ. Moreover, $A_\xi(\lambda)$ a is polynomial of degree $2m$ in λ, which is exemplarily shown in Example 3.5.1 below.

The operator $A_\xi(\lambda)$ realizes a continuous mapping

$$H^{2m}(I_\xi) \to L_2(I_\xi),$$

for every $\lambda \in \mathbb{C}$. Furthermore, $A_\xi(\lambda)$ is an isomorphism for all $\lambda \in \mathbb{C}$ with the possible exception of a denumerable set of isolated points, the *spectrum of* $A_\xi(\lambda)$, which consists of its eigenvalues with finite algebraic multiplicities: Here a complex number λ_0 is called an *eigenvalue of the pencil* $A_\xi(\lambda)$ if there exists a nonzero function $U \in H^{2m}(I_\xi)$ such that $A_\xi(\lambda_0)U = 0$. It is known that the '*energy line*' $\text{Re}\lambda = m - 1$ does not contain eigenvalues of the pencil $A_\xi(\lambda)$. We denote by $\delta_\pm^{(\xi)}$ the greatest positive real numbers such that the strip

$$m - 1 - \delta_-^{(\xi)} < \text{Re}\lambda < m - 1 + \delta_+^{(\xi)} \tag{3.5.4}$$

is free of eigenvalues of the pencil $A_\xi(\lambda)$. Furthermore, we put

$$\delta_\pm^{(k)} = \inf_{\xi \in M_k} \delta_\pm^{(\xi)}, \qquad k = 1, \ldots, l. \tag{3.5.5}$$

Example 3.5.1 (Dirichlet Problem for Poisson Equation on Dihedron) Consider the Dirichlet problem for the Poisson equation

$$-\Delta u = f \quad \text{in} \quad \mathcal{D}, \qquad u\big|_{\partial\mathcal{D}} = 0$$

on a dihedron $\mathcal{D} = K \times \mathbb{R} = \{x = (x', x_3) : x' = (x_1, x_2) \in K, \ x_3 \in \mathbb{R}\}$, where K is a 2-dimensional wedge given in polar coordinates via

$$K = \{x' = (x_1, x_2) : 0 < r < \infty, \ -\theta/2 < \varphi < \theta/2\}.$$

Then for the pencil $A(\lambda)$ introduced in (3.5.3) we see that

$$
\begin{aligned}
A(\lambda)U(\varphi) &= r^{2-\lambda}L_0(0, D_{x'})(r^{\lambda}U(\varphi)) \\
&= r^{2-\lambda}\Delta_{x'}(r^{\lambda}U(\varphi)) \\
&= r^{2-\lambda}\left[\partial_{r^2} + \frac{1}{r}\partial_r + \frac{1}{r^2}\partial_{\varphi^2}\right](r^{\lambda}U(\varphi)) \\
&= r^{2-\lambda}\left[\lambda(\lambda-1)r^{\lambda-2}U(\varphi) + \lambda r^{\lambda-2}U(\varphi) + r^{\lambda-2}\partial_{\varphi^2}U(\varphi)\right] \\
&= \lambda^2 U(\varphi) + \partial_{\varphi^2}U(\varphi),
\end{aligned}
$$

where in the third line we used the representation of the Laplacian in polar coordinates. The above calculation yields

$$
A(\lambda) = \lambda^2 + \partial_{\varphi^2}.
$$

Concerning eigenvalues of $A(\lambda)$ from Example 3.2.22 we see that a solution to $\Delta_{x'}(r^{\lambda}U(\varphi)) = 0$ on the segment of a circle with opening angle θ is given by $U(\varphi) = \sin\left(\frac{\pi}{\theta}\varphi\right)$. This gives

$$
A(\lambda)U(\varphi) = \left[\lambda^2 - \left(\frac{\pi}{\theta}\right)^2\right]\sin\left(\frac{\pi}{\theta}\varphi\right) = 0 \qquad \Longleftrightarrow \qquad \lambda = \pm\pi/\theta.
$$

Therefore, the first positive eigenvalue is $\lambda_1 = \pi/\theta$, where θ is the inner angle of the dihedron. Moreover, we obtain $\delta_{\pm} = \frac{\pi}{\theta}$ in (3.5.5).

(2) Operator Pencil $\mathfrak{A}_i(\lambda)$ for Corner Points
Let $x^{(i)}$ be a vertex of D. By the characterization of polyhedral type domains, cf. Remark 2.4.6(i), there exists a neighborhood U_i of $x^{(i)}$ and a diffeomorphism κ_i mapping $D \cap U_i$ onto $K_i \cap B_1(0)$, where

$$
K_i = \{x \in \mathbb{R}^3 : x/|x| \in \Omega_i\}
$$

is a polyhedral cone with edges and vertex at the origin. W.l.o.g. we may assume that the Jacobian matrix $\kappa_i'(x)$ is equal to the identity matrix at the point $x^{(i)}$. We introduce spherical coordinates $\rho = |x|$, $\omega = \frac{x}{|x|}$ in K_i and define the operator pencil

$$
\mathfrak{A}_i(\lambda)U(\omega) = \rho^{2m-\lambda}L_0(x^{(i)}, D_x)u, \tag{3.5.6}
$$

where $u(x) = \rho^\lambda U(\omega)$ and $U \in \mathring{H}^m(\Omega_i)$ is a function on Ω_i. An *eigenvalue of* $\mathfrak{A}_i(\lambda)$ is a complex number λ_0 such that $\mathfrak{A}_i(\lambda_0)U = 0$ for some nonzero function $U \in \mathring{H}^m(\Omega_i)$. The operator $\mathfrak{A}_i(\lambda)$ realizes a continuous mapping

$$\mathring{H}^m(\Omega_i) \to H^{-m}(\Omega_i).$$

Furthermore, it is known that $\mathfrak{A}_i(\lambda)$ is an isomorphism for all $\lambda \in \mathbb{C}$ with the possible exception of a denumerable set of isolated points. The mentioned enumerable set consists of eigenvalues with finite algebraic multiplicities.

Moreover, the eigenvalues of $\mathfrak{A}_i(\lambda)$ are situated, except for finitely many, outside a double sector $|\text{Re}\lambda| < \varepsilon|\text{Im}\lambda|$ containing the imaginary axis, cf. [KMR01, Thm. 10.1.1]. In Fig. 3.5 the situation is illustrated: Outside the yellow area there are only finitely many eigenvalues of the operator pencil $\mathfrak{A}_i(\lambda)$.

Dealing with regularity properties of solutions, we look for the widest strip in the λ-plane, free of eigenvalues and containing the '*energy line*'

$$\text{Re}\lambda = m - 3/2,$$

cf. Assumption 5.2.5. From what was outlined above, information on the width of this strip is obtained from lower estimates for real parts of the eigenvalues situated over the energy line.

Fig. 3.5 Eigenvalues of operator pencil $\mathfrak{A}_i(\lambda)$

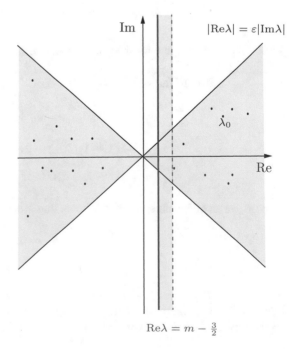

Example 3.5.2 (Dirichlet Problem for Poisson Equation on Polyhedral Type Domain) We consider the Dirichlet problem for the Poisson equation

$$-\Delta u = f \quad \text{in} \quad D, \qquad u\big|_{\partial D} = 0,$$

where $D \subset \mathbb{R}^3$ is a domain of polyhedral type. The eigenvalues of the pencil $A_\xi(\lambda)$ are given by

$$\lambda_k = k\pi/\theta_\xi, \qquad k = \pm 1, \pm 2, \ldots,$$

where θ_ξ is the inner angle at the edge point ξ. We refer to Example 3.5.1, where λ_1 was computed and [KMR01, Ch. 2] for the general case. Let Λ_k, $k \in \mathbb{N}$, be the eigenvalues of the Laplace-Beltrami operator $-\delta = -\Delta_\omega$ (with Dirichlet boundary condition on the subdomain Ω_j of the unit sphere). Then we can show that the eigenvalues of the pencils $\mathfrak{A}_i(\lambda)$ are given by

$$\tilde{\lambda}_k^\pm = -\frac{1}{2} \pm \sqrt{\Lambda_k + \frac{1}{4}}. \tag{3.5.7}$$

This can be seen as follows: The Laplace operator in d-dimensional spherical coordinates can be written as

$$\Delta u = \partial_{\rho^2} u + \frac{d-1}{\rho} \partial_\rho u + \frac{1}{\rho^2} \Delta_\omega u.$$

Since $d = 3$ we compute for the operator $\mathfrak{A}_i(\lambda)$,

$$\begin{aligned}
\mathfrak{A}_i(\lambda)U(\omega) &= \rho^{2-\lambda} L_0(0, \partial_\rho, \partial_\omega)(\rho^\lambda U(\omega)) \\
&= -\rho^{2-\lambda} \left[\partial_{\rho^2} + \frac{2}{\rho}\partial_\rho + \frac{1}{\rho^2}\Delta_\omega \right] (\rho^\lambda U(\omega)) \\
&= -\lambda(\lambda - 1)U(\omega) - 2\lambda U(\omega) - \delta U(\omega) \\
&= [-\lambda^2 - \lambda - \delta]U(\omega).
\end{aligned}$$

Since Λ_k is an eigenvalue of $-\delta$ it follows that λ is an eigenvalue of $\mathfrak{A}_i(\lambda)$ if

$$\lambda^2 + \lambda - \Lambda_k = 0,$$

which gives (3.5.7).

Remark 3.5.3 (Operator Pencils for Parabolic Problems) Since we study parabolic PDEs in Chap. 5, where the differential operator $L(t, x, D_x)$ additionally depends on the time t, we have to work with operator pencils $A_\xi(\lambda, t)$ and $\mathfrak{A}_i(\lambda, t)$ in this context. The philosophy is to fix $t \in [0, T]$ and define the pencils as above: We replace (3.5.3) by

$$A_\xi(\lambda, t)U(\varphi) = r^{2m-\lambda}L_0(t, 0, D_x)u,$$

and work with $\delta_\pm^{(\xi)}(t)$ and $\delta_\pm^{(k)}(t) = \inf_{\xi \in M_k} \delta_\pm^{(\xi)}(t)$ in (3.5.4) and (3.5.5), respectively. Moreover, we put

$$\delta_\pm^{(k)} = \inf_{t \in [0,T]} \delta_\pm^{(k)}(t), \qquad k = 1, \ldots, l. \tag{3.5.8}$$

Similar for $\mathfrak{A}_i(\lambda, t)$, where now (3.5.6) is replaced by

$$\mathfrak{A}_i(\lambda, t)U(\omega) = \rho^{2m-\lambda}L_0(t, x^{(i)}, D_x)u. \tag{3.5.9}$$

Chapter 4
Regularity Theory for Elliptic PDEs

In this chapter we are concerned with regularity estimates in Kondratiev and Besov spaces for the solutions of semilinear elliptic partial differential equations. The nonlinear term $g(x, \xi)$ we consider in this context is quite general and required to satisfy a growth condition w.r.t. its partial derivatives. In our considerations we pay special attention to monomials, i.e., when g is of the form $g(x, \xi) = \xi^n$, where $n \in \mathbb{N}$ with $n \geq 2$, and give precise estimates in this case.

In order to study the regularity of the solutions u we proceed as follows: We rewrite our elliptic problem as a fixed point problem of the form $(L \circ N)u = u$, for suitable (non-)linear operators L and N, respectively. Moreover, we study mapping properties of the nonlinear composition operator N in Kondratiev spaces.

Our central tool will then be to use the fixed point theorem stated in Proposition 4.2.7 in combination with regularity results in Kondratiev spaces for the corresponding linear problem from [BMNZ10, MN10]. However, in order to apply this fixed point theorem we have to show that the Kondratiev spaces and (non-)linear operators L and N under consideration satisfy the requirements of Proposition 4.2.7. In particular, we show that Kondratiev spaces are so-called admissible spaces and deal with complete continuity of $L \circ N$. The latter forces us to even work with fractional Kondratiev spaces (in order to obtain compactness and not loose too much information).

Our first fundamental regularity result is Theorem 4.2.18, which implies that regularity estimates for linear elliptic operators in Kondratiev spaces carry over to the semilinear equations. Note that from Proposition 4.2.7 we do only get existence but not uniqueness of the solution. Therefore, in Theorem 4.2.21 we show that for monomials it is also possible to apply Banach's fixed point theorem, which establishes also uniqueness of the solution in this context.

Finally, using embedding results between Kondratiev and Triebel-Lizorkin spaces, we obtain Besov regularity for our elliptic problem in Theorem 4.3.1. In Chap. 7 we discuss the impact of our findings in regard to N-term wavelet and finite element approximation.

© The Author(s), under exclusive license to Springer Nature Switzerland AG 2021   145
C. Schneider, *Beyond Sobolev and Besov*, Lecture Notes in Mathematics 2291,
https://doi.org/10.1007/978-3-030-75139-5_4

4.1   The Fundamental Semilinear Elliptic Problem

We consider the following semilinear elliptic PDE of the form

$$- \nabla\big(A(x) \cdot \nabla u(x)\big) + g(x, u(x)) = f(x) \quad \text{in} \quad D, \qquad u|_{\partial D} = 0, \qquad (4.1.1)$$

where $A = (a_{i,j})_{i,j=1}^{d}$ is symmetric and its coefficients satisfy certain smoothness and growth conditions as stated in Proposition 4.1.2 below. Moreover, the nonlinear term $g : D \times \mathbb{R} \to \mathbb{R}$ is assumed to be continuous with derivatives satisfying the growth conditions (4.2.4).

Unless otherwise stated, throughout this chapter we will always assume that the domain D under consideration satisfies the

Assumption 4.1.1 *Let $D \subset \mathbb{R}^d$, $d \in \{2, 3\}$, be a bounded Lipschitz domain of polyhedral type according to Definition 2.4.5.*

The following shift theorem for the corresponding linear elliptic problem of (4.1.1) from [BMNZ10, MN10] will form the basis for the investigations presented in this chapter. It demonstrates the relevance of Kondratiev spaces by showing that within the scale of these spaces one can prove shift theorems also on non-smooth domains analogously to those in the usual Sobolev scale on smooth domains.

Proposition 4.1.2 (Kondratiev Regularity of Linear Elliptic Problem) *Let D be some bounded polyhedral domain without cracks in $\mathbb{R}^d$, $d \in \{2, 3\}$ and $m \in \mathbb{N}$. Consider the problem*

$$- \nabla\big(A(x) \cdot \nabla u(x)\big) = f \quad \text{in} \quad D, \qquad u|_{\partial D} = 0, \qquad (4.1.2)$$

where $A = (a_{i,j})_{i,j=1}^{d}$ is symmetric and

$$a_{i,j} \in \mathcal{K}_{\infty,0}^{m}(D) = \big\{v : D \longrightarrow \mathbb{R} : \rho^{|\alpha|} \partial^{\alpha} v \in L_{\infty}(D), |\alpha| \leq m \big\}, \qquad 1 \leq i, j \leq d.$$

Let the associated bilinear form

$$B(v, w) = \int_{D} \sum_{i,j} a_{i,j}(x) \partial_i v(x) \partial_j w(x) dx$$

satisfy

$$|B(v, w)| \leq R \|v|H^1(D)\| \cdot \|w|H^1(D)\| \qquad and \qquad r \|v|H^1(D)\|^2 \leq B(v, v)$$

for all $v, w \in \mathring{H}^1(D)$ and some constants $0 < r \leq R < \infty$. Then there exists some
$\bar{a} > 0$ such that for any $m \in \mathbb{N}_0$, any $|a| < \bar{a}$, and any $f \in \mathcal{K}^{m-1}_{2,a-1}(D)$ the problem
(4.1.2) admits a uniquely determined solution $u \in \mathcal{K}^{m+1}_{2,a+1}(D)$, and it holds

$$\|u|\mathcal{K}^{m+1}_{2,a+1}(D)\| \leq C \|f|\mathcal{K}^{m-1}_{2,a-1}(D)\|$$

for some constant $C > 0$ independent of f.

Remark 4.1.3

(i) The conditions of Proposition 4.1.2 clearly imply that the bilinear form B is continuous and $\mathring{H}^1(D)$-elliptic according to (3.2.2) and (3.2.3), respectively. Hence, assuming $f \in H^{-1}(D)$, according to Theorem 3.2.14 there exists a unique weak solution $u \in \mathring{H}^1(D)$ to problem (4.1.2). Proposition 4.1.2 implies that under certain conditions on the coefficients $a_{i,j}, 1 \leq i, j \leq d$, and the right–hand side f this weak solution possesses additional regularity in the scale of Kondratiev spaces.

(ii) For a polygon $\Omega \subset \mathbb{R}^2$ without cracks we have $\bar{a} = \frac{\pi}{\alpha_{max}}$ in Proposition 4.1.2, where α_{max} is the largest angle of Ω, cf. [BNZ05, Sect. 2.1].

(iii) In the literature there are further results of this type, either treating different boundary conditions, or using slightly different scales of function spaces. We particularly refer to [KMR97, Ch. 6] and [MR10, Part 1, Ch. 4].

(iv) The domains D satisfying Assumption 4.1.1 are special polyhedral domains. Moreover, they are covered by the domains considered in Proposition 4.1.2.

4.2 Regularity of Semilinear Elliptic PDEs in Kondratiev Spaces

The goal of this section is to establish regularity estimates in Kondratiev spaces for the solution to semilinear problems of the form (4.1.1).

4.2.1 Composition Operators in Kondratiev Spaces

Let L be the solution operator associated to (4.1.2), i.e., $Lf = u$. Then, by Proposition 4.1.2, we know that L is well-defined on $\mathcal{K}^{m-1}_{2,a-1}(D)$ with values in the set

$$\mathcal{K}^{m+1}_{2,a+1,0}(D) := \left\{ u \in \mathcal{K}^{m+1}_{2,a+1}(D) : u|_{\partial D} = 0 \right\}. \tag{4.2.1}$$

Vice versa, to each $u \in \mathcal{K}^{m+1}_{2,a+1,0}(D)$ there exists an $f \in \mathcal{K}^{m-1}_{2,a-1}(D)$ such that (4.1.2) is satisfied and this operation is bounded as well. This can be seen as follows: Write (4.1.2) as

$$-\nabla(A \cdot \nabla u) = -\left(\sum_{i,j=1}^{d}(\partial_{x_i}a_{i,j})(\partial_{x_j}u) + \sum_{i,j=1}^{d}a_{i,j}(\partial_{x_i}\partial_{x_j}u)\right)$$

$$= -\left(\sum_{i,j=1}^{d}(\rho\partial_{x_i}a_{i,j})(\rho^{-1}\partial_{x_j}u) + \sum_{i,j=1}^{d}a_{i,j}(\partial_{x_i}\partial_{x_j}u)\right).$$

Since $u \in \mathcal{K}^{m+1}_{2,a+1}(D)$ Theorem 2.4.4 yields that $\partial_{x_i}\partial_{x_j}u \in \mathcal{K}^{m-1}_{2,a-1}(D)$ and $\partial_{x_j}u \in \mathcal{K}^{m}_{2,a}(D)$. Thus, $\rho^{-1}\partial_{x_j}u \in \mathcal{K}^{m}_{2,a-1}(D) \hookrightarrow \mathcal{K}^{m-1}_{2,a-1}(D)$, where we used Remark 2.4.3(ii),(iv). Moreover, since the coefficients satisfy $a_{i,j} \in \mathcal{K}^{m}_{\infty,0}(D)$, we deduce that $a_{i,j}$ as well as $\rho\partial_{x_i}a_{i,j}$ are bounded in L_∞. Therefore, $-\nabla(A \cdot \nabla u) \in \mathcal{K}^{m-1}_{2,a-1}(D)$ for $u \in \mathcal{K}^{m+1}_{2,a+1}(D)$.

Hence, L is an isomorphism. By defining the linear operator

$$L^{-1}u(x) := -\nabla(A(x) \cdot \nabla u(x)), \qquad x \in D, \qquad (4.2.2)$$

and the nonlinear map

$$N(u)(x) := f(x) - g(x, u(x)), \qquad x \in D, \qquad (4.2.3)$$

we see that Eq. (4.1.1) can be written as

$$u = (L \circ N)u,$$

which will enable us to apply suitable fixed point theorems in the regularity spaces we are interested in. Therefore, we have to study the mapping properties of $L \circ N$ and in this context derive bounds for the nonlinear map N in (4.2.3). For this purpose, we will directly estimate the weighted L_p-norms of partial derivatives of the function g, which requires imposing certain growth conditions on the classical partial derivatives of the function. Our result can be formulated as follows:

Theorem 4.2.1 *Let D be as in Definition 2.4.5. Let $1 < p < \infty$, $a \geq \frac{d}{p} - 1$, $m \in \mathbb{N}$, and $\delta \geq \max(1, m-1)$ such that the continuous function $g : D \times \mathbb{R} \to \mathbb{R}$ and its continuous classical derivatives fulfill the growth conditions*

$$|\partial_\xi^l \partial_x^\alpha g(x, \xi)| \leq c_{\alpha,l}|\xi|^{\delta - l}, \qquad l \in \mathbb{N}_0, \quad \alpha \in \mathbb{N}_0^d, \quad l + |\alpha| \leq m - 1. \qquad (4.2.4)$$

Moreover, let either

$$\min(m + 1, 3) - \frac{d}{p} > 0 \qquad (4.2.5)$$

or

$$0 > m + 1 - \frac{d}{p} \geq -\frac{2}{\delta - 1}$$

(no lower restriction in case $\delta = 1$).

Then the nonlinear operator $T_G : u \mapsto G(u) := g(\cdot, u(\cdot))$ maps functions from $\mathcal{K}^{m+1}_{p,a+1}(D)$ to functions in $\mathcal{K}^{m-1}_{p,a-1}(D)$. Moreover, it holds

$$\|G(u)|\mathcal{K}^{m-1}_{p,a-1}(D)\| \leq C \, \|u|\mathcal{K}^{m+1}_{p,a+1}(D)\|^{\delta}$$

with some constant $C > 0$ independent of u.

Proof

Step 1.   Preparations. As a first step we shall have a look at partial derivatives of the composed function $g(x, u(x))$. We first observe that, except for $\partial_x^{\alpha} g(x, \xi)|_{\xi = u(x)}$, all occurring terms are of the form

$$\partial_\xi^l \partial_x^{\alpha'} g(x, \xi)\big|_{\xi = u(x)} \partial_x^{\alpha - \alpha'}\left(u^l\right) \qquad \alpha' \leq \alpha, \quad 1 \leq l \leq |\alpha - \alpha'|. \tag{4.2.6}$$

This can be seen by induction. Here is the induction step. A further partial derivative ∂^{e_j} (e_j having entry 1 at position j, all the other entries being 0) gives

$$\partial_\xi^l \partial_x^{\alpha' + e_j} g(x, \xi)\big|_{\xi = u(x)} \partial_x^{\alpha - \alpha'}\left(u^l\right) + \partial_\xi^{l+1} \partial_x^{\alpha'} g(x, \xi)\big|_{\xi = u(x)} \partial_x^{\alpha - \alpha'}\left(u^l\right) \partial_x^{e_j} u$$

$$+ \partial_\xi^l \partial_x^{\alpha'} g(x, \xi)\big|_{\xi = u(x)} \partial_x^{\alpha - \alpha' + e_j}\left(u^l\right),$$

all three terms clearly being covered by (4.2.6) (with α being replaced by $\alpha + e_j$). Further, applying the Leibniz formula to the second factor, (4.2.6) results in pointwise estimates

$$\left|\partial^{\alpha}\big(g(x, u(x))\big)\right| \lesssim \left|\partial_x^{\alpha} g(x, \xi)|_{\xi = u(x)}\right| \tag{4.2.7}$$

$$+ \sum_{\alpha' \leq \alpha} \sum_{l=1}^{|\alpha - \alpha'|} \left|\partial_\xi^l \partial_x^{\alpha'} g(x, \xi)\big|_{\xi = u(x)}\right| \sum_{\beta_1 + \cdots + \beta_l = \alpha - \alpha'} \left|\partial_x^{\beta_1} u(x) \cdots \partial_x^{\beta_l} u(x)\right|.$$

A comparison with the general Faa di Bruno formula further yields that we can restrict the last sum to multiindices $|\beta_j| \geq 1$, $j = 1, \ldots, l$ (recall that we originally started with the chain rule in (4.2.6) therefore no terms of $u(x)$ without derivatives appear).

Secondly, let use mention that

$$\|\rho^{-a}u|L_p(D)\| + \sum_{|\alpha|=m} \|\rho^{m-a}\partial^\alpha u|L_p(D)\|$$

generates an equivalent norm for the Kondratiev space $\mathcal{K}^m_{p,a}(D)$, see [DHSS18b].
Below we shall work with this norm without further reference.

Step 2.   Now assume first $m + 1 - \frac{d}{p} < 0$. Then for a typical term in (4.2.7) we
can estimate

$$\int_D \left(\rho^{|\alpha|-a+1}(x)\left|\partial^l_\xi \partial^{\alpha'}_x g(x,\xi)\right|_{\xi=u(x)}\right|\left|\partial^{\beta_1}_x u(x)\cdots\partial^{\beta_l}_x u(x)\right|\right)^p dx$$

$$\lesssim \int_D \left(\rho^{|\alpha|-a+1}(x)|u(x)|^{\delta-l}\left|\partial^{\beta_1}_x u(x)\cdots\partial^{\beta_l}_x u(x)\right|\right)^p dx$$

$$\lesssim \left(\int_D \left(\rho^{\gamma_0}(x)|u(x)|\right)^{q_0} dx\right)^{\frac{p(\delta-l)}{q_0}} \prod_{j=1}^l \left(\int_D \left(\rho^{|\beta_j|+\gamma_j}(x)|\partial^{\beta_j}u(x)|\right)^{q_j} dx\right)^{\frac{p}{q_j}}$$

$$\tag{4.2.8}$$

$$\leq \|u|\mathcal{K}^0_{q_0,-\gamma_0}(D)\|^{p(\delta-l)} \prod_{j=1}^l \|u|\mathcal{K}^{|\beta_j|}_{q_j,-\gamma_j}(D)\|^p,$$

where we first used the growth condition (4.2.4), and then Hölder's inequality
assuming

$$\frac{\delta-l}{q_0} + \frac{1}{q_1} + \cdots + \frac{1}{q_l} \leq \frac{1}{p}$$

(since D is a bounded domain), as well as

$$\gamma_0(\delta-l) + \gamma_1 + \cdots + \gamma_l + |\alpha-\alpha'| \leq |\alpha| - a + 1.$$

Note that this step also required the condition $\delta - l \geq 0$ for all l, hence $\delta \geq m - 1$.
To satisfy these two conditions, we choose

$$\frac{d}{q_0} = \frac{d}{p} - m - 1 > 0 \qquad \text{and} \qquad \gamma_0 = -\frac{d}{q_0} + \frac{d}{p} - a - 1$$

as well as

$$\frac{d}{q_j} = |\beta_j| + \frac{d}{p} - m - 1 > 0 \qquad \text{and} \qquad \gamma_j = -\frac{d}{q_j} + \frac{d}{p} - a - 1.$$

This choice implies

$$\frac{d(\delta - l)}{q_0} + \frac{d}{q_1} + \cdots + \frac{d}{q_l}$$

$$= \delta\frac{d}{p} - \delta(m+1) + |\alpha - \alpha'| \le \delta\frac{d}{p} - \delta(m+1) + m - 1,$$

which is bounded by $\frac{d}{p}$ if, and only if, $\frac{d}{p} - m - 1 \le \frac{2}{\delta - 1}$ (in case $\delta = 1$ there is no extra condition). This reasoning particularly ensures $p < q_j < \infty$ for $j = 0, \ldots, l$. Similarly, we find

$$\gamma_0(\delta - l) + \gamma_1 + \cdots + \gamma_l + |\alpha - \alpha'|$$

$$= \delta\left(\frac{d}{p} - a - 1\right) - \frac{d(\delta - l)}{q_0} - \frac{d}{q_1} - \cdots - \frac{d}{q_l} + |\alpha - \alpha'|$$

$$= \delta\left(\frac{d}{p} - a - 1\right) - \delta\left(\frac{d}{p} - m - 1\right)$$

$$= \delta(m - a) \overset{!}{\le} |\alpha| - a + 1$$

which is fulfilled with $|\alpha| = m - 1$ (sufficient by Step 1) due to $a \ge m$, which follows from our assumptions in view of $a \ge \frac{d}{p} - 1 = \frac{d}{p} - 1 - m + m \ge m$ (again for $\delta = 1$ no assumption on the parameters is required).

With this choice of parameters γ_j and q_j we now can further argue using the Sobolev-embedding from Theorem 2.4.8, which yields

$$\int_D \left(\rho^{|\alpha| - a + 1}(x)\left|\partial_\xi^l \partial_x^{\alpha'} g(x, \xi)\right|_{\xi = u(x)}\right|\left|\partial_x^{\beta_1} u(x) \cdots \partial_x^{\beta_l} u(x)\right|\right)^p dx$$

$$\lesssim \|u|\mathcal{K}^0_{q_0, -\gamma_0}(D)\|^{p(\delta - l)} \prod_{j=1}^{l} \|u|\mathcal{K}^{|\beta_j|}_{q_j, -\gamma_j}(D)\|^p$$

$$\lesssim \|u|\mathcal{K}^{m+1}_{p, a+1}(D)\|^{p(\delta - l)} \prod_{j=1}^{l} \|u|\mathcal{K}^{m+1}_{p, a+1}(D)\|^p = \|u|\mathcal{K}^{m+1}_{p, a+1}(D)\|^{p\delta}.$$

This proves the claim in case $\frac{d}{p} - m - 1 > 0$.

Step 3.   Now assume $m + 1 - \frac{d}{p} > 0$.

Substep 3.1.   For the first term in (4.2.7), corresponding to $l = 0$, we use (4.2.4) and see that $|\partial_x^\alpha g(x, u(x))| \le c_{\alpha, 0}|u(x)|^\delta$. This yields

$$\max_{|\alpha| \le m - 1} \int_D |\rho^{|\alpha| - a + 1}(x) \left(\partial_x^\alpha g\right)(x, u(x))|^p dx$$

$$\lesssim \int_D (\rho^{-a+1}(x) |u(x)|^\delta)^p dx$$

$$\leq \left(\sup_{x \in D} \rho^{(\delta-1)\gamma}(x)|u(x)|^{\delta-1} \right)^p \int_D \left(\rho^{-a-1}(x)|u(x)| \right)^p dx$$

$$\leq \|u|\mathcal{K}^0_{\infty,-\gamma}(D)\|^{(\delta-1)p}\|u|\mathcal{K}^{m+1}_{p,a+1}(D)\|^p, \tag{4.2.9}$$

where the second but last step holds if $-a + 1 \geq (\delta - 1)\gamma - a - 1$, i.e., we may choose $\gamma := \frac{2}{\delta-1}$ if $\delta > 1$. Furthermore, by Theorem 2.4.8 we see that

$$\mathcal{K}^{m+1}_{p,a+1}(D) \hookrightarrow \mathcal{K}^0_{\infty,-\gamma}(D) \tag{4.2.10}$$

if

$$m + 1 - \frac{d}{p} > 0 \quad \text{and} \quad a + 1 - \frac{d}{p} \geq -\gamma,$$

which is satisfied by our assumptions. Now (4.2.9) and (4.2.10) give the desired estimate for the first term in (4.2.7). If $\delta = 1$ the same result follows from a slight modification of (4.2.9). In this case we have

$$\max_{|\alpha| \leq m-1} \int_D |\rho^{|\alpha|-a+1}(x) (\partial_x^\alpha g)(x, u(x))|^p dx$$

$$\leq \int_D (\rho^{-a+1}(x)|u(x)|)^p dx = \|u|\mathcal{K}^0_{p,a-1}(D)\|^p \leq \|u|\mathcal{K}^{m+1}_{p,a+1}(D)\|^p,$$

where the last step is a consequence of the elementary embeddings for Kondratiev spaces, cf. Remark 2.4.3.

Substep 3.2. Next we shall deal with the terms in (4.2.7) with $l = 1$, i.e., the terms $\left|\partial_\xi \partial_x^{\alpha'} g(x, \xi)\right|_{\xi=u(x)} \left| \left|\partial_x^{\alpha-\alpha'} u(x)\right|$. Note that this step is only relevant for $m \geq 2$, since we consider derivatives up to order $m - 1 \geq |\alpha| = |\alpha - \alpha'| + |\alpha'| \geq l$. Using the growth condition (4.2.4) we find

$$\int_D \left(\rho^{|\alpha|-a+1}(x) \left|\partial_\xi \partial_x^{\alpha'} g(x, \xi)\right|_{\xi=u(x)}\right| \left|\partial_x^{\alpha-\alpha'} u(x)\right| \right)^p dx$$

$$\lesssim \int_D \left(\rho^{|\alpha|-a+1}(x)|u(x)|^{\delta-1} \left|\partial_x^{\alpha-\alpha'} u(x)\right| \right)^p dx$$

$$\lesssim \left(\sup_{x \in D} \rho^{\gamma_0}(x)|u(x)| \right)^{(\delta-1)p} \int_D \left(\rho^{|\alpha-\alpha'|+\gamma_1}(x)|\partial^{\alpha-\alpha'} u(x)| \right)^p dx$$

$$\leq \|u|\mathcal{K}^0_{\infty,-\gamma_0}(D)\|^{p(\delta-1)}\|u|\mathcal{K}^{|\alpha-\alpha'|}_{p,-\gamma_1}(D)\|^p.$$

For this it needs to hold $\gamma_0(\delta-1)+|\alpha-\alpha'|+\gamma_1 \le |\alpha|-a+1$. In addition, we want the embedding $\mathcal{K}^{m+1}_{p,a+1}(D) \hookrightarrow \mathcal{K}^0_{\infty,-\gamma_0}(D)$ to be valid, which requires

$$m+1-\frac{d}{p} > 0 \qquad \text{and} \qquad a+1-\frac{d}{p} \ge -\gamma_0.$$

In view of our assumption $a \ge \frac{d}{p} - 1$ this is fulfilled for arbitrary $\gamma_0 \ge 0$. Hence choosing $-\gamma_1 = a+1$ and $\gamma_0(\delta-1) = |\alpha'|+2$ (if $\delta = 1$ we may choose $\gamma_0 \ge 0$ arbitrarily), the mentioned condition is satisfied and we conclude

$$\int_D \left(\rho^{|\alpha|-a+1}(x)\left|\partial_\xi \partial_x^{\alpha'} g(x,\xi)\big|_{\xi=u(x)}\right| \left|\partial_x^{\alpha-\alpha'} u(x)\right| \right)^p dx$$

$$\lesssim \|u|\mathcal{K}^0_{\infty,-\gamma_0}(D)\|^{p(\delta-1)} \|u|\mathcal{K}^{|\alpha-\alpha'|}_{p,a+1}(D)\|^p$$

$$\lesssim \|u|\mathcal{K}^{m+1}_{p,a+1}(D)\|^{p(\delta-1)} \|u|\mathcal{K}^{|\alpha-\alpha'|}_{p,a+1}(D)\|^p \le \|u|\mathcal{K}^{m+1}_{p,a+1}(D)\|^{p\delta}.$$

$$(4.2.11)$$

Substep 3.3.   Now consider the terms in (4.2.7) with $l \ge 2$. Note that this step is only relevant for $m \ge 3$ since as before $l \le m-1$. Once again we shall use the growth condition (4.2.4). We obtain this time

$$\max_{|\alpha|\le m-1} \int_D \left| \rho^{|\alpha|-a+1}(x)\, \partial_\xi^l \partial_x^{\alpha'} g(x,u(x))\, \partial_x^{\beta_1} u(x) \cdots \partial_x^{\beta_l} u(x) \right|^p dx$$

$$\lesssim \int_D \left(|\rho^{|\alpha|-a+1}(x)|u(x)|^{\delta-l} \prod_{j=1}^{l} |\partial^{\beta_j} u(x)| \right)^p dx$$

$$\lesssim \left(\sup_{x\in D} \rho^{\gamma(\delta-l)}(x)|u(x)|^{\delta-l} \right)^p \prod_{j=2}^{l} \left(\sup_{x\in D} \rho^{|\beta_j|+\gamma}(x)|\partial^{\beta_j} u(x)| \right)^p$$

$$\cdot \left(\int_D \rho^{p(|\beta_1|-a-1)}(x)|\partial^{\beta_1} u(x)|^p dx \right).$$

$$(4.2.12)$$

Here, in the last step, we have used that $\rho(x) \le 1$ and thus, in order to obtain an estimate from above, the exponents of ρ have to satisfy

$$|\alpha| - a + 1 \overset{!}{\ge} \gamma(\delta-l) + \sum_{j=2}^{l}(|\beta_j|+\gamma) + |\beta_1| - a - 1$$

$$= \gamma(\delta-1) + |\alpha| - |\alpha'| - a - 1,$$

which leads to $\gamma \leq \frac{2+|\alpha'|}{\delta-1}$ in case $\delta \neq 1$ and γ arbitrary in case $\delta = 1$. Therefore we may choose $\gamma = \frac{2}{\delta-1}$ if $\delta > 1$ and $\gamma = 0$ if $\delta = 1$. In addition, without loss of generality, we assume $|\beta_1| \geq |\beta_j|$ for all $2 \leq j \leq l$. Clearly,

$$\left(\sup_{x \in D} \rho^\gamma(x)\, |u(x)| \right)^{\delta-l} = \|u|\mathcal{K}^0_{\infty,-\gamma}(D)\|^{\delta-l},$$

$$\left(\sup_{x \in D} \rho^{|\beta_j|+\gamma}(x)|\partial^{\beta_j}u(x)| \right) \leq \|u|\mathcal{K}^{|\beta_j|}_{\infty,-\gamma}(D)\|,$$

$$\left(\int_D \rho^{p(|\beta_1|-a-1)}(x)|\partial^{\beta_1}u(x)|^p dx \right)^{1/p} \leq \|u|\mathcal{K}^{|\beta_1|}_{p,a+1}(D)\|.$$

From Theorem 2.4.8 we conclude

$$\mathcal{K}^{m+1}_{p,a+1}(D) \hookrightarrow \mathcal{K}^0_{\infty,-\gamma}(D) \qquad \text{if} \qquad m+1-\frac{d}{p} > 0, \quad a+1-\frac{d}{p} \geq -\gamma.$$

Furthermore, we have

$$\mathcal{K}^{m+1}_{p,a+1}(D) \hookrightarrow \mathcal{K}^{m-2}_{\infty,-\gamma}(D) \qquad \text{if} \qquad 3-\frac{d}{p} > 0, \quad a+1-\frac{d}{p} \geq -\gamma.$$

Observe that $|\beta_1|+|\beta_j| \leq |\alpha| - |\alpha'| \leq m-1$. Hence, by (4.2.9), (4.2.11) and (4.2.12) we get

$$\|G(u)|\mathcal{K}^{m-1}_{p,a-1}(D)\| \lesssim \|u|\mathcal{K}^{m+1}_{p,a+1}(D)\|^\delta$$

as claimed.

$\square$

Remark 4.2.2 Some more remarks concerning Theorem 4.2.1 seem to be in order. Estimates for Nemytskij operators $T_G : u \mapsto G(u) = g(\cdot, u(\cdot))$ in Sobolev spaces are a delicate topic. Even in the more simple case of composition operators $u \mapsto F(u) := f(u(\cdot))$ there are many open questions, we refer e.g. to [Bou91], [BM01], [MS02] or the survey [BS11]. The naive conjecture that F maps a Sobolev space into itself if f is sufficiently smooth is known to be true if $W_p^m(\mathbb{R}^d) \hookrightarrow L_\infty(\mathbb{R}^d)$. On the other hand, if the Sobolev space contains unbounded functions, such a statement is not true. One may think on composition operators related to $f(t) := t^n$ for some $n \geq 2$. Then, under the mapping F, the unboundedness is enhanced. Since we allow a shift in the smoothness from $m+1$ to $m-1$ (and from $a+1$ to $a-1$) we can deal also with Kondratiev spaces containing unbounded functions as long as the unboundedness is small enough. This is expressed by the restriction $0 > m+1-\frac{d}{p} \geq -\frac{2}{\delta-1}$.

We finish this subsection by taking a closer look at the mapping $u \mapsto u^n$, $n \geq 2$.

Corollary 4.2.3 *Let D be as in Definition 2.4.5. Let $1 < p < \infty$, $a \geq \frac{d}{p} - 1$, $m \in \mathbb{N}$, and $n \in \mathbb{N}$, $n \geq 2$. Moreover, assume either*

$$\min(m + 1, 3) - \frac{d}{p} > 0 \tag{4.2.13}$$

or

$$0 > m + 1 - \frac{d}{p} \geq -\frac{2}{n-1}. \tag{4.2.14}$$

Then the nonlinear operator $T_n : u \mapsto u^n$ maps functions from $\mathcal{K}^{m+1}_{p,a+1}(D)$ to functions in $\mathcal{K}^{m-1}_{p,a-1}(D)$. Moreover, it holds

$$\|u^n|\mathcal{K}^{m-1}_{p,a-1}(D)\| \leq C \, \|u|\mathcal{K}^{m+1}_{p,a+1}(D)\|^n \tag{4.2.15}$$

with some constant $C > 0$ independent of u.

Proof It will be enough to give a few comments. Clearly, monomials $g(x, \xi) = \xi^n$, where $n \in \mathbb{N}$ and $n \geq 2$, satisfy the growth condition (4.2.4) for $\delta = n$. Hence, we may follow the proof of Theorem 4.2.1 step by step. Observe that in formula (4.2.7) the summation with respect to l is limited by $\min(|\alpha|, n)$ in this case. If $l = n \leq |\alpha|$, then we drop the first factor in (4.2.8). All other arguments can be repeated, which proves the claim in case $0 > m + 1 - \frac{d}{p} \geq -\frac{2}{n-1}$. Also the proof under the restriction (4.2.13) follows along the lines of the previous arguments in Step 3, where again $\delta = n$. □

Remark 4.2.4 Under the stronger assumption $d/p < 2$ instead of (4.2.13), an alternative proof can be given applying Theorem 2.4.9(ii) together with an obvious induction argument.

4.2.2 Fixed Points of Nonlinear Operators in Banach Spaces

As already outlined above, our regularity results will be established by means of suitable fixed point theorems. In particular, we will use the fixed point theorem as stated in Proposition 4.2.7 below that works for admissible spaces.

Definition 4.2.5 (Admissible Space) A quasi-normed space A is said to be *admissible*, if for every compact subset $K \subset A$ and for every $\varepsilon > 0$ there exists a continuous map $\tilde{T} : K \to A$ such that $\tilde{T}(K)$ is contained in a finite-dimensional subset of A and $x \in K$ implies

$$\|\tilde{T}x - x|A\| \leq \varepsilon.$$

Remark 4.2.6 We show in Proposition 4.2.9 that Kondratiev spaces are admissible.

Let X and Y be admissible quasi-Banach spaces. Furthermore, we assume that $\tilde{L} : Y \to X$ is a linear and continuous operator and $\tilde{N} : X \to Y$ is (in general) a nonlinear map. We are looking for a fixed point of the problem

$$u = (\tilde{L} \circ \tilde{N})u. \qquad (4.2.16)$$

For this we make use of the following result.

Proposition 4.2.7 (Fixed Point Theorem) *Let X, Y, $\tilde{L}$, and $\tilde{N}$ be as above. Suppose that there exist $\eta \geq 0$, $\vartheta \geq 0$, and $\delta \geq 0$ such that*

$$\| \tilde{N}u|Y \| \leq \eta + \vartheta \, \|u|X\|^{\delta} \qquad (4.2.17)$$

holds for all $u \in X$. Furthermore, we assume that the mapping $\tilde{L} \circ \tilde{N} : X \to X$ is completely continuous. Then there exists at least one solution $u \in X$ of (4.2.16) provided one of the following conditions is satisfied:

(a) $\delta \in [0, 1)$,

(b) $\delta = 1$, $\vartheta < \| \tilde{L} \|^{-1}$,   (4.2.18)

(c) $\delta > 1$ *and* $\eta \, \| \tilde{L} \| < \left[\dfrac{1}{\vartheta \, \| \tilde{L} \|} \right]^{\frac{1}{\delta-1}} \left[\left(\dfrac{1}{\delta} \right)^{\frac{1}{\delta-1}} - \left(\dfrac{1}{\delta} \right)^{\frac{\delta}{\delta-1}} \right].$

Clearly, we will apply Proposition 4.2.7 to the case when $\tilde{L} = L$ and $\tilde{N} = N$ as defined in (4.2.2) and (4.2.3), respectively.

Remark 4.2.8

(i) A proof of this proposition, which is based on Schauder's fixed point theorem, can be found in [Fuc80]. Later on, in [RS96], by means of the Leray-Schauder principle, these results have also been generalized to admissible quasi-Banach spaces.

(ii) In [DS13] semilinear problems associated with the Poisson equation have been studied. There, the authors considered nonlinear terms that lead to bounds with $\delta \leq 1$. For this reason, in this manuscript we particularly discuss the case $\delta > 1$.

We want to establish our regularity results by applying Proposition 4.2.7 to Kondratiev spaces. Therefore, we have to clarify that all the necessary assumptions are satisfied in this case.

Admissibility of Kondratiev Spaces

Before we come to the existence of solutions of problem (4.1.1), we need to discuss the admissibility of Kondratiev spaces, this being one of the requirements of our main tool Proposition 4.2.7.

Proposition 4.2.9 (Admissibility of Kondratiev Spaces) *Let the domain D satisfy Assumption 4.1.1. Then the spaces $\mathcal{K}^m_{p,a}(D)$ and $\mathcal{K}^{m+1}_{2,a+1,0}(D)$ are admissible for all $1 < p < \infty$, $m \in \mathbb{N}$, and $a \in \mathbb{R}$.*

Proof The admissibility of the spaces $\mathcal{K}^m_{p,a,0}(D)$ is an immediate consequence of the admissibility of the Kondratiev spaces $\mathcal{K}^m_{p,a}(D)$ themselves.

Furthermore, the admissibility of the Kondratiev spaces $\mathcal{K}^m_{p,a}(D)$ can be traced back to the one of the so-called refined localization spaces $F^{s,\mathrm{rloc}}_{p,q}(D)$ and their relation to the spaces $\mathcal{K}^m_{p,m}(D)$. In turn, the admissibility of the spaces $F^{s,\mathrm{rloc}}_{p,q}(D)$ follows from the existence of wavelet bases.

Step 1. Concerning the definition and further properties of refined localization spaces $F^{s,\mathrm{rloc}}_{p,q}(D)$ we refer to [Tri06, Ch. 4], [Tri08a, Ch. 2], and [Han17]. In particular, the latter reference provides the following equivalent characterization of their norm: With δ being the distance to the boundary, i.e., $\delta(x) = \mathrm{dist}(x, \partial D)$, we have

$$\|u|F^{s,\mathrm{rloc}}_{p,q}(D)\| \sim \|u|F^s_{p,q}(D)\| + \|\delta^{-s}u|L_p(D)\|.$$

This equivalence holds for $s > \sigma_{p,q}$, $0 < p < \infty$, and $0 < q \leq \infty$. One of the key properties of these refined localization spaces is their characterization by suitable wavelet systems. Moreover, these wavelet systems then form a basis in case $q < \infty$. From the existence of such a basis, the admissibility now follows by standard arguments: Expanding $x \in K$ as a series $x = \sum_{j=1}^{\infty} \lambda_j(x)w_j$, where $(w_j)_{j=1}^{\infty}$ is the wavelet system and $\lambda_j \in (F^{s,\mathrm{rloc}}_{p,q}(D))'$, we can find $j_0(x)$ such that $\|\sum_{j=1}^{j_0(x,\varepsilon)} \lambda_j(x)w_j - x|F^{s,\mathrm{rloc}}_{p,q}(D)\| \leq \varepsilon$. Now a standard compactness argument ensures that we can choose $j_0(x, \varepsilon)$ independent of $x \in K$, so that defining $T_\varepsilon x = \sum_{j=1}^{j_0(\varepsilon)} \lambda_j(x)w_j$ satisfies the requirements of the definition for admissibility of $F^{s,\mathrm{rloc}}_{p,q}(D)$.

Step 2. In [Han14] it was shown that $\mathcal{K}^m_{p,m}(\mathbb{R}^d \setminus E) = F^{m,\mathrm{rloc}}_{p,2}(\mathbb{R}^d \setminus E)$, where E is an arbitrary closed set with Lebesgue measure $|E| = 0$. This particularly applies to the case where E is the singular set S of a bounded Lipschitz domain D of polyhedral type. Moreover, the spaces $\mathcal{K}^m_{p,m}(D)$ and $\mathcal{K}^m_{p,m}(\mathbb{R}^d \setminus S)$ are related via the boundedness of Stein's extension operator, $\mathcal{E} : \mathcal{K}^m_{p,m}(D) \to \mathcal{K}^m_{p,m}(\mathbb{R}^d \setminus S)$, which was proven in [Han15, Lem. 5.1]. For general a, the admissibility of $\mathcal{K}^m_{p,a}(D)$ now follows from the admissibility of $\mathcal{K}^m_{p,m}(D)$ since both spaces are isomorphic, cf. Remark 2.4.3(iv).

$\square$

Remark 4.2.10 (Admissibility of Fractional Kondratiev Spaces) The fractional Kondratiev spaces $\mathfrak{K}_{p,a}^s(D)$ from Definition 2.4.18 are admissible as well for all $1 < p < \infty$, $s \geq 0$, and $a \in \mathbb{R}$. We sketch the proof which follows from the observations in Proposition 4.2.9 together with the results presented in [Han17]. There, for $s \geq 0$ and $1 < p < \infty$, alternative fractional Kondratiev spaces

$$\tilde{\mathfrak{K}}_{p,s}^s(\mathbb{R}^d \setminus S) = F_{p,2}^{s,\mathrm{rloc}}(\mathbb{R}^d \setminus S)$$

were introduced, where S is an arbitrary closed set with $|S| = 0$ (e.g. the exceptional set of a bounded Lipschitz domain of polyhedral type). Moreover, for $a \in \mathbb{R}$ put

$$\tilde{\mathfrak{K}}_{p,a}^s(\mathbb{R}^d \setminus S) := T_{a-s}\tilde{\mathfrak{K}}_{p,s}^s(\mathbb{R}^d \setminus S),$$

where $T_{a-s}(u) = \rho^{a-s}u$. Since refined localization spaces are admissible and T_{a-s} is an isomorphism, cf. Remark 2.4.3(iv), we conclude that the spaces $\tilde{\mathfrak{K}}_{p,a}^s(\mathbb{R}^d \setminus S)$ are admissible as well. Furthermore, it was shown in [Han17] that $\tilde{\mathfrak{K}}_{p,a}^s(D)$ and $\tilde{\mathfrak{K}}_{p,a}^s(\mathbb{R}^d \setminus S)$ are related via the boundedness of Stein's extension operator. This implies admissibility of $\tilde{\mathfrak{K}}_{p,a}^s(D)$. Finally, in [Han17] it was proven that for $1 < p < \infty$, $s \geq 0$, and $a \in \mathbb{R}$, we have the coincidence

$$\mathfrak{K}_{p,a}^s(D) = \tilde{\mathfrak{K}}_{p,a}^s(D), \tag{4.2.19}$$

which shows admissibility of the spaces $\mathfrak{K}_{p,a}^s(D)$.

Complete Continuity of $L \circ N$

Now we want to prove another technical aspect for the application of the fixed point result (Proposition 4.2.7), namely the complete continuity of the composed mapping $L \circ N$. For an operator T to be *completely continous* it is sufficient to know that it is continuous and compact. Hence, we shall prove compactness of the linear solution map L for appropriate pairs of spaces, together with continuity of N, which will ultimately yield complete continuity of the composed map.

A natural idea to prove compactness of L would be to use compact embeddings of Kondratiev spaces. For classical non-weighted Sobolev spaces, this strategy has successfully been worked out in [DS13]. Moreover, in [MR10, Lem. 4.1.4] and [DHSS18b, Thm. 4] compact embeddings of Kondratiev spaces have already been established.

If $p \leq q$, then $\mathcal{K}_{p,a}^m(D)$ is compactly embedded into $\mathcal{K}_{q,a'}^{m'}(D)$ if, and only if,

$$m - \frac{d}{p} > m' - \frac{d}{q} \quad \text{and} \quad a - \frac{d}{p} > a' - \frac{d}{q}.$$

However, by using this result with $p = q$ directly, we would loose at least one order in the regularity in the end. To avoid this problem, our strategy is to use the fractional Kondratiev spaces from Definition 2.4.18 in order to only loose an arbitrarily small ε of smoothness: From Remark 2.4.23 it follows that the embedding

$$\mathfrak{K}^{s'}_{p,a'}(D) \hookrightarrow \mathfrak{K}^{s}_{p,a}(D)$$

is compact for $s' > s$ and $a' > a$.

We prodeed as follows: In order to deal with the complete continuity of $L \circ N$, we first show that

$$N : \mathfrak{K}^{m+1-\varepsilon}_{p,a+1-\varepsilon}(D) \to \mathfrak{K}^{m-1-\varepsilon'}_{p,a-1-\varepsilon'}(D)$$

is continuous for sufficiently small $\varepsilon > \varepsilon' > 0$. For this we require a slight strengthening of Theorem 4.2.1, which is given in Propostion 4.2.11 below and yields the boundedness N in the respective Kondratiev spaces. However, since N is nonlinear, continuity does not follow automatically from this. The fact that N is continuous nevertheless will be established in Lemma 4.2.13. Finally, using the continuity of N, the complete continuity of $L \circ N$ will be proven in Theorem 4.2.15.

Proposition 4.2.11 (Boundedness of N) *Let D be a domain which satisfies Assumption 4.1.1, $d/2 < p < \infty$, $a \geq \frac{d}{p} - 1$, and $m \in \mathbb{N}$ with $m > \frac{d}{p}$. Moreover, assume $\delta \geq \max(1, m - 1)$ and let the function g satisfy the growth-condition (4.2.4). Then for sufficiently small $\varepsilon > 0$ the nonlinear operator $G(u) = g(\cdot, u(\cdot))$ maps functions from $\mathfrak{K}^{m+1-\varepsilon}_{p,a+1-\varepsilon}(D)$ to functions in $\mathcal{K}^{m-1}_{p,a-1}(D)$ and it holds*

$$\|G(u)|\mathcal{K}^{m-1}_{p,a-1}(D)\| \leq C \, \|u|\mathfrak{K}^{m+1-\varepsilon}_{p,a+1-\varepsilon}(D)\|^{\delta},$$

for some constant $C > 0$ independent of u.

Proof

Step 1. For now let $0 < \varepsilon < 1$. In what follows we will have to choose ε small enough to suit our needs (from the proof $\varepsilon < \frac{1}{\delta}$ will turn out to be sufficient, but below we won't further comment on the specific choice). We reuse the estimate

$$\left|\partial^{\alpha}\big(g(x, u(x))\big)\right| \lesssim \left|\partial^{\alpha}_x g(x, \xi)|_{\xi=u(x)}\right| \tag{4.2.20}$$

$$+ \sum_{\alpha' \leq \alpha} \sum_{l=1}^{|\alpha-\alpha'|} \left|\partial^{l}_{\xi}\partial^{\alpha'}_x g(x, \xi)\big|_{\xi=u(x)}\right| \sum_{\substack{\beta_1+\cdots+\beta_l=\alpha-\alpha', \\ |\beta_1|, \ldots, |\beta_l| \geq 1}} \left|\partial^{\beta_1}_x u(x) \cdots \partial^{\beta_l}_x u(x)\right|,$$

obtained in Step 1 of the proof of Theorem 4.2.1. Again we start dealing with
the first term in (4.2.20), corresponding to $l = 0$. From (4.2.4) it follows that
$|\partial_x^\alpha g(x, u(x))| \leq c_{\alpha,0}|u(x)|^\delta$, which now yields

$$
\max_{|\alpha| \leq m-1} \int_D \left| \rho^{|\alpha|-a+1}(x) \, (\partial_x^\alpha g)(x, u(x)) \right|^p dx
$$

$$
\lesssim \int_D \left(\rho^{-a+1}(x) \, |u(x)|^\delta \right)^p dx
$$

$$
\leq \left(\sup_{x \in D} \rho^{\gamma(\delta-1)}(x) u(x)^{\delta-1} \right)^p \left(\int_D \left(\rho^{-a-1+\varepsilon}(x)|u(x)| \right)^p dx \right)
$$

$$
\lesssim \|u|\mathcal{K}^0_{\infty,-\gamma}\|^{(\delta-1)p} \|u|\mathcal{K}^0_{p,a+1-\varepsilon}\|^p
$$

$$
\lesssim \|u|\mathcal{K}^m_{p,a+1-\varepsilon}\|^{(\delta-1)p} \|u|\mathcal{K}^m_{p,a+1-\varepsilon}\|^p, \qquad (4.2.21)
$$

where the second step holds if

$$
-a + 1 \geq \gamma(\delta - 1) - a - 1 + \varepsilon,
$$

i.e., we choose $\gamma := \frac{2-\varepsilon}{\delta-1}$ if $\delta > 1$ and the 4th step is a consequence
of the elementary embeddings for Kondratiev spaces, cf. Remark 2.4.3 and
Theorem 2.4.8(ii), which holds if

$$
m - \frac{d}{p} > 0 \qquad \text{and} \qquad a + 1 - \varepsilon - \frac{d}{p} \geq -\gamma,
$$

and is satisfied by our assumptions upon choosing ε small enough. Note that for
$\delta = 1$ estimate (4.2.21) is just a consequence of the elementary embeddings of
Kondratiev spaces, i.e., we have

$$
\max_{|\alpha| \leq m-1} \int_D \left| \rho^{|\alpha|-a+1}(x) \, (\partial_x^\alpha g)(x, u(x)) \right|^p dx
$$

$$
\lesssim \int_D \left(\rho^{-a+1}(x) \, |u(x)| \right)^p dx = \|u|\mathcal{K}^0_{p,a-1}(D)\|^p \leq \|u|\mathcal{K}^m_{p,a+1-\varepsilon}(D)\|^p.
$$

$$
\qquad (4.2.22)
$$

From the definition of fractional Kondratiev spaces and the properties of complex
interpolation it follows that

$$
\mathfrak{K}^{m+1-\varepsilon}_{p,a+1-\varepsilon}(D) = \left[\mathcal{K}^m_{p,a+1-\varepsilon}(D), \mathcal{K}^{m+1}_{p,a+1-\varepsilon}(D) \right]_{1-\varepsilon}
$$

$$
\hookrightarrow \left[\mathcal{K}^m_{p,a+1-\varepsilon}(D), \mathcal{K}^m_{p,a+1-\varepsilon}(D) \right]_{1-\varepsilon} = \mathcal{K}^m_{p,a+1-\varepsilon}(D).
$$

$$
\qquad (4.2.23)
$$

This together with (4.2.21) and (4.2.22) yields for $\delta \geq 1$,

$$\max_{|\alpha| \leq m-1} \int_D \left| \rho^{|\alpha|-a+1}(x) \, (\partial_x^\alpha g)(x, u(x)) \right|^p dx \lesssim \|u|\mathfrak{K}_{p,a+1-\varepsilon}^{m+1-\varepsilon}(D)\|^{\delta p}.$$

$$(4.2.24)$$

Step 2.   Next we shall deal with the terms in (4.2.7) with $l = 1$, i.e., the terms $\left| \partial_\xi \partial_x^{\alpha'} g(x, \xi) \right|_{\xi=u(x)} \left| \left| \partial_x^{\alpha-\alpha'} u(x) \right|$. Note that this step is only relevant for $m \geq 2$ since we consider derivatives up to order $m - 1 \geq |\alpha| = |\alpha - \alpha'| + |\alpha'| \geq l$. Using the growth condition (4.2.4) we find

$$\int_D \left(\rho^{|\alpha|-a+1}(x) \left| \partial_\xi \partial_x^{\alpha'} g(x, \xi) \right|_{\xi=u(x)} \left| \left| \partial_x^{\alpha-\alpha'} u(x) \right| \right)^p dx$$

$$\lesssim \int_D \left(\rho^{|\alpha|-a+1}(x) |u(x)|^{\delta-1} \left| \partial_x^{\alpha-\alpha'} u(x) \right| \right)^p dx$$

$$\lesssim \left(\sup_{x \in D} \rho^{\gamma_0}(x) |u(x)| \right)^{(\delta-1)p} \int_D \left(\rho^{|\alpha-\alpha'|+\gamma_1}(x) |\partial^{\alpha-\alpha'} u(x)| \right)^p dx$$

$$\leq \|u|\mathcal{K}_{\infty,-\gamma_0}^0(D)\|^{p(\delta-1)} \|u|\mathcal{K}_{p,-\gamma_1}^{|\alpha-\alpha'|}(D)\|^p.$$

For this to hold we need $\gamma_0(\delta - 1) + |\alpha - \alpha'| + \gamma_1 \leq |\alpha| - a + 1$. In addition we want the embedding $\mathcal{K}_{p,a+1-\varepsilon}^m(D) \hookrightarrow \mathcal{K}_{\infty,-\gamma_0}^0(D)$ to be valid, which requires

$$m - \frac{d}{p} > 0 \qquad \text{and} \qquad a + 1 - \varepsilon - \frac{d}{p} \geq -\gamma_0.$$

In view of our assumption $a \geq \frac{d}{p} - 1$ this is fulfilled for ε small (i.e., $0 < \varepsilon < \gamma_0$). Hence choosing $-\gamma_1 = a + 1 - \varepsilon$ and $\gamma_0(\delta - 1) = |\alpha'| + 2 - \varepsilon$ (if $\delta = 1$ we may choose $\gamma_0 > 0$ arbitrarily), the mentioned condition is satisfied and we conclude

$$\int_D \left(\rho^{|\alpha|-a+1}(x) \left| \partial_\xi \partial_x^{\alpha'} g(x, \xi) \right|_{\xi=u(x)} \left| \left| \partial_x^{\alpha-\alpha'} u(x) \right| \right)^p dx$$

$$\lesssim \|u|\mathcal{K}_{\infty,-\gamma_0}^0(D)\|^{p(\delta-1)} \|u|\mathcal{K}_{p,a+1-\varepsilon}^{|\alpha-\alpha'|}(D)\|^p$$

$$\lesssim \|u|\mathcal{K}_{p,a+1-\varepsilon}^m(D)\|^{p(\delta-1)} \|u|\mathcal{K}_{p,a+1-\varepsilon}^{|\alpha-\alpha'|}(D)\|^p$$

$$\leq \|u|\mathcal{K}_{p,a+1-\varepsilon}^m(D)\|^{p\delta} \lesssim \|u|\mathfrak{K}_{p,a+1-\varepsilon}^{m+1-\varepsilon}(D)\|^{p\delta}, \qquad (4.2.25)$$

where the last embedding is a consequence of (4.2.23).

Step 3.   We estimate of the terms in (4.2.20) with $l \geq 2$ (this case only occurs for $m \geq 3$). Using the inequality (4.2.4) in case $2 \leq l \leq |\alpha| - |\alpha'|$ one obtains

$$\max_{|\alpha| \leq m-1} \int_D \left| \rho^{|\alpha|-a+1}(x) \, \partial_\xi^l \partial_x^{\alpha'} g(x, u(x)) \, \partial_x^{\beta_1} u(x) \cdots \partial_x^{\beta_j} u(x) \right|^p dx$$

$$\lesssim \int_D \left| \rho^{|\alpha|-a+1}(x) |u(x)|^{\delta-l} \prod_{j=1}^{l} |\partial^{\beta_j} u(x)| \right|^p dx$$

$$\lesssim \left(\sup_{x \in D} \rho^{\gamma(\delta-l)}(x) |u(x)|^{\delta-l} \right)^p \prod_{j=2}^{l} \left(\sup_{x \in D} \rho^{|\beta_j|+\gamma-\varepsilon}(x) |\partial^{\beta_r} u(x)| \right)^p$$

$$\cdot \left(\int_D \rho^{p(|\beta_1|-a-1+\varepsilon)}(x) |\partial^{\beta_1} u(x)|^p dx \right). \tag{4.2.26}$$

Here, in the last step, we have used that $\rho(x) \leq 1$ and thus, in order to obtain an estimate from above, the exponents of ρ have to satisfy

$$|\alpha| - a + 1 \overset{!}{\geq} \gamma(\delta-l) + \sum_{j=2}^{l} (|\beta_j| + \gamma - \varepsilon) + |\beta_1| - a - 1 + \varepsilon$$

$$= \gamma(\delta-1) + |\alpha| - |\alpha'| - a - 1 - \varepsilon(l-1) + \varepsilon ,$$

which leads to $\gamma \leq \frac{(2-\varepsilon)+|\alpha'|+\varepsilon(l-1)}{\delta-1}$ in case $\delta \neq 1$ and γ arbitrary in case $\delta = 1$. Therefore, we may choose $\gamma = \frac{1}{\delta-1}$ if $\delta > 1$ and $\gamma = 1$ if $\delta = 1$. In addition, without loss of generality, we assume $|\beta_1| \geq |\beta_j|$ for all $2 \leq j \leq l$. Clearly,

$$\left(\sup_{x \in D} \rho^\gamma(x) |u(x)| \right)^{(\delta-l)} \leq \|u| \mathcal{K}^0_{\infty, -\gamma}(D)\|^{\delta-l},$$

$$\left(\sup_{x \in D} \rho^{|\beta_j|+\gamma-\varepsilon}(x) |\partial^{\beta_r} u(x)| \right) \leq \|u| \mathcal{K}^{|\beta_j|}_{\infty, -\gamma+\varepsilon}(D)\|,$$

$$\left(\int_D \rho^{p(|\beta_1|-a-1+\varepsilon)}(x) |\partial^{\beta_1} u(x)|^p dx \right)^{1/p} \leq \|u| \mathcal{K}^{|\beta_1|}_{p, a+1-\varepsilon}(D)\|.$$

From (4.2.23) and Theorem 2.4.8 we conclude

$$\mathfrak{K}^{m+1-\varepsilon}_{p, a+1-\varepsilon}(D) \hookrightarrow \mathcal{K}^m_{p, a+1-\varepsilon}(D) \hookrightarrow \mathcal{K}^0_{\infty, -\gamma}(D)$$

if $m - \frac{d}{p} > 0$ and $a + 1 - \varepsilon - \frac{d}{p} \geq -\gamma$. Both inequalities are guaranteed by our assumptions on m and a (if ε is chosen small enough such that $\varepsilon < \gamma$). Furthermore, we have

$$\mathscr{K}^{m+1-\varepsilon}_{p,a+1-\varepsilon}(D) \hookrightarrow \mathcal{K}^{m}_{p,a+1-\varepsilon}(D) \hookrightarrow \mathcal{K}^{m-2}_{\infty,-\gamma+\varepsilon}(D)$$

if $m - \frac{d}{p} > m - 2$ and $a + 1 - \varepsilon - \frac{d}{p} \geq -\gamma + \varepsilon$. Both inequalities are satisfied by our assumptions $d/2 < p$ and $a \geq \frac{d}{p} - 1$ (if ε is chosen small enough such that $\varepsilon < \gamma/2$).

Observe that $|\beta_1| + |\beta_j| \leq |\alpha| - |\alpha'| \leq m - 1$. Hence, by (4.2.24), (4.2.25), and (4.2.26) we get

$$\|G(u)|\mathcal{K}^{m-1}_{p,a-1}(D)\| \lesssim \|u|\mathscr{K}^{m+1-\varepsilon}_{p,a+1-\varepsilon}(D)\|^{\delta}$$
$$+ \|u|\mathcal{K}^{0}_{\infty,-\gamma}(D)\|^{\delta-l}\|u|\mathcal{K}^{m-2}_{\infty,-\gamma+\varepsilon}(D)$$
$$\times \|^{l-1}\|u|\mathcal{K}^{m-1}_{p,a+1-\varepsilon}(D)\|$$
$$\lesssim \|u|\mathscr{K}^{m+1-\varepsilon}_{p,a+1-\varepsilon}(D)\|^{\delta}$$

as claimed.

$\square$

Remark 4.2.12

(i) Since we assume $p > \frac{d}{2}$ in Proposition 4.2.11 the condition $m > \frac{d}{p}$ is always satisfied if $m \geq 2$.

(ii) Moreover, using the fact that $\mathcal{K}^{m-1}_{p,a-1}(D) \hookrightarrow \mathscr{K}^{m-1-\varepsilon'}_{p,a-1-\varepsilon'}(D)$ for some $\varepsilon' > 0$, we see from Proposition 4.2.11 that

$$G : \mathscr{K}^{m+1-\varepsilon}_{p,a+1-\varepsilon}(D) \hookrightarrow \mathscr{K}^{m-1-\varepsilon'}_{p,a-1-\varepsilon'}(D) \tag{4.2.27}$$

is bounded as well.

So far Proposition 4.2.11 and Remark 4.2.12(ii) give boundedness of $N :$ $\mathscr{K}^{m+1-\varepsilon}_{p,a+1-\varepsilon}(D) \to \mathscr{K}^{m-1-\varepsilon'}_{p,a-1-\varepsilon'}(D)$. Since N is nonlinear, continuity does not follow automatically but is proven in the following Lemma.

Lemma 4.2.13 (Continuity of N) *Let D be a domain which satisfies Assumption 4.1.1, $d/2 < p < \infty$, $a \geq \frac{d}{p} - 1$, and $m \in \mathbb{N}$ with $m > \frac{d}{p}$. Moreover, assume $\delta \geq \max(1, m-1)$ and let the function g satisfy the growth-condition (4.2.4). Then for sufficiently small $\varepsilon > \varepsilon' > 0$ the nonlinear mapping*

$$N : \mathscr{K}^{m+1-\varepsilon}_{p,a+1-\varepsilon}(D) \to \mathscr{K}^{m-1-\varepsilon'}_{p,a-1-\varepsilon'}(D)$$

is continuous.

Proof Since $N(u)(x) = f(x) - g(x, u(x))$ Proposition 4.2.11 implies

$$\|N(u)|\mathcal{K}^{m-1}_{p,a-1}(D)\| \le \|f|\mathcal{K}^{m-1}_{p,a-1}(D)\| + c\|u|\mathfrak{K}^{m+1-\varepsilon}_{p,a+1-\varepsilon}(D)\|^{\delta}. \qquad (4.2.28)$$

Making use of (4.2.4), we have $|G(u)(x)| = |g(x, u(x))| \le c|u(x)|^{\delta}$ for some $\delta \ge 1$. From this we see that

$$\|Gu|L_p(D)\| \le c\left(\int_D |u(x)|^{\delta p} dx\right)^{1/p} = c\|u|L_{\delta p}(D)\|^{\delta},$$

therefore, $G : L_{\delta p}(D) \to L_p(D)$ is bounded. Continuity of G follows from [AZ90, Thm. 3.7], since g is at least continuous in both variables. Hence,

$$N : L_{\delta p}(D) \to L_p(D) \qquad (4.2.29)$$

is continuous as well. By (4.2.23) and Theorem 2.4.8 we have

$$\mathfrak{K}^{m+1-\varepsilon}_{p,a+1-\varepsilon}(D) \hookrightarrow \mathcal{K}^{m}_{p,a+1-\varepsilon}(D) \hookrightarrow L_{\delta p}(D) = \mathcal{K}^{0}_{\delta p,0}(D), \qquad (4.2.30)$$

if, and only if,

$$m \ge \frac{d}{p} - \frac{d}{\delta p} \quad \text{and} \quad a \ge \frac{d}{p} - 1 - \frac{d}{\delta p} + \varepsilon.$$

This is satisfied by our assumptions $m > \frac{d}{p}$ and $a \ge \frac{d}{p} - 1$ if $\varepsilon > 0$ is chosen small enough. We now make use of the interpolation result

$$\left[L_p(D), \mathcal{K}^{m-1}_{p,a-1}(D)\right]_\theta = \left[\mathcal{K}^{0}_{p,0}(D), \mathcal{K}^{m-1}_{p,a-1}(D)\right]_\theta = \mathfrak{K}^{\theta(m-1)}_{p,\theta(a-1)}(D), \quad 0 < \theta < 1,$$

which follows from formula (2.4.22). By employing the associated interpolation inequality we have for all $u \in \mathcal{K}^{m-1}_{p,a-1}(D) \cap L_p(D)$,

$$\|u|\mathfrak{K}^{\theta(m-1)}_{p,\theta(a-1)}(D)\| \le \|u|\mathcal{K}^{m-1}_{p,a-1}(D)\|^{\theta}\|u|L_p(D)\|^{1-\theta}, \qquad (4.2.31)$$

cf. [BL76, Thm. 4.1.4]. Replacing u by $Nu_1 - Nu_2 = Gu_1 - Gu_2$ and using (4.2.28)–(4.2.31) we find

$$\|Nu_1 - Nu_2|\mathfrak{K}^{\theta(m-1)}_{p,\theta(a-1)}(D)\|$$

$$\le \|Gu_1 - Gu_2|\mathcal{K}^{m-1}_{p,a-1}(D)\|^{\theta}\|Nu_1 - Nu_2|L_p(D)\|^{1-\theta}$$

$$\le c^{\theta}\left(\|u_1|\mathfrak{K}^{m+1-\varepsilon}_{p,a+1-\varepsilon}(D)\|^{\delta} + \|u_2|\mathfrak{K}^{m+1-\varepsilon}_{p,a+1-\varepsilon}(D)\|^{\delta}\right)^{\theta}\|Nu_1 - Nu_2|L_p(D)\|^{1-\theta}.$$

Observe that ε and θ can be chosen independent from each other. Since we have
$\|Nu_1 - Nu_2|L_p(D)\| \to 0$ if $u_1 \to u_2$ in $\mathfrak{K}^{m+1-\varepsilon}_{p,a+1-\varepsilon}(D) \hookrightarrow L_{\delta p}(D)$ as well as
$\mathfrak{K}^{\theta(m-1)}_{p,\theta(a-1)}(D) \hookrightarrow \mathfrak{K}^{m-1-\varepsilon'}_{p,a-1-\varepsilon'}(D)$ for some $\varepsilon' < \varepsilon$ (choosing θ close enough to 1),
the above calculations show that

$$N : \mathfrak{K}^{m+1-\varepsilon}_{p,a+1-\varepsilon}(D) \to \mathfrak{K}^{m-1-\varepsilon'}_{p,a-1-\varepsilon'}(D)$$

is continuous.   □

Remark 4.2.14 Continuity of Nemytskij operators (composition operators) is even
more delicate than boundedness. In the framework of Sobolev spaces we refer to
[AZ90], [BL08], [BM01] and [MS02].

Now we are finally in a position to prove complete continuity of $L \circ N$ on
the fractional Kondratiev space $\mathfrak{K}^{m+1-\varepsilon}_{p,a+1-\varepsilon}(D)$. Recall that $\bar{a}$ has been defined in
Proposition 4.1.2.

Theorem 4.2.15 (Complete Continuity of $L \circ N$) *Let D be a domain which
satisfies Assumption 4.1.1, $d/2 < p < \infty$, $\frac{d}{p} - 1 \le a < \bar{a}$, and $m \in \mathbb{N}$ with $m > \frac{d}{p}$.
Moreover, assume $\delta \ge \max(1, m - 1)$ and let the function g satisfy the growth-
condition (4.2.4). Then for sufficiently small $\varepsilon > 0$ it follows that the operator*

$$(L \circ N) : \mathfrak{K}^{m+1-\varepsilon}_{p,a+1-\varepsilon}(D) \hookrightarrow \mathfrak{K}^{m+1-\varepsilon}_{p,a+1-\varepsilon}(D)$$

is completely continuous.

Proof From Proposition 4.1.2 we know that the linear operator L maps $\mathcal{K}^{m-1}_{p,a-1}(D)$
into $\mathcal{K}^{m+1}_{p,a+1}(D)$ if $m \in \mathbb{N}_0$ and $|a| < \bar{a}$. Using complex interpolation and $m \ge 1$ we
obtain

$$L : \underbrace{\left[\mathcal{K}^{m-2}_{p,a-1-\varepsilon'}(D), \mathcal{K}^{m-1}_{p,a-1-\varepsilon'}(D)\right]_{1-\varepsilon'}}_{=\mathfrak{K}^{m-1-\varepsilon'}_{p,a-1-\varepsilon'}(D)} \to \underbrace{\left[\mathcal{K}^{m}_{p,a+1-\varepsilon'}(D), \mathcal{K}^{m+1}_{p,a+1-\varepsilon'}(D)\right]_{1-\varepsilon'}}_{=\mathfrak{K}^{m+1-\varepsilon'}_{p,a+1-\varepsilon'}(D)},$$

which together with the compact embedding $\mathfrak{K}^{m+1-\varepsilon'}_{p,a+1-\varepsilon'}(D) \hookrightarrow \mathfrak{K}^{m+1-\varepsilon}_{p,a+1-\varepsilon}(D)$ for
$0 < \varepsilon' < \varepsilon$ (cf. Remark 2.4.23) shows that

$$L : \mathfrak{K}^{m-1-\varepsilon'}_{p,a-1-\varepsilon'}(D) \to \mathfrak{K}^{m+1-\varepsilon}_{p,a+1-\varepsilon}(D), \tag{4.2.32}$$

is a compact operator. Now using (4.2.32) together with Lemma 4.2.13 yields
complete continuity of

$$(L \circ N) : \mathfrak{K}^{m+1-\varepsilon}_{p,a+1-\varepsilon}(D) \hookrightarrow \mathfrak{K}^{m+1-\varepsilon}_{p,a+1-\varepsilon}(D).$$

□

4.2.3   Existence of Solutions in Kondratiev Spaces

We are now in a position to formulate existence results for solutions to problem (4.1.1) within Kondratiev spaces.

Proposition 4.2.16 *Let D be as in Definition 2.4.5 with singular set S. Consider the mapping $N(u)(x) = f(x) - g(x, u(x))$ and let $f \in \mathcal{K}^{m-1}_{p,a-1}(D)$ and $u \in \mathcal{K}^{m+1}_{p,a+1}(D)$. Further, let us assume $m \in \mathbb{N}$, $a \geq \frac{d}{p} - 1$ and $d/2 < p < \infty$.*

(i) Let $g(x, \xi) = \xi^n$ for some natural number $n > 1$. Then $N(u)$ satisfies the estimate

$$\|N(u)|\mathcal{K}^{m-1}_{p,a-1}(D)\| \leq \|f|\mathcal{K}^{m-1}_{p,a-1}(D)\| + C_n \|u|\mathcal{K}^{m+1}_{p,a+1}(D)\|^n,$$

where $C_n = \tilde{c}c^{n-1}$ and c denotes the constant in Theorem 2.4.9 (ii).
(ii) Now let $g(x, \xi)$ satisfy the growth condition (4.2.4) and, additionally, assume m satisfies the restrictions of Theorem 4.2.1 (in particular, $m > d/p$ is sufficient). Then it holds

$$\|N(u)|\mathcal{K}^{m-1}_{p,a-1}(D)\| \leq \|f|\mathcal{K}^{m-1}_{p,a-1}(D)\| + C_{m-1,g} \|u|\mathcal{K}^{m+1}_{p,a+1}(D)\|^\delta.$$

Proof Part (i) follows simply by repeatedly applying Theorem 2.4.9. This results in an estimate

$$\|f - u^n|\mathcal{K}^{m-1}_{p,a-1}(D)\| \leq \|f|\mathcal{K}^{m-1}_{p,a-1}(D)\| + c^{n-1}\|u|\mathcal{K}^{m+1}_{p,a+1}(D)\|^{n-1}\|u|\mathcal{K}^{m-1}_{p,a-1}(D)\|,$$

and in view of the embedding properties of Kondratiev spaces as stated in Remark 2.4.3(ii) we have $\|u|\mathcal{K}^{m-1}_{p,a-1}(D)\| \leq \tilde{c}\|u|\mathcal{K}^{m+1}_{p,a+1}(D)\|$.
   Part (ii) follows immediately from Theorem 4.2.1.   □

Remark 4.2.17 Proposition 4.2.16 can further be strengthened with the help of our results from Proposition 4.2.11. The result then reads as follows. Let D be a domain which satisfies Assumption 4.1.1, $d/2 < p < \infty$, $a \geq \frac{d}{p} - 1$, and $m \in \mathbb{N}$ with $m > \frac{d}{p}$. Moreover, assume $\delta \geq \max(1, m - 1)$ and let the function g satisfy the growth-condition (4.2.4). Then $N(u)$ satisfies the estimate

$$\|N(u)|\mathcal{K}^{m-1}_{p,a-1}(D)\| \leq \|f|\mathcal{K}^{m-1}_{p,a-1}(D)\| + C \|u|\mathcal{R}^{m+1-\varepsilon}_{p,a+1-\varepsilon}(D)\|^\delta \qquad (4.2.33)$$

for $f \in \mathcal{K}^{m-1}_{p,a-1}(D)$, $u \in \mathcal{K}^{m+1}_{p,a+1}(D)$, and sufficiently small $\varepsilon > 0$. In the particular case of $g(x, \xi) = \xi^n$ for $n \in \mathbb{N}$, $n > 1$, in view of Corollary 4.2.3, we have $\delta = n$ in (4.2.33).

   With this we now obtain the following existence and regularity result for problem (4.1.1). Since we rely on Proposition 4.1.2 in the sequel, we restrict our considerations to the case when $p = 2$.

Theorem 4.2.18 (Existence of Solution in Kondratiev Spaces) *Let D be a domain which satisfies Assumption 4.1.1. Assume $m \in \mathbb{N}$ with $m > \frac{d}{2}$ and $\frac{d}{2} - 1 \le a < \bar{a}$, where $\bar{a}$ is the constant from Proposition 4.1.2. Let the function g satisfy the growth condition (4.2.4) and assume that the condition (4.2.18)(c) is satisfied for $\eta = \|f|\mathcal{K}_{2,a-1}^{m-1}(D)\|$ and $\vartheta = C$ from (4.2.33). Then there exists a solution $u \in \mathcal{K}_{2,a+1}^{m+1}(D)$ of problem (4.1.1).*

Proof This is almost an immediate consequence of Proposition 4.2.7 and (4.2.33). The admissibility of the fractional Kondratiev space $\mathfrak{K}_{2,a+1-\varepsilon}^{m+1-\varepsilon}(D)$ follows from the explanations given in Remark 4.2.10. Furthermore, the complete continuity of the composition operator $L \circ N$ in $\mathfrak{K}_{2,a+1-\varepsilon}^{m+1-\varepsilon}(D)$ was proven in Theorem 4.2.15. Hence we may apply Proposition 4.2.7 with $X := \mathfrak{K}_{2,a+1-\varepsilon}^{m+1-\varepsilon}(D)$ and $Y := \mathcal{K}_{2,a-1}^{m-1}(D)$ for sufficiently small $\varepsilon > 0$. This yields the existence of a function $u \in \mathfrak{K}_{2,a+1-\varepsilon}^{m+1-\varepsilon}(D)$ which solves the partial differential equation but does not necessarily fulfill the boundary condition. In order to see that $u|_{\partial D} = 0$ holds, we argue as follows. Because u is a fixed point, i.e., $u = (L \circ N)u$, and $N(u) \in \mathcal{K}_{2,a-1}^{m-1}(D)$, see Proposition 4.2.11, we conlude that $(L \circ N)u \in \mathcal{K}_{2,a+1,0}^{m+1}(D)$ by using Proposition 4.1.2. □

Remark 4.2.19 Clearly, C from (4.2.33) depends on the nonlinearity and the Kondratiev space. By choosing $\eta = \|f|\mathcal{K}_{2,a-1}^{m-1}(D)\|$ small enough we can always apply Theorem 4.2.18.

Remark 4.2.20 The reader might wonder why the results in this subsection can be established without any essential restriction on the power δ in the nonlinear term. The reason is that so far we have studied distributional solutions to (4.1.1). Our solutions are regular distributions contained in Kondratiev spaces, but until now we have not claimed that they are 'classical weak solutions' contained in $\overset{\circ}{H}{}^1(D)$. We will come back to this problem in the following subsection.

4.2.4   Uniqueness of Solutions

Unfortunately, Proposition 4.2.7 doesn't make any assertions concerning the uniqueness of the fixed point. In the simplified setting where the nonlinearity is just of monomial type, it turns out that we may also use Banach's fixed point theorem instead of Proposition 4.2.7. Then we additionally obtain (under the same assumptions) uniqueness of the solution to the semilinear problem in a (sufficiently small) ball around the solution of the corresponding linear problem in $\mathcal{K}_{2,a+1}^{m+1}(D)$. The precise result reads as follows.

Theorem 4.2.21 (Unique Solution in Kondratiev Spaces) *Let D be a domain which satisfies Assumption 4.1.1, $m \in \mathbb{N}$ and $\frac{d}{2} - 1 \le a < \bar{a}$, where $\bar{a}$ is the constant from Proposition 4.1.2. Furthermore, let $g(x, \xi) = \xi^n$ for some natural*

*number $n > 1$ and let $f \in \mathcal{K}^{m-1}_{2,a-1}(D)$. Assume the condition (4.2.18)(c) to be
satisfied for $\delta = n$, $\eta = \|f|\mathcal{K}^{m-1}_{2,a-1}(D)\|$, and $\vartheta = C_n = \tilde{c}c^{n-1}$ (c taken from
Theorem 2.4.9).*

*Then there exists a unique solution $u \in K_0 \subset \mathcal{K}^{m+1}_{2,a+1}(D)$ of problem (4.1.1),
where K_0 denotes a small closed ball in $\mathcal{K}^{m+1}_{2,a+1}(D)$ with center Lf (the solution of
the corresponding linear problem) and radius $r := \frac{\|L\|\eta}{n-1}$.*

Proof We wish to apply Banach's fixed point theorem, which guarantees uniqueness of the solution if we can show that $T := (L \circ N) : K_0 \to K_0$ is a contraction
mapping, i.e., there exists some $q \in [0, 1)$ such that

$$\|T(u) - T(v)|\mathcal{K}^{m+1}_{2,a+1}(D)\| \le q\|u - v|\mathcal{K}^{m+1}_{2,a+1}(D)\| \quad \text{for all} \quad u, v \in K_0,$$

where K_0 is a closed ball in $\mathcal{K}^{m+1}_{2,a+1}(D)$ with center Lf and suitable radius r.
Observe that

$$(L \circ N)(u) - (L \circ N)(v) = L(f - G(u)) - L(f - G(v)) = (L \circ G)(v) - (L \circ G)(u),$$

thus, $L \circ N$ is a contraction if, and only if, $L \circ G$ is. Let us analyze the resulting
scaling condition in the monomial case $g(x, \xi) = \xi^n$. Since we have $G(u) -
G(v) = u^n - v^n = (u - v) \sum_{j=0}^{n-1} u^j v^{n-1-j}$, we can apply Proposition 4.1.2 and
Theorem 2.4.9 to obtain the estimate

$$\|(L \circ G)(u) - (L \circ G)(v)|\mathcal{K}^{m+1}_{2,a+1}(D)\|$$

$$\le \|L\| \, \|G(u) - G(v)|\mathcal{K}^{m-1}_{2,a-1}(D)\|$$

$$= \|L\| \, \|u^n - v^n|\mathcal{K}^{m-1}_{2,a-1}(D)\|$$

$$\le \|L\| \, c \, \|u - v|\mathcal{K}^{m+1}_{2,a+1}(D)\| \sum_{j=0}^{n-1} \tilde{c}c^{n-2} \, \|u|\mathcal{K}^{m+1}_{2,a+1}(D)$$

$$\times \|^j \, \|v|\mathcal{K}^{m+1}_{2,a+1}(D)\|^{n-1-j}$$

$$\le \|L\| \, n \, C_n \, (r + \|L\|\eta)^{n-1} \, \|u - v|\mathcal{K}^{m+1}_{2,a+1}(D)\| \tag{4.2.34}$$

for all $u, v \in K_0$, the closed ball in $\mathcal{K}^{m+1}_{2,a+1}(D)$ with center Lf and radius r. With
$r = \frac{\|L\|\eta}{n-1}$ we conlude that $L \circ G$ is a contraction if

$$n \left(1 + \frac{1}{n-1}\right)^{n-1} C_n \, \eta^{n-1} \, \|L\|^n < 1 \,. \tag{4.2.35}$$

Elementary calculations yield that this equivalent to (4.2.18)(c).

Moreover, we need $(L \circ N)(B_r(Lf)) \subset B_r(Lf)$. Since $(L \circ N)(0) = Lf$, due to $G(0) = 0$, we see that

$$\|(L \circ N)(u) - Lf|\mathcal{K}^{m+1}_{2,a+1}(D)\|$$

$$= \|(L \circ N)(u) - (L \circ N)(0)|\mathcal{K}^{m+1}_{2,a+1}(D)\|$$

$$\leq \|L\| \, \|u^n - 0|\mathcal{K}^{m-1}_{2,a-1}(D)\|$$

$$\leq \|L\| \, C_n \, (r + \|L\|\eta)^n \overset{!}{\leq} r = \frac{\|L\|\eta}{n-1} \qquad (4.2.36)$$

with $u \in K_0$. The claimed inequality is equivalent to

$$\left(1 + \frac{1}{n-1}\right)^n C_n \, \eta^{n-1} \, \|L\|^n \leq \frac{1}{n-1} .$$

Using (4.2.35) we conclude

$$\left(1 + \frac{1}{n-1}\right)^n C_n \, \eta^{n-1} \, \|L\|^n < \left(1 + \frac{1}{n-1}\right)^n \frac{1}{n(1 + \frac{1}{n-1})^{n-1}} = \frac{1}{n-1} .$$

$\square$

Remark 4.2.22 The main restriction in Theorem 4.2.21 stems from condition (4.2.18)(c), which upon inserting $\eta = \|f|\mathcal{K}^{m-1}_{2,a-1}(D)\|$ turns into a scaling condition for the right-hand side. In other words: By our method of proof a unique solution only exists in case of "sufficiently small" right-hand sides.

This observation is in accordance with what is known from the classical theory for semilinear elliptic problems in Sobolev spaces, as can be found e.g. in [Str08]. Particularly for semilinear problems with monomial nonlinearities $\pm|u|^{p-2}u$, much is known about existence and (non-)uniqueness of solutions. The delicate dependence on the sign of the nonlinearity is eliminated in our setting by the usage of the simple growth-condition (4.2.4).

More precisely, for the problem

$$-\Delta u = u|u|^{p-2} + f, \quad u|_{\partial D} = 0$$

it is known that, for a certain range of parameters $p > 2$, for arbitrary $f \in L_2(D)$ we have an unbounded sequence of solutions in $\mathring{H}^1(D)$; we refer to [Str08, Thm. 7.2, Rem. 7.3]. Therefore, in order to have any notion of uniqueness, additional restrictions (usually taking the form of scaling conditions) become necessary.

In order to obtain classical weak solutions contained in $\mathring{H}^1$ we can strengthen Theorem 4.2.21 in the following way.

Corollary 4.2.23 (Unique Weak Solution in Kondratiev Spaces) *Let D be a domain which satisfies Assumption 4.1.1, $m \in \mathbb{N}$, and $\frac{d}{2} - 1 \leq a < \bar{a}$, where $\bar{a}$ is the constant from Proposition 4.1.2. Furthermore, let $g(x, \xi) = \xi^n$ where $n \in \mathbb{N}$ with $n \geq 2$ if $d = 2$ and $n \in \{2, \ldots, 5\}$ if $d = 3$. Let $f \in \mathcal{K}^{m-1}_{2,a-1}(D) \cap H^{-1}(D)$ and assume the condition (4.2.18)(c) to be satisfied for $\delta = n$,*

$$\eta := \|f|\mathcal{K}^{m-1}_{2,a-1}(D) \cap H^{-1}(D)\| := \max\left(\|f|\mathcal{K}^{m-1}_{2,a-1}(D)\|, \|f|H^{-1}(D)\|\right),$$

$$\|L\| := \max(\|L|\mathcal{L}(\mathcal{K}^{m-1}_{2,a-1}(D), \mathcal{K}^{m+1}_{2,a+1}(D))\|, \|L|\mathcal{L}(H^{-1}(D), \mathring{H}^1(D))\|),$$

and

$$\vartheta := \max\left(C_n, 3^{n-1}\|I_D\|^{n+1}_{n+1}\right)$$

where $C_n = \tilde{c}c^{n-1}$ and c is the constant from Theorem 2.4.9. Moreover, $\|I_D\|_{n+1}$ denotes the operator norm $\|I_D|H^1(D) \to L_{n+1}(D)\|$ of the embedding operator I_D. Then there exists a unique solution $u \in \tilde{K}_0 \subset \mathcal{K}^{m+1}_{2,a+1}(D) \cap \mathring{H}^1(D)$ of problem (4.1.1), where $\tilde{K}_0$ denotes the small closed ball in this space with center Lf (the solution of the corresponding linear problem) and radius $r = \frac{\|L\|\eta}{n-1}$.

Proof

Step 1.   In this step we show that

$$L \circ N : \mathcal{K}^{m+1}_{2,a+1}(D) \cap \mathring{H}^1(D) \to \mathcal{K}^{m+1}_{2,a+1}(D) \cap \mathring{H}^1(D)$$

is a bounded and continuous operator. For this we first show that already $N : \mathring{H}^1(D) \to H^{-1}(D)$ is a bounded and continuous operator, where $N(u) = f - u^n$. Since $f \in H^{-1}(D)$ this is the case if $G : \mathring{H}^1(D) \to H^{-1}(D)$, $G(u) = u^n$, is bounded and continuous. Let $2 < q < \infty$. By Sobolev's embedding theorem we have

$$H^1(D) \hookrightarrow L_q(D) \quad \text{if} \quad 1 - \frac{d}{2} \geq -\frac{d}{q}, \tag{4.2.37}$$

i.e., $q \leq \frac{2d}{d-2}$ if $d = 3$ and no extra condition for $d = 2$. The operator norm $\|I_D|H^1(D) \to L_q(D)\|$ of the embedding operator I_D will be abbreviated by $\|I_D\|_q$. Because of (4.2.37), if $u^n \in L_{q'}(D)$ then

$$\left|\int_D u^n(x)\,\varphi(x)\,dx\right| \leq \|\varphi|L_q(D)\| \cdot \|u^n|L_{q'}(D)\| < \infty \quad \text{for all} \quad \varphi \in \mathring{H}^1(D)$$

$$\tag{4.2.38}$$

and we conclude $u^n \in H^{-1}(D)$, where $1/q' := 1 - 1/q$. This holds if $nq' \le q$, i.e., $n \le \frac{q}{q'} = q - 1 \le \frac{d+2}{d-2}$, which for $d = 3$ yields $n \le 5$. Thus, $G : \mathring{H}^1(D) \to H^{-1}(D)$, $G(u) = u^n$, is a bounded operator for $n \le 5$ if $d = 3$ and for all n if $d = 2$. Concerning continuity we make use of the formula $u^n - v^n = (u - v) \sum_{j=0}^{n-1} u^j v^{n-1-j}$. Since $q > 2$ applying Hölder's inequality twice with $r := \frac{q}{2}$, $r' = \frac{q}{q-2}$, i.e., $\frac{1}{r} + \frac{1}{r'} = 1$, and afterwards with $r = r' = 2$, we obtain for $\varphi \in \mathring{H}^1(D)$,

$$\left| \int_D (u^n(x) - v^n(x))\, \varphi(x)\, dx \right| \tag{4.2.39}$$

$$\le \sum_{j=0}^{n-1} \int_D |u(x) - v(x)|\, |u^j(x) v^{n-1-j}(x)|\, |\varphi(x)|\, dx$$

$$\le \sum_{j=0}^{n-1} \left(\int_D |(u(x) - v(x))\varphi(x)|^{\frac{q}{2}}\, dx \right)^{\frac{2}{q}} \left(\int_D |u^j(x) v^{n-1-j}(x)|^{\frac{q}{q-2}}\, dx \right)^{\frac{q-2}{q}}$$

$$\le \sum_{j=0}^{n-1} \left(\int_D |u(x) - v(x)|^q\, dx \right)^{\frac{1}{q}} \times \left(\int_D |\varphi(x)|^q\, dx \right)^{\frac{1}{q}}$$

$$\times \left(\int_D |u^j(x) v^{n-1-j}(x)|^{\frac{q}{q-2}}\, dx \right)^{\frac{q-2}{q}}$$

$$\le n \cdot \|I_D\|_q^2 \cdot \|u - v|H^1(D)\| \cdot \|\varphi|H^1(D)\|$$

$$\cdot \left(\int_D \left(\max(|u(x)|, |v(x)|)^{n-1} \right)^{\frac{q}{q-2}}\, dx \right)^{\frac{q-2}{q}}.$$

If $u, v \in H^1(D)$ it follows from [GT01, Lem. 7.6, p. 152] that $u^+, |u| \in H^1(D)$, where $u^+ := \max(u, 0)$, which together with the formula

$$\max(|u|, |v|) = \frac{1}{2} \left((|u| - |v|)^+ + (|v| - |u|)^+ + |u| + |v| \right)$$

shows that $\max(|u|, |v|) \in H^1(D)$. Hence, in order for the integral in the last line of (4.2.39) to be bounded, we require from Sobolev's embedding theorem that $(n - 1)\frac{q}{q-2} \overset{!}{\le} q$, i.e., $n \le q - 1 \le \frac{d+2}{d-2}$, which for $d = 3$ gives $n \le 5$ and for $d = 2$ no extra condition. Under the given restrictions (4.2.39) yields

$$\|u^n - v^n|H^{-1}(D)\| \lesssim \|u - v|H^1(D)\|,$$

where the suppressed constant depends on n and R for $u, v \in B_R(0) \subset \mathring{H}^1(D)$. Hence, $G : \mathring{H}^1(D) \to H^{-1}(D)$ is locally Lipschitz continuous. This, together with Remark 4.1.3(i), shows that

$$L \circ N : \mathring{H}^1(D) \to \mathring{H}^1(D)$$

is a bounded and continuous operator. From the proof of Theorem 4.2.21 (in particular the calculations in (4.2.34)) we already know that

$$L \circ N : \mathcal{K}^{m+1}_{2,a+1}(D) \to \mathcal{K}^{m+1}_{2,a+1}(D)$$

is a bounded and continuous operator. Thus,

$$L \circ N : \mathcal{K}^{m+1}_{2,a+1}(D) \cap \mathring{H}^1(D) \to \mathcal{K}^{m+1}_{2,a+1}(D) \cap \mathring{H}^1(D)$$

is bounded and continuous as well.

Step 2.   Now we wish to apply Banach's fixed point theorem. Applying again [GT01, Lem. 7.6, p. 152] we see that

$$\max(\|u^+|H^1(D)\|, \| |u| \,|H^1(D)\|) \le \|u|H^1(D)\|, \qquad u \in H^1(D).$$

Let $u, v \in \tilde{K}_0$. A close inspection of (4.2.39) gives that

$$\|u^n - v^n|H^{-1}(D)\| \le n \, \|I_D\|_q^{n+1} \left(\frac{3}{2} \|u|H^1(D)\| + \frac{3}{2} \|v|H^1(D)\|\right)^{n-1}$$

$$\times \|u - v|H^1(D)\|$$

$$\le n \, \|I_D\|_q^{n+1} \, 3^{n-1} \, (r + \eta\|L\|)^{n-1} \, \|u - v|H^1(D)\|$$

$$\le n \, \|I_D\|_q^{n+1} \, 3^{n-1} \, \|L\|^{n-1} \, \eta^{n-1} \left(1 + \frac{1}{n-1}\right)^{n-1}$$

$$\times \|u - v|H^1(D)\|.$$

This implies

$$\|(L \circ N)(u) - (L \circ N)(v)|\mathring{H}^1(D)\|$$

$$= \|(L \circ G)(u) - (L \circ G)(v)|\mathring{H}^1(D)\|$$

$$\le \|L\|\|G(u) - G(v)|H^{-1}(D)\|$$

$$\le n \, \|I_D\|_q^{n+1} \, 3^{n-1} \, \|L\|^n \, \eta^{n-1} \left(1 + \frac{1}{n-1}\right)^{n-1} \|u - v|H^1(D)\|. \qquad (4.2.40)$$

The structure of this estimate is exactly as in (4.2.34) except that C_n is replaced by $\|I_D\|_q^{n+1}\, 3^{n-1}$. Hence, by our assumptions,

$$n\,\|I_D\|_q^{n+1}\,3^{n-1}\,\|L\|^n\,\eta^{n-1}\left(1+\frac{1}{n-1}\right)^{n-1}<1 \qquad (4.2.41)$$

and $L\circ N$ is a contraction. We also need $(L\circ N)(B_r(Lf))\subset B_r(Lf)$. Therefore we apply (4.2.38) and find

$$\left|\int_D u^n(x)\,\varphi(x)\,dx\right|\le \|\varphi|L_q(D)\|\cdot\|u|L_{q'n}(D)\|^n$$

$$\le \|I_D\|_q^{n+1}\,\|\varphi|H^1(D)\|\cdot(r+\|L\|\eta)^n$$

Hence, since $u\in\tilde{K}_0$ we obtain

$$\|(L\circ N)(u)-Lf|\mathring{H}^1(D)\|=\|(L\circ G)(u)-(L\circ G)(0)|\mathring{H}^1(D)\|$$

$$\le \|L\|\|G(u)|H^{-1}(D)\|$$

$$\le \|I_D\|_q^{n+1}\,\|L\|^{n+1}\,\eta^n\left(1+\frac{1}{n-1}\right)^n. \qquad (4.2.42)$$

Because of (4.2.41) it follows

$$\|I_D\|_q^{n+1}\,\|L\|^{n+1}\,\eta^n\left(1+\frac{1}{n-1}\right)^n\le r=\frac{\|L\|\,\eta}{n-1}.$$

Combining these arguments with those used in the proof of Theorem 4.2.21 and taking into account that the smallest possible q is given by $q=n+1$, see Step 1, the claim follows.

□

Remark 4.2.24 If $m\ge 1$ and $a\ge 1$ we see from the embedding

$$\mathcal{K}_{2,a-1}^{m-1}(D)\hookrightarrow\mathcal{K}_{2,0}^0(D)=L_2(D)\hookrightarrow H^{-1}(D)$$

that no additional restrictions on the right hand side f are needed in Corollary 4.2.23 compared to Theorem 4.2.21.

4.3   Regularity of Semilinear Elliptic PDEs in Besov Spaces

The Kondratiev regularity results in Theorem 4.2.18 of the last section are the basis for assertions on Besov regularity of solutions to (4.1.1). More precisely, we use the embedding results between Kondratiev and Triebel-Lizorkin spaces from Theorem 2.4.16, which enable us to show that the solutions belong to spaces $F_{\tau,2}^{m+1}(D) \hookrightarrow L_2(D)$ for a suitable parameter $0 < \tau < 2$.

We specialize to the case $p = 2$. Our main result reads as follows.

Theorem 4.3.1 (Semilinear Elliptic Besov Regularity) *Let $D \subset \mathbb{R}^d$, $d \in \{2, 3\}$, be a bounded Lipschitz domain of polyhedral type with singular set S of dimension l. Let $\bar{a}, m, g, f, \eta, C$ be as in Theorem 4.2.18. Let a and $0 < \tau < 2$ be such that*

$$\frac{d}{2} - 1 \leq a < \bar{a} \qquad \text{and} \qquad \frac{m-a}{d-l} + \frac{1}{2} < \frac{1}{\tau} \leq \frac{2m+d}{2d}.$$

Then there exists a solution $u \in F_{\tau,2}^{m+1}(D) \hookrightarrow H^1(D)$ of problem (4.1.1).

Proof The claim is an immediate consequence of Theorems 4.2.18 and 2.4.16. Let us mention that Proposition 2.3.4 yields

$$F_{\tau,2}^{m+1}(D) \hookrightarrow F_{2,2}^1(D) = H^1(D),$$

if $m + 1 - \frac{d}{\tau} \geq 1 - \frac{d}{2}$, i.e., if $\tau \geq \frac{2d}{2m+d}$.   □

Remark 4.3.2 F-spaces are closely related to B-spaces: Under the assumptions of Theorem 4.3.1, using the embedding (2.3.5) gives

$$F_{\tau,2}^{m+1}(D) \hookrightarrow B_{\tau,2}^{m+1}(D),$$

which provides the Besov regularity for the solution of problem 4.1.1. In Chap. 7, Sects. 7.1 and 7.2, we further discuss the impact of these findings with regard to approximation spaces for N-term wavelet approximation and adaptive finite element approximation.

Chapter 5
Regularity Theory for Parabolic PDEs

The present chapter is the heart of Part I of this manuscript dealing with the regularity theory of PDEs. In contrast to Chap. 4 we now consider parabolic problems and the (spacial) fractional Sobolev and Besov regularity of their solutions.

Let us explain the structure of Chap. 5 in a nutshell: In Sect. 5.1 we introduce the various parabolic settings we want to study, which are collected in Problems I–IV. All of our problems are considered on Lipschitz domains. In particular, Problems I–III are treated on domains of polyhedral type and generalized wedges, respectively. These restrictions to special domains are due to the techniques we use. On the other hand, the fact that these specific domains are not as bad as general Lipschitz domains, leads to an improvement in the Besov regularity of the solutions. This becomes clear when comparing the regularity results in Besov spaces of Problems I and II with corresponding results for Problem IV (which is considered on a general Lipschitz domain).

To achieve our goal we rely on regularity results in Sobolev and Kondratiev spaces for Problems I–IV, which we present in Sect. 5.2. Our presentations for Problems I and II are essentially new and not published elsewhere so far, whereas for Problems III and IV we merely rely on already existing results in the literature which we reformulate in terms of our Kondratiev spaces from Sect. 2.4.

In Sect. 5.3 we then proceed as follows: We make use of the embedding from Theorem 2.4.16 which states that subject to some restrictions on the parameters,

$$\mathcal{K}^m_{p,a}(D) \hookrightarrow F^m_{\tau,2}(D), \qquad m - a < (d - \delta)\left(\frac{1}{\tau} - \frac{1}{p}\right),$$

where D is some Lipschitz domain of polyhedral type with singular set of dimension δ. Using the fact that the fractional Sobolev spaces are included in the scale of Triebel-Lizorkin spaces via $F^s_{p,2} = H^s_p$, we obtain fractional Sobolev regularity of Problem III.

C. Schneider, *Beyond Sobolev and Besov*, Lecture Notes in Mathematics 2291, https://doi.org/10.1007/978-3-030-75139-5_5

Finally, in Sect. 5.4 we deal with the Besov regularity of Problems I, II, and IV. Here we make use of Theorem 2.4.10 and related embeddings. To demonstrate the main idea a slight reformulation of Theorem 2.4.10 yields

$$\mathcal{K}^{\gamma}_{p,a}(D) \cap H^s(D) \hookrightarrow B^{\eta}_{\tau,\tau}(D), \qquad \eta < \min\left(\gamma, \frac{d}{\delta}s\right), \qquad \frac{1}{\tau} = \frac{\eta}{d} + \frac{1}{p},$$

$$(5.0.1)$$

subject to some restrictions on the parameters (e.g. for a large enough). Thus, we see from (5.0.1) that the knowledge of Kondratiev and Sobolev regularity of our solutions gives regularity in the specific scale of Besov spaces (1.0.3).

As an important example we discuss the heat equation as the prototype of parabolic PDEs and give precise upper bounds for its Besov and fractional Sobolev regularity in Sects. 5.3 and 5.4. Also the role of the weight parameter a appearing in the Kondratiev spaces and its restrictions will be discussed several times. Comparision of our findings with related results in the literature (and further references) will be given at adequate places within Chap. 5.

5.1   The Fundamental Parabolic Problems

In this section we present the different parabolic settings, Problems I–IV, that will be studied throughout Chap. 5. Precisely, Problems I and II are of general order and defined on domains of polyhedral type according to Definition 2.4.5. In particular, Problem II is the nonlinear version of Problem I and we investigate the spacial Besov regularity of the solutions of these two problems and to some extend also the Hölder regularity with respect to the time variable of Problem I.

Moreover, we also study the spacial Besov regularity of the solutions of Problem IV, which is only of second order but defined on general Lipschitz domains. Due to the fact that the underlying domain is more general in this context, our regularity results in the Besov spaces are not as good as those for Problems I and II.

In contrast to this, we investigate the fractional Sobolev regularity of the solution of Problem III, which is of second order and defined on generalized wedges acccording to Definition 2.1.6.

5.1.1   Problems I and II: Parabolic PDEs on Domains of Polyhedral Type

Let D denote some domain of polyhedral type in $\mathbb{R}^d$ according to Definition 2.4.5 with faces Γ_j, $j = 1, \ldots, n$.

For $0 < T < \infty$ put $D_T = (0, T] \times D$ and $\Gamma_{j,T} = [0, T] \times \Gamma_j$.

Fig. 5.1 Polyhedron

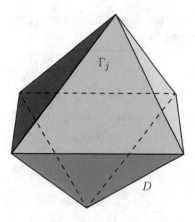

Figure 5.1 gives an example of a polyhedron in $\mathbb{R}^3$ which is of polyhedral type.
We will investigate the Besov regularity of the following linear parabolic problem.

Problem I (Linear Parabolic Problem in Divergence Form) *Let* $m \in \mathbb{N}$. *We consider the following first initial-boundary value problem*

$$
\left\{
\begin{array}{rll}
\frac{\partial}{\partial t}u + (-1)^m L(t, x, D_x)u & = f & in \; D_T, \\
\left.\frac{\partial^{k-1}u}{\partial \nu^{k-1}}\right|_{\Gamma_{j,T}} & = 0, & k = 1, \ldots, m, \; j = 1, \ldots, n, \\
u\big|_{t=0} & = 0 & in \; D.
\end{array}
\right\} \tag{5.1.1}
$$

Here f is a function given on D_T, ν denotes the exterior normal to $\Gamma_{j,T}$, and the partial differential operator L is given by

$$
L(t, x, D_x) = \sum_{|\alpha|,|\beta|=0}^{m} D_x^\alpha (a_{\alpha\beta}(t, x)D_x^\beta),
$$

where $a_{\alpha\beta}$ are bounded real-valued functions from $C^\infty(D_T)$ with $a_{\alpha\beta} = (-1)^{|\alpha|+|\beta|}a_{\beta\alpha}$. Furthermore, the operator L is assumed to be uniformly elliptic with respect to $t \in [0, T]$, i.e.,

$$
\sum_{|\alpha|,|\beta|=m} a_{\alpha\beta}\xi^\alpha\xi^\beta \geq c|\xi|^{2m} \qquad \text{for all} \quad (t, x) \in D_T, \quad \xi \in \mathbb{R}^d. \tag{5.1.2}
$$

Let us denote by

$$
B(t, u, v) = \int_D \sum_{|\alpha|,|\beta|=0}^{m} a_{\alpha\beta}(t, x)(D_x^\beta u)(D_x^\alpha v)dx \tag{5.1.3}
$$

the time-dependent bilinear form. In the sequel, w.l.o.g. we suppose that $B(t, \cdot, \cdot)$ satisfies

$$B(t, u, u) \geq \mu \|u | H^m(D)\|^2 \tag{5.1.4}$$

for all $u \in \mathring{H}^m(D)$ and a.e. $t \in [0, T]$. We refer to the explanation in Remark 3.3.5.
    Moreover, for simplicity we set

$$B_{\partial_{t,k}}(t, u, v) = \sum_{|\alpha|, |\beta| \leq m} \int_D \frac{\partial a_{\alpha\beta}(t, x)}{\partial t^k} D^\beta u(t, x) D^\alpha v(t, x) dx. \tag{5.1.5}$$

It is our intention to also study nonlinear versions of Problem I. Therefore, we modify (5.1.1) as follows.

Problem II (Nonlinear Parabolic Problem in Divergence Form) *Let* $m, M \in \mathbb{N}$ *and* $\varepsilon > 0$. *We consider the following nonlinear parabolic problem*

$$\left\{ \begin{aligned} \frac{\partial}{\partial t} u + (-1)^m L(t, x, D_x) u + \varepsilon u^M &= f \ \ in \ D_T, \\ \left. \frac{\partial^{k-1} u}{\partial v^{k-1}} \right|_{\Gamma_{j,T}} &= 0, \ \ k = 1, \ldots, m, \ \ j = 1, \ldots, n, \\ u \big|_{t=0} &= 0 \ \ in \ D. \end{aligned} \right\} \tag{5.1.6}$$

The assumptions on f and the operator L are as in Problem I. In order to establish Besov regularity results for Problem II we interpret (5.1.6) as a fixed point problem and show that the regularity estimates for Problem I carry over to Problem II, provided that ε is sufficiently small.

5.1.2   Problem III: Linear Parabolic PDEs of Second Order on Generalized Wedges

When studying the fractional Sobolev regularity of parabolic PDEs we deal with a slightly different setting. In particular, we restrict ourselves to differential operators of second order now. Moreover, instead of having PDEs with underlying domains of polyhedral type we now deal with generalized wedges (only). The precise parabolic problem is stated below.

Problem III (Linear Parabolic Problem in Non-divergence Form) *Consider the following parabolic problem of second order*

$$\mathcal{L}u = f \ \ in \ \ \mathbb{R} \times \mathcal{K}, \qquad u \big|_{\mathbb{R} \times \partial \mathcal{K}} = 0. \tag{5.1.7}$$

Here $\mathcal{K}$ denotes a generalized wedge according to Definition 2.1.6. Moreover, the differential operator $\mathcal{L}$ is of second order and given by

$$\mathcal{L}u(t, x) := \frac{\partial}{\partial t}u(t, x) - \sum_{i,j=1}^{d} A_{ij}(t)\frac{\partial^2 u}{\partial x_i \partial x_j}(t, x). \tag{5.1.8}$$

We only assume that the coefficients A_{ij} are real valued measurable functions of t satisfying $A_{ij} = A_{ji}$ and that for some constant $\nu > 0$ we have

$$\nu|\xi|^2 \le A_{ij}(t)\xi_i\xi_j \le \nu^{-1}|\xi|^2 \quad \text{for all} \quad \xi \in \mathbb{R}^d. \tag{5.1.9}$$

Remark 5.1.1 (Generalized Solution) Equation (5.1.7) is understood in the distributional sense (i.e., in the sense of generalized functions) only with resrpect to x. By a solution of (5.1.7) we mean a function $u(t)$, $t \in \mathbb{R}$, taking values in the set of generalized functions on $\mathcal{K}$ (i.e., in $\mathcal{D}'(\mathcal{K})$) such that, for any $t, s \in \mathbb{R}$ satisfying $t \ge s$ and $\varphi \in \mathcal{D}(\mathcal{K})$, we have

$$\langle u(t), \varphi \rangle = \langle u(s), \varphi \rangle + \int_s^t \left[\sum_{i,j=1}^{d} A_{ij}(r)\left\langle u(r), \frac{\partial^2 \varphi}{\partial x_i \partial x_j} \right\rangle + \langle f(r), \varphi \rangle \right] dr.$$

5.1.3   Problem IV: Linear Parabolic PDEs of Second Order on General Lipschitz Domains

We also want to study regularity in Besov spaces for parabolic PDEs on general Lipschitz domains. Compared to Problems I and II, where we restricted our investigations to domains of polyhedral type, in case of general Lipschitz domains all points on the boundary are equally bad. Therefore, the singularities of the solution induced by the boundary have a much stronger influence. As a consequence, the regularity results on domains of polyhedral type turn out to be much stronger compared to the general Lipschitz case.

Let $\mathcal{O} \subset \mathbb{R}^d$ be a bounded Lipschitz domain and as before put $\mathcal{O}_T = (0, T] \times \mathcal{O}$. Moreover, let $\varrho(x) = \text{dist}(x, \partial\mathcal{O})$.

Problem IV (Second Order Parabolic PDE in Non-divergence Form) *We consider the following class of parabolic equations*

$$\begin{cases} \frac{\partial}{\partial t}u = \sum_{i,j=1}^{d} a_{ij}\frac{\partial^2}{\partial x_i \partial x_j}u + \sum_{i=1}^{d} b_i\frac{\partial}{\partial x_i}u + cu + f & \text{on } \mathcal{O}_T, \\ u(0, \cdot) = u_0 & \text{on } \mathcal{O}, \end{cases} \tag{5.1.10}$$

where the coefficients are assumed to satisfy the assumptions listed below.

We need to introduce some more notation. Put $\varrho(x, y) = \min(\varrho(x), \varrho(y))$. For $\alpha \in \mathbb{R}$, $\delta \in (0, 1]$, and $m \in \mathbb{N}_0$, we set

$$[f]_m^{(\alpha)} := \sup_{x \in \mathcal{O}} \varrho^{m+\alpha}(x) |D^m f(x)|,$$

$$[f]_{m+\delta}^{(\alpha)} := \sup_{\substack{x, y \in \mathcal{O} \\ |\beta| = m}} \varrho^{m+\alpha}(x, y) \frac{|D^\beta f(x) - D^\beta f(y)|}{|x - y|^\delta},$$

$$|f|_m^{(\alpha)} := \sum_{l=0}^{m} [f]_l^{(\alpha)} \quad \text{and} \quad |f|_{m+\delta}^{(\alpha)} := |f|_m^{(\alpha)} + [f]_{m+\delta}^{(\alpha)},$$

whenever it makes sense.

Assumption 5.1.2 (Assumptions on the Coefficients)

 (i) *Parabolicity: There are constants δ_0, $K \in (0, \infty)$ such that for all $\lambda \in \mathbb{R}^d$ it holds*

$$\delta_0 |\lambda|^2 \le a_{ij}(t, x) \lambda_i \lambda_j \le K |\lambda|^2.$$

 (ii) *The behaviour of the coefficients b_i and c can be controlled near the boundary of $\mathcal{O}$:*

$$\lim_{\substack{\varrho(x) \to 0, \\ x \in \mathcal{O}}} \sup_t \left(\varrho(x) |b_i(t, x)| + \varrho^2(x) |c(t, x)| \right) = 0.$$

 (iii) *The coefficients $a_{ij}(t, \cdot)$ are uniformly continuous in x, i.e., for any $\varepsilon > 0$ there is a $\delta = \delta(\varepsilon) > 0$ such that*

$$\left| a_{ij}(t, x) - a_{ij}(t, y) \right| < \varepsilon$$

for all $x, y \in \mathcal{O}$ with $|x - y| < \delta$.

 (iv) *For any $t > 0$,*

$$\left| a_{ij}(t, \cdot) \right|_{\gamma_{(+\varepsilon_0)}}^{(0)} + \left| \varrho(x) b_i(t, \cdot) \right|_{\gamma_{(+\varepsilon_0)}}^{(0)} + \left| \varrho^2(x) c(t, \cdot) \right|_{\gamma_{(+\varepsilon_0)}}^{(0)} \le K,$$

where for $\gamma \ge 0$ and a fixed constant $\varepsilon_0 > 0$ we put

$$\gamma_{(+\varepsilon_0)} = \begin{cases} \gamma, & \text{if } \gamma \in \mathbb{N}_0, \\ \gamma + \varepsilon_0, & \text{otherwise.} \end{cases}$$

Remark 5.1.3 Observe that Problems I and II are more general compared to Problem IV in the sense that there differential operators of arbitrary order are considered (but with smooth C^∞ coefficients), whereas the analysis on general Lipschitz domains is restricted to second order operators.

5.2   Regularity Results in Sobolev and Kondratiev Spaces

This section presents regularity results for Problems I–IV in Sobolev and Kondratiev spaces. They will form the basis for obtaining regularity results in Besov and fractional Sobolev spaces in Sects. 5.4 and 5.3 via the embeddings established in Sect. 2.4.4.

The results in Sobolev and Kondratiev spaces for Problems I and II on domains of polyhedral type $D \subset \mathbb{R}^d$, $d = 3$, are are essentially new and not published elsewhere so far: In [DS19] we restricted our investigations to polyhedral cones $K \subset \mathbb{R}^3$ relying on the results from [LL15].

For Problems III and IV we merely collect what is known concerning estimates in weighted Sobolev spaces from [KN14] and [Kim09], respectively, and formulate the results from those papers in terms of our Kondratiev spaces introduced in Chap. 2.

5.2.1   Regularity Results in Sobolev Spaces for Problem I

In this subsection we are concerned with the Sobolev regularity of the weak solution of Problem I. Our main result (which also involves regularity of time derivatives of the solution) is formulated in Theorem 5.2.3. Forerunners can be found in [LL15, Thm. 2.1., Lem. 3.1], where the underlying domain is a polyhedral cone $K \subset \mathbb{R}^3$. We generalize this setting to bounded domains of polyhedral type $D \subset \mathbb{R}^d$. A clarification of this problem is crucial for our purposes: In order to study the Besov regularity of Problem I with the help of the embedding from Theorem 2.4.10, one needs to know the regularity of the solution not only in Kondratiev but also in Sobolev spaces.

We start with the following lemma.

Lemma 5.2.1 (Continuity of Bilinear Form) *Assume that for each* $t \in [0, T]$, $F(t, \cdot, \cdot) : \mathring{H}^m(D) \times \mathring{H}^m(D) \to \mathbb{R}$ *is a bilinear map satisfying*

$$|F(t, u, v)| \leq C \|u|\mathring{H}^m(D)\| \|v|\mathring{H}^m(D)\| \tag{5.2.1}$$

for all $t \in [0, T]$ *and all* $u, v \in \mathring{H}^m(D)$, *where* C *is a constant independent of* $u, v,$ *and* t. *Assume further that* $F(\cdot, u, v)$ *is measurable on* $[0, T]$ *for each pair* $u, v \in \mathring{H}^m(D)$. *Assume that* $u \in H^{m,1*}(D_T)$ *satisfies* $u(0) \equiv 0$ *and*

$$\langle \partial_t u(t), v \rangle + B(t, u(t), v) = \int_0^t F(\tau, u(\tau), v) d\tau \tag{5.2.2}$$

for a.e. $t \in [0, T]$ *and all* $v \in \mathring{H}^m(D)$. *Then* $u \equiv 0$ *on* $[0, T] \times D$.

Proof The proof is similar to [AH08, Lem. 4.1]. Put $v = u(t)$ in (5.2.2). Integrating both sides w.r.t. t from 0 to b ($b > 0$) and using (5.2.1) we arrive at

$$\int_0^b \langle \partial_t u, u \rangle dt + \int_0^b B(t, u(t), u(t)) dt \qquad (5.2.3)$$

$$= \int_0^b \int_0^t F(\tau, u(\tau), u(t)) d\tau dt$$

$$\leq C \int_0^b \int_0^t \|u(\tau)|H^m(D)\| \|u(t)|H^m(D)\| d\tau dt$$

$$\leq \frac{C}{2} \int_0^b \int_0^t \left(\|u(\tau)|H^m(D)\|^2 + \|u(t)|H^m(D)\|^2 \right) d\tau dt$$

$$\leq Cb \|u|L_2((0, b), H^m(D))\|^2,$$

where we used that fact that $0 \leq (a - b)^2 = a^2 - 2ab + b^2$, i.e., $2ab \leq a^2 + b^2$ for $a, b \in \mathbb{R}$. The left hand side of (5.2.3) can be estimated from below using (5.1.4) and the fact that $2 \int_0^b \langle \partial_t u, u \rangle dt = \|u(b)|L_2(D)\|^2 - \|u(0)|L_2(D)\|^2$ (a proof of this assertion for $m = 1$ can be found in [Eva10, Sect. 5.9.2, Thm. 3] and holds for general $m \in \mathbb{N}$ by using $|\langle \partial_t u, u \rangle| \leq \|\partial_t u|H^{-m}(D)\| \|u|H^m(D)\|$). This gives

$$\int_0^b \langle \partial_t u, u \rangle dt + \int_0^b B(t, u(t), u(t)) dt$$

$$\geq \frac{1}{2} \left(\|u(b)|L_2(D)\|^2 - \|u(0)|L_2(D)\|^2 \right) + \mu \|u|L_2((0, b), H^m(D))\|^2$$

$$= \frac{1}{2} \|u(b)|L_2(D)\|^2 + \mu \|u|L_2((0, b), H^m(D))\|^2. \qquad (5.2.4)$$

Choosing $b \leq \frac{\mu}{2C}$ we see from (5.2.3) and (5.2.4) that

$$\frac{1}{2} \left(\|u(b)|L_2(D)\|^2 + \mu \|u|L_2((0, b), H^m(D))\|^2 \right) \leq 0.$$

This implies $u \equiv 0$ on $\left[0, \frac{\mu}{2C} \right] \times D$. Repeating this argument we can show that $u \equiv 0$ on intervals $\left[\frac{\mu}{2C}, \frac{\mu}{C} \right] \times D$, $\left[\frac{\mu}{C}, \frac{3\mu}{2C} \right] \times D$, ..., and therefore, $u \equiv 0$ on $[0, T] \times D$. $\square$

Using the spectral method we are now able to prove the following regularity result.

Theorem 5.2.2 (Sobolev Regularity Without Time Derivatives) *Let $f \in L_2([0, T], H^{-m}(D))$. Then Problem I has a unique weak solution u in the space*

$H^{m,1*}(D_T)$ and the following estimate holds

$$\|u|H^{m,1*}(D_T)\| \le C\|f|L_2([0, T], H^{-m}(D))\|, \tag{5.2.5}$$

where C is a constant independent of f and u.

Proof This proof follows [AH08, Lem. 4.2], which in turn is based on [Eva10, Sect. 7.1.2]. The uniqueness of the solution can be deduced directly from Lemma 5.2.1: For two weak solutions $u_1, u_2 \in H^{m,1*}(D_T)$ we have

$$\langle \partial_t u_1 - \partial_t u_2, v \rangle + B(t, u_1(t) - u_2(t), v) = 0, \qquad v \in \mathring{H}^m(D),$$

where the bilinear form $F \equiv 0$ on $\mathring{H}^m(D) \times \mathring{H}^m(D)$ satisfies the assumptions of Lemma 5.2.1. But this implies $u_1 - u_2 \equiv 0$ on $[0, T] \times D$. Now we turn towards proving the existence of a solution. By the assumption $a_{\alpha\beta} = (-1)^{|\alpha|+|\beta|} a_{\beta\alpha}$ for $|\alpha|, |\beta| \le m$, L is formally a self-adjoint operator. Moreover, the Rellich-Kondrachov Theorem tells us that the embedding $H^m(D) \hookrightarrow L_2(D)$ is compact (holds for bounded Lipschitz domains Ω and therefore for bounded domains of polyhedral type). From this we deduce that the operator $L(0, x, D)$ possesses a set $\{\psi_k\}_{k=1}^\infty$ consisting of all its eigenfunctions, which is not only an orthogonal basis of $\mathring{H}^m(D)$ but also an orthonormal basis of $L_2(D)$. We refer to [Dob10, Thm. 8.39] in this context (the fact that the embedding $H^m(D) \hookrightarrow L_2(D)$ is compact is crucial here). For each positive integer N, we consider the function

$$u^N(t, x) = \sum_{k=1}^N C_k^N(t)\psi_k(x),$$

where $\{C_k^N(t)\}_{k=1}^N$ is the solution of the ordinary differential system

$$(\partial_t u^N, \psi_l) + B(t, u^N, \psi_l) = \langle f, \psi_l \rangle, \qquad l = 1, \ldots, N, \tag{5.2.6}$$

$$C_k^N(0) = 0, \qquad k = 1, \ldots, N. \tag{5.2.7}$$

After multiplying both sides of (5.2.6) by $C_l^N(t)$, taking the sum with respect to l from 1 to N, and integrating with respect to t from 0 to T ($T > 0$), we arrive at

$$\int_0^T (\partial_t u^N, u^N)dt + \int_0^T B(t, u^N, u^N)dt = \int_0^T \langle f, u^N \rangle dt.$$

From this we deduce

$$\int_0^T \frac{\partial}{\partial t}\|u^N(t)|L_2(D)\|^2 dt + 2\int_0^T B(t, u^N, u^N)dt = 2\int_0^T \langle f, u^N \rangle dt. \tag{5.2.8}$$

Since $u(0) = 0$ this yields

$$\|u^N(T)|L_2(D)\|^2 + 2\int_0^T B(t, u^N, u^N)dt = 2\int_0^T \langle f, u^N\rangle dt.$$

Using the fact that $\left(\sqrt{\varepsilon}a - \frac{1}{\sqrt{\varepsilon}}b\right)^2 = \varepsilon a^2 - 2ab + \frac{b^2}{\varepsilon} > 0$, we can estimate the right hand side by

$$\left|2\int_0^T \langle f, u^N\rangle dt\right| \leq 2\int_0^T \|f|H^{-m}(D)\| \|u^N|H^m(D)\| dt$$

$$\leq \varepsilon\|u^N|L_2([0, T], H^m(D))\|^2 + \frac{1}{\varepsilon}\|f|L_2([0, T], H^{-m}(D))\|^2.$$

Therefore, choosing $0 < \varepsilon < 2\mu$ and using assumption (5.1.4), we see from (5.2.8) that

$$\|u^N|L_2([0, T], H^m(D))\|^2 \leq C\|f|L_2([0, T], H^{-m}(D))\|^2. \qquad (5.2.9)$$

Now fix any $v \in \mathring{H}^m(D)$ with $\|v|\mathring{H}^m(D)\| \leq 1$, and write $v = v_1 + v_2$, where $v_1 \in$ span$(\psi_1, \ldots, \psi_N)$ and $(v_2, \psi_l)_{L_2(D)} = 0$, for $l = 1, \ldots, N$. Since the functions $\{\psi_l\}_{l=1}^N$ are orthogonal in $\mathring{H}^m(D)$, $\|v_1|H^m(D)\| \leq \|v|H^m(D)\| \leq 1$. We obtain from (5.2.6) that

$$(\partial_t u^N, v_1) + B(t, u^N, v_1) = \langle f, v_1\rangle.$$

Therefore,

$$\langle \partial_t u^N, v\rangle = (\partial_t u^N, v) = (\partial_t u^N, v_1) = \langle f, v_1\rangle - B(t, u^N, v_1).$$

Hence, we get

$$|\langle \partial_t u^N, v\rangle| \leq C\left(\|f|H^{-m}(D)\| + \|u^N|H^m(D)\|\right),$$

since $\|v_1|H^m(D)\| \leq 1$. Thus,

$$\|\partial_t u^N|H^{-m}(D)\| \leq C\left(\|f|H^{-m}(D)\| + \|u^N|H^m(D)\|\right),$$

and therefore, by (5.2.9),

$$\|\partial_t u^N|L_2([0, T], H^{-m}(D))\|$$

$$\leq c_1\left(\|f|L_2([0, T], H^{-m}(D))\|^2 + \|u^N|L_2([0, T], H^m(D))\|^2\right)$$

$$\leq c_2\|f|L_2([0, T], H^{-m}(D))\|^2. \qquad (5.2.10)$$

Combining (5.2.9) and (5.2.10) we get

$$\|u^N|H^{m,1*}(D_T)\|^2 \leq C\|f|L_2([0, T], H^{-m}(D))\|^2,$$

where C is a constant independent of f and N. By the same arguments as in [Eva10, Ch. 7, Thm. 3], we conclude that there exists a subsequence of $\{u^N\}$ which weakly converges to a weak solution $u \in H^{m,1*}(D)$ of Problem I.   □

By an application of Theorem 5.2.2 and induction we can proof the following regularity result.

Theorem 5.2.3 (Sobolev Regularity with Time Derivatives) *Let $l \in \mathbb{N}_0$ and assume that the right hand side f of Problem I satisfies*

$$f \in H^l([0, T], H^{-m}(D)) \quad and \quad \partial_{t^k} f(x, 0) = 0 \quad for \quad k = 0, \ldots, l-1.$$

Then the weak solution u in the space $H^{m,1*}(D_T)$ of Problem I in fact belongs to $H^{m,l+1*}(D_T)$, i.e., has derivatives with respect to t up to order l satisfying

$$\partial_{t^k} u \in H^{m,1*}(D_T) \quad for \quad k = 0, \ldots, l,$$

and

$$\sum_{k=0}^{l} \|\partial_{t^k} u | H^{m,1*}(D_T)\| \leq C \sum_{k=0}^{l} \|\partial_{t^k} f | L_2([0, T], H^{-m}(D))\|,$$

where C is a constant independent of u and f.

Proof Our proof is similar to [ALL16, Thm. 2]. We will show the theorem by induction on l and additionally prove that the following equalities hold:

$$\partial_{t^k} u(0, \cdot) = 0, \quad k = 0, \ldots, l, \tag{5.2.11}$$

and

$$\langle \partial_{t^{l+1}} u, \eta \rangle + \sum_{k=0}^{l} \binom{l}{k} B_{\partial_{t^{l-k}}}(t, \partial_{t^k} u, \eta) = \langle \partial_{t^l} f, \eta \rangle \quad \text{for all} \quad \eta \in \mathring{H}^m(D). \tag{5.2.12}$$

The case $l = 0$ follows from Theorem 5.2.2. Assuming now that the assumptions hold for $l - 1$, we will prove them for l ($l \geq 1$). We consider first the following problem: find a function $v \in H^{m,1*}(D_T)$ satisfying $v(0, \cdot) = 0$ and

$$\langle \partial_t v, \eta \rangle + B(t, v, \eta) = \langle \partial_{t^l} f, \eta \rangle - \sum_{k=0}^{l-1} \binom{l}{k} B_{\partial_{t^{l-k}}}(t, \partial_{t^k} u, \eta) \tag{5.2.13}$$

for all $\eta \in \mathring{H}^m(D)$ and a.e. $t \in [0, T]$. Let $F : [0, T] \to H^{-m}(D)$ be a function defined by

$$\langle F(t), \eta \rangle = \langle \partial_{t^l} f, \eta \rangle - \sum_{k=0}^{l-1} \binom{l}{k} B_{\partial_{t^{l-k}}}(t, \partial_{t^k} u, \eta), \qquad \eta \in \mathring{H}^m(D).$$

From the inductive hypothesis, the fact that

$$\|F(t)|H^{-m}(D)\| = \sup_{\eta \in \mathring{H}^m(D), \|\eta|\mathring{H}^m(D)\| \leq 1} |\langle F(t), \eta \rangle|,$$

and (5.1.5), we see that $F \in L_2([0, T], H^{-m}(D))$ with

$$\|F|L_2([0, T], H^{-m}(D))\|$$

$$\leq \|\partial_{t^l} f|L_2([0, T], H^{-m}(D))\| + C \sum_{k=0}^{l-1} \|\partial_{t^k} u|L_2([0, T], H^m(D))\|$$

$$\leq C \sum_{k=0}^{l} \|\partial_{t^k} f|L_2([0, T], H^{-m}(D))\|,$$

where C is a constant independent of f. Hence, according to Theorem 5.2.2 the problem (5.2.13) has a solution $v \in H^{m,1*}(D_T)$ with

$$\|v|H^{m,1*}(D_T)\| \leq C \sum_{k=0}^{l} \|\partial_{t^k} f|L_2([0, T], H^{-m}(D))\|,$$

where C is a constant independent of f. We put now

$$w(t, x) = \int_0^t v(\tau, x)d\tau, \qquad t \in [0, T], \ x \in D.$$

Then, we have $w(0, \cdot) = 0$, $\partial_t w = v$, $\partial_t w(0, \cdot) = 0$. We rewrite (5.2.13) as follows

$$\langle \partial_{t^2} w, \eta \rangle + B(t, \partial_t w, \eta) = \langle \partial_{t^l} f, \eta \rangle - \sum_{k=0}^{l-1} \binom{l}{k} B_{\partial_{t^{l-k}}}(t, \partial_{t^k} u, \eta). \qquad (5.2.14)$$

Note that

$$B(t, \partial_t w, \eta) = \frac{\partial}{\partial t} B(t, w, \eta) - B_{\partial_t}(t, w, \eta), \qquad (5.2.15)$$

and

$$\frac{\partial}{\partial t}\left(\sum_{k=0}^{l-2}\binom{l-1}{k}B_{\partial_{t^{l-1-k}}}(t,\partial_{t^k}u,\eta)\right)$$

$$=\sum_{k=0}^{l-2}\binom{l-1}{k}\left(B_{\partial_{t^{l-k}}}(t,\partial_{t^k}u,\eta)+B_{\partial_{t^{l-1-k}}}(t,\partial_{t^{k+1}}u,\eta)\right)$$

$$=B_{\partial_{t^l}}(t,u,\eta)+\sum_{k=1}^{l-2}\binom{l-1}{k}B_{\partial_{t^{l-k}}}(t,\partial_{t^k}u,\eta)+\sum_{k=0}^{l-2}\binom{l-1}{k}B_{\partial_{t^{l-1-k}}}(t,\partial_{t^{k+1}}u,\eta)$$

$$=B_{\partial_{t^l}}(t,u,\eta)+\sum_{k=1}^{l-2}\binom{l-1}{k}B_{\partial_{t^{l-k}}}(t,\partial_{t^k}u,\eta)+\sum_{h=1}^{l-1}\binom{l-1}{h-1}B_{\partial_{t^{l-h}}}(t,\partial_{t^h}u,\eta)$$

$$=B_{\partial_{t^l}}(t,u,\eta)+\sum_{k=1}^{l-2}\left(\binom{l-1}{k}+\binom{l-1}{k-1}\right)B_{\partial_{t^{l-k}}}(t,\partial_{t^k}u,\eta)+(l-1)B_{\partial_t}(t,u_{t^{l-1}},\eta)$$

$$=B_{\partial_{t^l}}(t,u,\eta)+\sum_{k=1}^{l-2}\binom{l}{k}B_{\partial_{t^{l-k}}}(t,\partial_{t^k}u,\eta)+\left(\binom{l}{l-1}-1\right)B_{\partial_t}(t,\partial_{t^{l-1}}u,\eta)$$

$$=\sum_{k=0}^{l-1}\binom{l}{k}B_{\partial_{t^{l-k}}}(t,\partial_{t^k}u,\eta)-B_{\partial_t}(t,\partial_{t^{l-1}}u,\eta). \tag{5.2.16}$$

Hence, we get from (5.2.14), (5.2.15), and (5.2.16) that

$$\langle\partial_{t^2}w,\eta\rangle+\frac{\partial}{\partial t}B(t,w,\eta)=\langle\partial_{t^l}f,\eta\rangle+B_{\partial_t}(t,w-\partial_{t^{l-1}}u,\eta)$$

$$-\frac{\partial}{\partial t}\sum_{k=0}^{l-2}\binom{l-1}{k}\partial_{t^{l-1-k}}B(t,\partial_{t^k}u,\eta). \tag{5.2.17}$$

Now by integrating equality (5.2.17) with respect to t from 0 to t and using the inductive hypothesis (5.2.11), we arrive at

$$\langle\partial_t w,\eta\rangle+B(t,w,\eta)=\langle\partial_{t^{l-1}}f,\eta\rangle+\int_0^t B_{\partial_t}(\tau,w-\partial_{t^{l-1}}u,\eta)d\tau$$

$$-\sum_{k=0}^{l-2}\binom{l-1}{k}B_{\partial_{t^{l-1-k}}}(t,\partial_{t^k}u,\eta). \tag{5.2.18}$$

Put $z = w - \partial_{t^{l-1}} u$. Then $z(0, \cdot) = 0$. It follows from the inductive assumption (5.2.12) with l replaced by $l - 1$ and (5.2.18) that

$$\langle \partial_t w - \partial_{t^l} u, \eta \rangle + B(t, w - \partial_{t^{l-1}} u, \eta)$$

$$= \sum_{k=0}^{l-1} \binom{l-1}{k} B_{\partial_{t^{l-1-k}}}(t, \partial_{t^k} u, \eta) - \langle \partial_{t^{l-1}} f, \eta \rangle$$

$$+ \langle \partial_{t^{l-1}} f, \eta \rangle + \int_0^t B_{\partial_t}(\tau, w - \partial_{t^{l-1}} u, \eta) d\tau$$

$$- \sum_{k=0}^{l-2} \binom{l-1}{k} B_{\partial_{t^{l-1-k}}}(t, \partial_{t^k} u, \eta) - B(t, \partial_{t^{l-1}} u, \eta)$$

$$= \int_0^t B_{\partial_t}(\tau, z, \eta) d\tau + B(t, \partial_{t^{l-1}} u, \eta) - B(t, \partial_{t^{l-1}} u, \eta),$$

i.e.,

$$\langle \partial_t z(t), \eta \rangle + B(t, z(t), \eta) = \int_0^t B_{\partial_t}(\tau, z(\tau, \cdot), \eta) d\tau, \qquad (5.2.19)$$

for all $\eta \in \mathring{H}^m(D)$. Now by applying Lemma 5.2.1, we can see from (5.2.19) that $z \equiv 0$ on $[0, T] \times D$. This implies $\partial_{t^l} u = \partial_t w = v \in H^{m,1*}(D_T)$. The proof is complete. $\square$

Remark 5.2.4 Note that the regularity results for the solution u in [LL15, Thm. 2.1., Lem. 3.1] are slightly stronger than the ones obtained in Theorem 5.2.3 above (with the cost of also assuming more regularity on the right hand side f). By using similar arguments as in [AH08, Lem. 4.3] we are probably able to also show in our context that Theorem 5.2.2 can be strengthened in the sense that if $f \in L_2([0, T], L_2(D))$ then the weak solution u of Problem I belongs in fact to $L_2([0, T], \mathring{H}^m) \cap H^1([0, T], L_2(D))$. A corresponding generalization of Theorem 5.2.3 should also be possible in the spirit of [AH08, Thm. 3.1]. However, for our purposes the above results on the Sobolev regularity are sufficient, so these investigations are postphoned for the time being.

5.2.2   Regularity Results in Kondratiev Spaces for Problem I

Concerning weighted Sobolev regularity of Problem I first fundamental results on polyhedral cones $K \subset \mathbb{R}^3$ can be found in [LL15, Thm. 3.3, 3.4], which form the starting point for our investigations. However, for our purposes we slightly modify and generalize the results to domains of polyhedral type $D \subset \mathbb{R}^d$, $d = 3$, and give a detailed proof. The restriction to the 3-dimensional setting in this subsection is due to the operator pencils from Sect. 3.5 (which are not available in higher dimensions), that we use in order to state the regularity results in Kondratiev spaces below. In particular, we obtain an *a priori* estimate for the derivatives of the solution *u* within our scale of Kondratiev spaces, which in turn is needed in Theorem 5.2.14 for proving the existence of a solution of Problem II in Kondratiev spaces.

For our regularity assertions we rely on known results for elliptic equations. Therefore, we consider first the following Dirichlet problem for elliptic equations

$$\left\{ \begin{array}{ll} Lu = F & \text{on} \quad D, \\ \frac{\partial^k u}{\partial \nu^k}\big|_{\Gamma_j} = 0, & k = 1, \ldots, m, \quad j = 1, \ldots, n, \end{array} \right\} \tag{5.2.20}$$

where $D \subset \mathbb{R}^3$ is a domain of polyhedral type according to Definition 2.4.5 with faces Γ_j. Moreover, we assume that

$$L(x, D_x) = \sum_{|\alpha| \leq 2m} A_\alpha(x) D_x^\alpha$$

is a uniformly elliptic differential operator of order $2m$ with smooth coefficients A_α. We need the following technical assumptions in order to state the Kondratiev regularity of (5.2.20).

Assumption 5.2.5 (Assumptions on Operator Pencils) *Consider the operator pencils $\mathfrak{A}_i(\lambda, t)$, $i = 1, \ldots, l'$ for the vertices and $A_\xi(\lambda, t)$ with $\xi \in M_k$, $k = 1, \ldots, l$ for the edges of the polyhedral type domain $D \subset \mathbb{R}^3$ introduced in Sect. 3.5. For the elliptic problem (5.2.20) we may drop t from the notation of the pencils, otherwise (for our parabolic problems) we assume $t \in [0, T]$ is fixed.*

Let $\mathcal{K}_{p,b}^\gamma(D)$ and $\mathcal{K}_{p,b'}^{\gamma'}(D)$ be two Kondratiev spaces according to Definition 2.4.1, where the singularity set S of D is given by $S = M_1 \cup \ldots \cup M_l \cup \{x^{(1)}, \ldots, x^{(l')}\}$ and weight parameters $b, b' \in \mathbb{R}$. Then we assume that the closed strip between the lines

$$\text{Re}\lambda = b + 2m - \frac{3}{2} \quad \text{and} \quad \text{Re}\lambda = b' + 2m - \frac{3}{2} \tag{5.2.21}$$

does not contain eigenvalues of $\mathfrak{A}_i(\lambda, t)$. Moreover, b and b' satisfy

$$-\delta_-^{(k)} < b + m < \delta_+^{(k)}, \qquad -\delta_-^{(k)} < b' + m < \delta_+^{(k)}, \quad k = 1, \ldots, l, \tag{5.2.22}$$

where $\delta_\pm^{(k)}$ are defined in (3.5.5) (replaced by (3.5.8) for parabolic problems).

Remark 5.2.6 (Discussion of Assumption 5.2.5)

(i) Some remarks concerning the above assumption seem to be in order. If $l' = 0$ we have an edge domain without vertices, cf. Figure 3.3. In this case condition

(5.2.21) is empty. Moreover, if $l = 0$, we have a corner domain without edges, in which case condition (5.2.22) is empty.

(ii) The values of $\delta_\pm^{(k)}$ determine the range of b and b'. To keep our discussion simple, assume that the operator L does not depend on the time t and that the polyhedral type domain $D \subset \mathbb{R}^3$ is in fact a polyhedron and has straight edges, i.e, near a fixed vertex $x^{(i)}$ it looks like a polyhedral cone K_i and near a point $\xi \in M_k$ it looks like a dihedron with angle θ_k, respectively. In this case we can write $A_k(\lambda) = A_\xi(\lambda)$ for all $\xi \in M_k$ and put $\delta_\pm^{(k)} = \delta_\pm^{(\xi)}$. Then the results mentioned in [Koz91, p. 1] indicate that the strip $|\text{Re}\lambda - m + 1| \leq \frac{1}{2}$ contains no eigenvalues of $A_k(\lambda)$. Thus, we obtain $m - \frac{3}{2} \leq \text{Re}\lambda \leq m - \frac{1}{2}$, i.e., it follows from (3.5.4) that $\delta_\pm^{(k)} = \frac{1}{2}$, which yields

$$-\frac{1}{2} < b + m < \frac{1}{2}.$$

If we additionally assume that our domain D is convex, i.e., $\theta_k \in (0, \pi)$, $k = 1, \ldots, l$, then the above results can be improved: As is stated in [Koz91, p. 1] in this case we even know that the strip $|\text{Re}\lambda - m + 1| \leq 1$ does not contain eigenvalues of $A_k(\lambda)$. Hence, $-2 + m \leq \text{Re}\lambda \leq m$, which gives $\delta_\pm^{(k)} = 1$. Therefore, our restriction (5.2.22) then reads as

$$-1 < b + m < 1.$$

In particular, for $m = 1$ we see that $b \in (-2, 0)$. Finally, let us consider the heat equation ($L = \Delta$ and $m = 1$) on the polyhedron D. In good agreement with the results above from Example 3.5.2 we know in this case that $\delta_\pm^{(k)} = \frac{\pi}{\theta_k}$ (i.e., $\delta_\pm^{(k)} \geq 1$ for angles $\theta_k \in (0, \pi)$ and worst case $\delta_\pm^{(k)} = \frac{1}{2}$ if $\theta_k = 2\pi$). Therefore, (5.2.22) now reads as

$$-\frac{\pi}{\theta_k} < b + 1 < \frac{\pi}{\theta_k}.$$

(iii) We also require in Assumption 5.2.5 that the closed strip between the lines $\text{Re}\lambda = b + 2m - \frac{3}{2}$ and $\text{Re}\lambda = b' + 2m - \frac{3}{2}$ does not contain eigenvalues of the operator pencils $\mathfrak{A}_i(\lambda)$. Later on, in Theorem 5.2.9 we choose $b = a + 2m\gamma_m$ and $b' = -m$, leading to the condition that the strip

$$\left[m - \frac{3}{2}, a + 2m(\gamma_m + 1) - \frac{3}{2} \right], \qquad -m \leq a \leq m,$$

is free of eigenvalues. We see that if the spectrum is real and discrete without cluster points, then even for $\gamma_m = 0$ it is always possible to find some $a \in (-m, -m + \varepsilon)$ satisfying our condition as long as $m - \frac{3}{2}$ is not an eigenvalue of $\mathfrak{A}_i(\lambda)$. But this is known to be true, cf. the explanations given besides Fig. 3.5.

However, if we look at our nonlinear results established in Theorem 5.2.14 the situation becomes more delicate. There (for $d = 3$) we have the additional restrictions $\gamma_m \geq 1$, $m \geq 2$, and $m \geq a \geq -\frac{1}{2}$. This gives for $\gamma_m = 1$ and $m = 2$ the condition that the strip

$$\left[\frac{1}{2}, a + \frac{13}{2}\right], \qquad -\frac{1}{2} \leq a \leq 2,$$

is free of eigenvalues. Let us finally turn our attention once again to the heat equation and w.l.o.g. consider the vertex $x^{(i)}$. Then the eigenvalues of $\mathfrak{A}_1(\lambda) = \mathfrak{A}(\lambda)$ are given by

$$\lambda_k^{\pm} = -\frac{1}{2} \pm \sqrt{\Lambda_k + \frac{1}{4}}, \qquad k \in \mathbb{N},$$

where Λ_k denote the eigenvalues of the Laplace-Beltrami operator on $\Omega_1 = K \cap S^2$, cf. Example 3.5.2. It is well known that the spectrum of the Laplace-Beltrami operator is a countable set of positive eigenvalues, cf. [KMR01, Sect. 2.2.1]. Hence, the interval $[-1, 0]$ is free of eigenvalues of the pencil $\mathfrak{A}(\lambda)$. We denote the smallest positive eigenvalue of the pencil $\mathfrak{A}(\lambda)$ by λ_1^{+}.

In general, it is not possible to determine precise values for λ_1^{+} and arbitrary Ω. However, for the special case that the cone K_1 (describing the neighbourhood of the vertex $x^{(1)}$) is a rotationally invariant cone with opening angle θ_0 and $\Omega = \Omega_{\theta_0} = \{(y_1, y_2, y_3) \in S^2 : \theta \in (0, \theta_0), \varphi \in [0, 2\pi]\}$ a spherical cap, cf. Figs. 5.4 and 5.5 in Chap. 5, one finds different values of λ_1^{+} depending on $\theta_0 \in (0, \pi)$ in Table 5.1. In particular, it can be seen that for small θ_0 the eigenvalues may become quite large, e.g. we have $\lambda_1^{+} = 1$ for $\theta_0 = 90°$ and even $\lambda_1^{+} > 27$ for angle $\theta_0 = 5°$.

Table 5.1 Values for $\lambda_1^{+} = \nu_{\min}$ when $\Omega_{\theta_0} = K \cap S$ is a spherical cap

θ_0	λ_1^{+}	θ_0	λ_1^{+}	θ_0	λ_1^{+}
5°	27.0558	65°	1.5988	125°	0.5523
10°	13.2756	70°	1.4456	130°	0.5063
15°	8.6812	75°	1.3124	135°	0.4631
20°	6.3832	80°	1.1956	140°	0.4223
25°	5.0038	85°	1.0922	145°	0.3834
30°	4.0837	90°	1.000	150°	0.3462
35°	3.4260	95°	0.9172	155°	0.3101
40°	2.9323	100°	0.8423	160°	0.2745
45°	2.5479	105°	0.7741	165°	0.2387
50°	2.2400	110°	0.7118	170°	0.2012
55°	1.9878	115°	0.6545	175°	0.1581
60°	1.7773	120°	0.6015		

The following lemma on the regularity of solutions to elliptic boundary value problems in domains of polyhedral type is taken from [MR10, Cor. 4.1.10, Thm. 4.1.11]. We rewrite it for our scale of Kondratiev spaces.

Lemma 5.2.7 (Kondratiev Regularity for Elliptic PDEs) *Let* $D \subset \mathbb{R}^d$, $d = 3$, *be a domain of polyhedral type. Moreover, let* $u \in \mathcal{K}^{\gamma}_{2,a+2m}(D)$ *be a solution of* (5.2.20), *where*

$$F \in \mathcal{K}^{\gamma-2m}_{2,a}(D) \cap \mathcal{K}^{\gamma'-2m}_{2,a'}(D), \qquad \gamma \geq m, \quad \gamma' \geq m.$$

Suppose that $\mathcal{K}^{\gamma}_{2,a}(D)$ *and* $\mathcal{K}^{\gamma'}_{2,a'}(D)$ *satisfy Assumption 5.2.5. Then* $u \in \mathcal{K}^{\gamma'}_{2,a'+2m}(D)$ *and*

$$\|u|\mathcal{K}^{\gamma'}_{2,a'+2m}(D)\| \leq C\|F|\mathcal{K}^{\gamma'-2m}_{2,a'}(D)\|,$$

where C *is a constant independent of* u *and* F.

Remark 5.2.8 In particular, if in Theorem 5.2.3 we use the stronger assumption $\partial_{t^k} f(t) \in L_2(D)$ instead of $\partial_{t^k} f(t) \in H^{-m}(D)$ for $k = 0, \ldots, l$, then it follows that

$$\partial_{t^k} f(t) \in L_2(D) = \mathcal{K}^0_{2,0}(D) \hookrightarrow \mathcal{K}^{-m}_{2,-m}(D), \tag{5.2.23}$$

where the latter embedding follows from the corresponding duality assertion, i.e., we have $\mathcal{K}^m_{2,m}(D) \hookrightarrow \mathcal{K}^0_{2,0}(D)$ since $m \geq 0$. In this case the solution u of Problem I satisfies

$$\partial_{t^k} u(t) \in \mathring{H}^m(D) \hookrightarrow \mathring{\mathcal{K}}^m_{2,m}(D) \hookrightarrow \mathcal{K}^0_{2,a}(D), \qquad a \leq m, \tag{5.2.24}$$

where the first embedding is taken from [MR10, Lem. 3.1.6] and the second embedding for Kondratiev spaces holds whenever $m \geq a$. We additionally require in our later considerations that

$$\partial_{t^k} u(t) \in \mathcal{K}^0_{2,a}(D) \hookrightarrow \mathcal{K}^{-m}_{2,-m}(D), \tag{5.2.25}$$

which holds for $a \geq -m$. From (5.2.23) and (5.2.24) we see that it is possible to take $\gamma = m$ and $a = -m$ in Lemma 5.2.7, i.e., if $f(t) \in \mathcal{K}^{-m}_{2,-m}(D)$ then $u(t) \in \mathcal{K}^m_{2,m}(D)$. Note that all our arguments with $u(t)$ and $f(t)$, respectively, hold for a.e. $t \in [0, T]$. However, since lower order time derivatives are continuous w.r.t. suitable spaces (but not necessarily the highest one, cf. the proof of Theorem 5.4.8), we will suppress this distinction in the sequel.

Using similar arguments as in [LL15, Thm. 3.3] we are now able to show the following regularity result in Kondratiev spaces.

Theorem 5.2.9 (Kondratiev Regularity A) *Let $D \subset \mathbb{R}^d$, $d = 3$, be a domain of polyhedral type. Let $\gamma \in \mathbb{N}$ with $\gamma \geq 2m$ and put $\gamma_m := \left[\frac{\gamma-1}{2m}\right]$. Furthermore, let $a \in \mathbb{R}$ with $a \in [-m, m]$. Assume that the right hand side f of Problem I satisfies*

(i) $\partial_{t^k} f \in L_2(D_T) \cap L_2([0, T], \mathcal{K}^{2m(\gamma_m-k)}_{2,a+2m(\gamma_m-k)}(D))$, $k = 0, \ldots, \gamma_m$, *and* $\partial_{t^{\gamma_m+1}} f \in L_2(D_T)$.
(ii) $\partial_{t^k} f(x, 0) = 0$, $k = 0, 1, \ldots, \gamma_m$.

Furthermore, let Assumption 5.2.5 hold for weight parameters $b = a + 2m(\gamma_m - i)$, where $i = 0, \ldots, \gamma_m$, and $b' = -m$. Then for the weak solution $u \in H^{m, \gamma_m+2*}(D_T)$ of Problem I we have

$$\partial_{t^{l+1}} u \in L_2([0, T], \mathcal{K}^{2m(\gamma_m-l)}_{2,a+2m(\gamma_m-l)}(D))$$

for $l = -1, 0, \ldots, \gamma_m$. In particular, for the derivatives $\partial_{t^{l+1}} u$ up to order $\gamma_m + 1$ we have the a priori estimate

$$\sum_{l=-1}^{\gamma_m} \|\partial_{t^{l+1}} u | L_2([0, T], \mathcal{K}^{2m(\gamma_m-l)}_{2,a+2m(\gamma_m-l)}(D))\|$$

$$\lesssim \sum_{k=0}^{\gamma_m} \|\partial_{t^k} f | L_2([0, T], \mathcal{K}^{2m(\gamma_m-k)}_{2,a+2m(\gamma_m-k)}(D))\| + \sum_{k=0}^{\gamma_m+1} \|\partial_{t^k} f | L_2(D_T)\|,$$

$$(5.2.26)$$

where the constant is independent of u and f.

Proof We proof the theorem by induction. Let $\gamma = 2m$, then we have $\gamma_m = 0$. Since by our assumptions $f, \partial_t f \in L_2(D_T) \hookrightarrow L_2([0, T], H^{-m}(D))$ it follows from Theorem 5.2.3 that $\partial_t u(t) \in \mathring{H}^m(K) \subset \mathcal{K}^0_{2,a}(K)$ if $a \leq m$. In this case Theorem 5.2.3 gives the estimate

$$\|\partial_t u | L_2([0, T], \mathcal{K}^0_{2,a}(D))\| \lesssim \|\partial_t u | L_2([0, T], \mathring{H}^m_2(D))\|$$

$$\lesssim \sum_{k=0}^{1} \|\partial_{t^k} f | L_2([0, T], H^{-m}(D))\|$$

$$\lesssim \sum_{k=0}^{1} \|\partial_{t^k} f | L_2(D_T)\|. \qquad (5.2.27)$$

Moreover, since

$$(-1)^m L u = f - \partial_t u,$$

where for fixed t the right hand side belongs to $\mathcal{K}^0_{2,a}(D)$ an application of Lemma 5.2.7 (with $\gamma = m$, $a = -m$, $\gamma' = 2m$, $a' = a$) gives $u(t) \in \mathcal{K}^{2m}_{2,a+2m}(D)$ with the a priori estimate

$$\|u(t)|\mathcal{K}^{2m}_{2,a+2m}(D)\| \lesssim \|f(t)|\mathcal{K}^0_{2,a}(D)\| + \|\partial_t u|\mathcal{K}^0_{2,a}(D)\|.$$

Now integration w.r.t. the parameter t together with (5.2.27) proves the claim for $\gamma = 2m$, i.e.,

$$\|u|L_2([0,T], \mathcal{K}^{2m}_{2,a+2m}(D)\| \lesssim \|f|L_2([0,T], \mathcal{K}^0_{2,a}(D))\| + \sum_{k=0}^{1} \|\partial_{t^k} f|L_2(D_T)\|.$$

Assume inductively that our assumption holds for $\gamma - 1$. This means, in particular, that we have the following a priori estimate

$$\sum_{l=-1}^{(\gamma-1)_m} \|\partial_{t^{l+1}} u|L_2([0,T], \mathcal{K}^{2m((\gamma-1)_m - l)}_{2,a+2m((\gamma-1)_m - l)}(D))\|$$

$$\lesssim \sum_{k=0}^{(\gamma-1)_m} \|\partial_{t^k} f|L_2([0,T], \mathcal{K}^{2m((\gamma-1)_m - k)}_{2,a+2m((\gamma-1)_m - k)}(D))\| + \sum_{k=0}^{(\gamma-1)_m+1} \|\partial_{t^k} f|L_2(D_T)\|.$$

$$(5.2.28)$$

We are going to show that the claim then holds for γ as well. If $l = \gamma_m$ by our assumptions on f we have $\partial_{t^k} f \in L_2(D_T) \hookrightarrow L_2([0,T], H^{-m}(D))$ for $k = 0, \dots, \gamma_m + 1$, and by Theorem 5.2.3 together with (5.2.24) we have

$$\partial_{t^{\gamma_m+1}} u(t) \in \mathring{H}^m(D) \subset \mathcal{K}^0_{2,a}(D), \quad a \leq m.$$

In particular, Theorem 5.2.3 provides us with the a priori estimate

$$\|\partial_{t^{\gamma_m+1}} u|L_2([0,T], \mathcal{K}^0_{2,a}(D))\| \lesssim \|\partial_{t^{\gamma_m+1}} u|L_2([0,T], \mathring{H}^m(D))\|$$

$$\lesssim \sum_{k=0}^{\gamma_m+1} \|\partial_{t^k} f|L_2([0,T], H^{-m}(D))\|$$

$$\lesssim \sum_{k=0}^{\gamma_m+1} \|\partial_{t^k} f|L_2(D_T)\|, \qquad (5.2.29)$$

which shows the claim for $l = \gamma_m$ and arbitrary γ. Hence, the claim holds for $l = \gamma_m$. We proceed by backwards induction. Suppose now the result holds for $l = \gamma_m, \gamma_m - 1, \dots, i$ where $0 \leq i \leq \gamma_m$. We show that it then also holds for $i - 1$.

Differentiating (5.1.1) i-times gives

$$(-1)^m L(\partial_{t^i} u) = \partial_{t^i} f - \partial_{t^{i+1}} u - (-1)^m \sum_{k=0}^{i-1} \binom{i}{k} \partial_{t^{i-k}} L(\partial_{t^k} u). \tag{5.2.30}$$

From our initial assumptions on f we see that $\partial_{t^i} f(t) \in \mathcal{K}_{2,a+2m(\gamma_m-i)}^{2m(\gamma_m-i)}(D)$ and from the inductive assumptions it follows that $\partial_{t^{i+1}} u(t) \in \mathcal{K}_{2,a+2m(\gamma_m-i)}^{2m(\gamma_m-i)}(D)$ and

$$\partial_{t^{(k-1)+1}} u \in L_2([0, T], \mathcal{K}_{2,a+2m((\gamma-1)_m-(k-1))}^{2m((\gamma-1)_m-(k-1))}(D))$$

$$\hookrightarrow L_2([0, T], \mathcal{K}_{2,a+2m((\gamma_m-1-(k-1)))}^{2m(\gamma_m-1-(k-1))}(D))$$

$$\hookrightarrow L_2([0, T], \mathcal{K}_{2,a+2m((\gamma_m-k))}^{2m(\gamma_m-k)}(D))$$

$$\hookrightarrow L_2([0, T], \mathcal{K}_{2,a+2m((\gamma_m-i+1))}^{2m(\gamma_m-i+1)}(D)), \tag{5.2.31}$$

where we used $(\gamma - 1)_m = \left[\frac{\gamma-2}{2m}\right] \geq \gamma_m - 1$ in the second step and the fact that $k = 0, \ldots, i - 1$ in the last step. From (5.2.31) we see that

$$\partial_{t^{i-k}} L(\partial_{t^k} u) \in L_2([0, T], \mathcal{K}_{2,a+2m((\gamma_m-i))}^{2m(\gamma_m-i)}(D)),$$

hence, the right hand side of (5.2.30) belongs to $L_2([0, T], \mathcal{K}_{2,a+2m((\gamma_m-i))}^{2m(\gamma_m-i)}(D))$. An application of Lemma 5.2.7 (now with $\gamma' = 2m(\gamma_m - (i - 1))$, $a' = a + 2m(\gamma_m - i)$ and again taking $\gamma = m$, $a = -m$ according to Remark 5.2.8) yields

$$\partial_{t^{(i-1)+1}} u(t) \in \mathcal{K}_{2,a+2m(\gamma_m-(i-1))}^{2m(\gamma_m-(i-1))}(D).$$

Moreover, we have the a priori estimate

$$\|\partial_{t^{(i-1)+1}} u(t) | \mathcal{K}_{2,a+2m(\gamma_m-(i-1))}^{2m(\gamma_m-(i-1))}(D)\|$$

$$\lesssim \|\partial_{t^i} f(t) | \mathcal{K}_{2,a+2m(\gamma_m-i)}^{2m(\gamma_m-i)}(D)\| + \|\partial_{t^{i+1}} u(t) | \mathcal{K}_{2,a+2m(\gamma_m-i)}^{2m(\gamma_m-i)}(D)\|$$

$$+ \sum_{k=0}^{i-1} \|\partial_{t^{i-k}} L(\partial_{t^k} u)(t) | \mathcal{K}_{2,a+2m(\gamma_m-i)}^{2m(\gamma_m-i)}(D)\|$$

$$\lesssim \|\partial_{t^i} f(t) | \mathcal{K}_{2,a+2m(\gamma_m-i)}^{2m(\gamma_m-i)}(D)\| + \|\partial_{t^{i+1}} u(t) | \mathcal{K}_{2,a+2m(\gamma_m-i)}^{2m(\gamma_m-i)}(D)\|$$

$$+ \sum_{k=0}^{i-1} \|(\partial_{t^k} u)(t) | \mathcal{K}_{2,a+2m(\gamma_m-i+1)}^{2m(\gamma_m-i+1)}(D)\|$$

$$\lesssim \|\partial_{t^i} f(t)|\mathcal{K}_{2,a+2m(\gamma_m-i)}^{2m(\gamma_m-i)}(D)\| + \|\partial_{t^{i+1}} u(t)|\mathcal{K}_{2,a+2m(\gamma_m-i)}^{2m(\gamma_m-i)}(D)\|$$

$$+ \sum_{k=0}^{i-1} \|(\partial_{t^k} u)(t)|\mathcal{K}_{2,a+2m((\gamma-1)_m-(k-1))}^{2m((\gamma-1)_m-(k-1))}(D)\|$$

where we used (5.2.31) in the last step. Integration w.r.t. the parameter t together with the inductive assumptions on $\partial_{t^{i+1}} u$ and $\partial_{t^k} u$ (cf. (5.2.28)) gives the a priori estimate

$$\|\partial_{t^{(i-1)+1}} u|L_2([0, T], \mathcal{K}_{2,a+2m(\gamma_m-(i-1))}^{2m(\gamma_m-(i-1))}(D))\|$$

$$\lesssim \|\partial_{t^i} f|L_2([0, T], \mathcal{K}_{2,a+2m(\gamma_m-i)}^{2m(\gamma_m-i)}(K))\| + \|\partial_{t^{i+1}} u|L_2([0, T], \mathcal{K}_{2,a+2m(\gamma_m-i)}^{2m(\gamma_m-i)}(D))\|$$

$$+ \sum_{k=0}^{i-1} \|(\partial_{t^k} u)|L_2([0, T], \mathcal{K}_{2,a+2m((\gamma-1)_m-(k-1))}^{2m((\gamma-1)_m-(k-1))}(D))\|$$

$$\lesssim \sum_{k=0}^{\gamma_m} \|\partial_{t^k} f|L_2([0, T], \mathcal{K}_{2,a+2m(\gamma_m-k)}^{2m(\gamma_m-k)}(D))\| + \sum_{k=0}^{\gamma_m+1} \|\partial_{t^k} f|L_2(D_T)\|$$

$$+ \sum_{k=0}^{(\gamma-1)_m} \|\partial_{t^k} f|L_2([0, T], \mathcal{K}_{2,a+2m((\gamma-1)_m-k)}^{2m((\gamma-1)_m-k)}(D))\| + \sum_{k=0}^{(\gamma-1)_m+1} \|\partial_{t^k} f|L_2(D_T)\|$$

$$\lesssim \sum_{k=0}^{\gamma_m} \|\partial_{t^k} f|L_2([0, T], \mathcal{K}_{2,a+2m(\gamma_m-k)}^{2m(\gamma_m-k)}(D))\| + \sum_{k=0}^{\gamma_m+1} \|\partial_{t^k} f|L_2(D_T)\|,$$

where in the last step we used the fact that $(\gamma - 1)_m \le \gamma_m$. This shows that the claim is true for $i - 1$ and completes the proof.   □

Remark 5.2.10 The existence of the solution $u \in H^{m,\gamma_m+2*}(D_T)$ follows from Theorem 5.2.3 using $l = \gamma_m + 1$.

Remark 5.2.11 (The Parameter a)   We discuss the restrictions on the parameter a appearing in Theorem 5.2.9.

(i) For simplicity, let the domain of polyhedral type D be a polyhedron with straight edges and faces where θ_k denotes the angle at the edge M_k. Then we require

$$(1) \quad -\delta_-^{(k)} < a + 2m(\gamma_m - i) + m < \delta_+^{(k)}, \quad i = 0, \ldots, \gamma_m,$$

$$(2) \quad -m \le a \le m.$$

If we consider the heat equation, then $m = 1$ and $\delta_{\pm}^{(k)} = \frac{\pi}{\theta_k}$, cf. [LL15, Sect. 4]. For the extremal case that $\theta_k = 2\pi$ we obtain

$$(1) \quad -\frac{3}{2} < a + 2(\gamma_m - i) < -\frac{1}{2}, \quad i = 0, \dots, \gamma_m,$$

$$(2) \quad -1 \leq a \leq 1.$$

We see that (1) is only satisfied for $\gamma_m - i = 0$. Thus, the theorem holds in this case only for $\gamma_m = 0$ and $a \in \left[-1, -\frac{1}{2}\right)$, i.e., regularity $\gamma \leq 2m = 2$.

The results improve for smaller angles: By choosing $\theta_k = \frac{\pi}{4}$ we obtain the following conditions

$$1) \quad -5 < a + 2(\gamma_m - i) < 3, \quad i = 0, \dots, \gamma_m,$$

$$2) \quad -1 \leq a \leq 1.$$

We see that $-1 \leq a$ and $a < 3 - 2(\gamma_m - i)$. However,

$$-1 \leq a < 3 - 2(\gamma_m - i)$$

is satisfied for $\gamma_m - i < 2$, i.e., we take $\gamma_m = 1$ ($\gamma \leq 4m$) in this case and $a \in [-1, 1)$. The same phenomenon is known for elliptic problems: The shape of the domain limits the regularity of the solutions even for smooth data. For general angles θ_k the restrictions on a for the heat equation read as

$$1) \quad -\frac{\pi}{\theta_k} - 1 < a + 2(\gamma_m - i) < \frac{\pi}{\theta_k} - 1, \quad i = 0, \dots, \gamma_m,$$

$$2) \quad -1 \leq a \leq 1.$$

(1) and (2) together yield

$$-1 \leq a < \min\left(1, \frac{\pi}{\theta_k} - 1 - 2(\gamma_m - i)\right). \tag{5.2.32}$$

In particular, for the upper bound we see that

$$\min\left(1, \frac{\pi}{\theta_k} - 1 - 2(\gamma_m - i)\right) = \begin{cases} \min(1, \frac{\pi}{\theta_k} - 1), & \gamma_m = 0, \\ \min(1, \frac{\pi}{\theta_k} - 3), & \gamma_m = 1. \end{cases}$$

Therefore, when $\gamma_m = 0$, since it holds that $-1 < \frac{\pi}{\theta_k} - 1$, it is always possible to find suitable parameters a. On the other hand for $\gamma_m = 1$ we require $-1 < \frac{\pi}{\theta_k} - 3$, which is true for angles $\theta_k < \frac{\pi}{2}$ only.

(ii) Later on, in Sect. 5.4.2 we want to study the Besov regularity of the Problem II. This will be performed by using embedding results of Kondratiev spaces into Besov spaces. Since all functions in the adaptivity scale (1.0.3) of Besov spaces are locally integrable, the same must hold for the Kondratiev spaces which requires $a > 0$. For Problem I this is no restriction since from Theorem 5.2.9 it follows that our solution satisfies $u \in L_2([0, T], \mathcal{K}^{\gamma}_{2,a'}(D))$, where $a' = a + 2m(\gamma_m + 1) > 0$. Thus, we always have a locally integrable solution in this case.

However, the above calculations in (i) tell us that for non-convex polyhedrons, i.e., $\theta_k > \pi$ for some $k = 1, \ldots, l$, we can only choose $\gamma_m = 0$ in Theorem 5.2.9 and the restriction on a becomes

$$-1 \le a < \min\left(1, \frac{\pi}{\theta_k} - 1\right) < 0.$$

From this we see that it is possible to choose $a > 0$ only as long as the domain D is convex.

The regularity results obtained in Theorem 5.2.9 only hold under certain restrictions on the parameter a we are able to choose. We refer to the previous discussion in Remark 5.2.11, where the heat equation is treated in detail. In particular, it turns out that we are not able to choose $\gamma_m > 0$ if we have a non-convex polyhedral type domain D, since there is no suitable a satisfying all of our requirements in this case. Therefore, in order to be able to treat non-convex domains as well, stronger assumptions are needed: It turns out that we obtain a positive result if the right-hand side f is arbitrarily smooth with respect to time. With this additional assumption we are able to weaken the restrictions on the parameter a and allow a larger range. However, as a drawback, these results are hard to apply to nonlinear equations since the right-hand sides are not taken from a Banach or quasi-Banach space.

Theorem 5.2.12 (Kondratiev Regularity B) *Let $D \subset \mathbb{R}^d$, $d = 3$, be a domain of polyhedral type and $\eta \in \mathbb{N}$ with $\eta \ge 2m$. Moreover, let $a \in \mathbb{R}$ with $a \in [-m, m]$. Assume that the right hand side f of Problem I satisfies*

(i) $f \in \bigcap_{l=0}^{\infty} H^l([0, T], L_2(D) \cap \mathcal{K}^{\eta-2m}_{2,a}(D))$.
(ii) $\partial_{t^l} f(x, 0) = 0$, $l \in \mathbb{N}_0$.

Furthermore, let Assumption 5.2.5 hold for weight parameters $b = a$ and $b' = -m$. Then for the weak solution $u \in \bigcap_{l=0}^{\infty} H^{m,l+1*}(D_T)$ of Problem I we have

$$\partial_{t^l} u \in L_2([0, T], \mathcal{K}^{\eta}_{2,a+2m}(D)) \qquad \text{for all} \quad l \in \mathbb{N}_0.$$

In particular, for the derivative $\partial_{t^l} u$ we have the a priori estimate

$$\sum_{k=0}^{l} \|\partial_{t^k} u | L_2([0, T], \mathcal{K}_{2,a+2m}^{\eta}(D))\|$$

$$\lesssim \sum_{k=0}^{l+(\eta-2m)} \|\partial_{t^k} f | L_2([0, T], \mathcal{K}_{2,a}^{\eta-2m}(D))\| + \sum_{k=0}^{l+1+(\eta-2m)} \|\partial_{t^k} f | L_2(D_T)\|,$$

$$(5.2.33)$$

where the constant is independent of u and f.

Proof We proof the theorem by induction. Let $\eta = 2m$. Since by our assumptions $f, \partial_t f \in L_2(D_T) \hookrightarrow L_2([0, T], H^{-m}(D))$ it follows from Theorem 5.2.3 that $\partial_t u(t) \in \mathring{H}^m(D) \subset \mathcal{K}_{2,a}^0(D)$ if $a \leq m$. In this case Theorem 5.2.3 gives the estimate

$$\|\partial_t u | L_2([0, T], \mathcal{K}_{2,a}^0(D))\| \lesssim \|\partial_t u | L_2([0, T], \mathring{H}^m(D))\|$$

$$\lesssim \sum_{k=0}^{1} \|\partial_{t^k} f | L_2([0, T], H^{-m}(D))\|$$

$$\lesssim \sum_{k=0}^{1} \|\partial_{t^k} f | L_2(D_T)\|. \qquad (5.2.34)$$

Moreover, since

$$Lu = (-1)^m (f - \partial_t u) =: F,$$

where for fixed t the right hand side $F(t)$ belongs to $\mathcal{K}_{2,a}^0(D)$ (and due to the fact that $a \geq -m$, we have $\partial_t u(t) \in \mathring{H}^m(D) \hookrightarrow \mathcal{K}_{2,a}^0(D) \hookrightarrow \mathcal{K}_{2,-m}^{-m}(D)$, thus, also $F(t) \in \mathcal{K}_{2,-m}^{-m}(D))$ an application of Lemma 5.2.7 (with $\gamma = m, a = -m, \gamma' = 2m$, $a' = a$) gives $u(t) \in \mathcal{K}_{2,a+2m}^{2m}(D)$ with the a priori estimate

$$\|u(t) | \mathcal{K}_{2,a+2m}^{2m}(D)\| \lesssim \|f(t) | \mathcal{K}_{2,a}^0(D)\| + \|\partial_t u(t) | \mathcal{K}_{2,a}^0(D)\|.$$

Now integration w.r.t. the parameter t and (5.2.34) prove the claim for $\eta = 2m$ and $l = 0$, i.e.,

$$\|u | L_2([0, T], \mathcal{K}_{2,a+2m}^{2m}(D))\| \lesssim \|f | L_2([0, T], \mathcal{K}_{2,a}^0(D))\| + \sum_{k=0}^{1} \|\partial_{t^k} f | L_2(D_T)\|.$$

We now assume that the claim is true for $\eta = 2m$ and $k = 0, \dots, l - 1$. Then differentiating (5.1.1) l-times gives

$$L(\partial_{t^l} u) = (-1)^m \left(\partial_{t^l} f - \partial_{t^{l+1}} u - (-1)^m \sum_{k=0}^{l-1} \binom{l}{k} \partial_{t^{l-k}} L(\partial_{t^k} u) \right) =: F.$$

(5.2.35)

By our initial assumptions $\partial_{t^l} f(t) \in \mathcal{K}_{2,a}^0(D)$ and $\partial_{t^{l+1}} u(t) \in \mathring{H}^m(D) \hookrightarrow \mathcal{K}_{2,a}^0(D)$. Moreover, the inductive assumptions provide us with $\partial_{t^k} u(t) \in \mathcal{K}_{2,a+2m}^{2m}(D)$, thus, $\partial_{t^{l-k}} L(\partial_{t^k} u)(t) \in \mathcal{K}_{2,a}^0(D)$ and we see that $F(t) \in \mathcal{K}_{2,a}^0(D)$. Applying Lemma 5.2.7 (with $\gamma = m$, $a = -m$, $\gamma' = 2m$, $a' = a$) gives $\partial_{t^l} u(t) \in \mathcal{K}_{2,a+2m}^{2m}(D)$ and we have the following a priori estimate

$$\| \partial_{t^l} u(t) | \mathcal{K}_{2,a+2m}^{2m}(D) \|$$

$$\lesssim \| \partial_{t^l} f(t) | \mathcal{K}_{2,a}^0(D) \| + \| \partial_{t^{l+1}} u(t) | \mathcal{K}_{2,a}^0(D) \|$$

$$+ \sum_{k=0}^{l-1} \| \partial_{t^{l-k}} L(\partial_{t^k} u)(t) | \mathcal{K}_{2,a}^0(D) \|$$

$$\lesssim \| \partial_{t^l} f(t) | \mathcal{K}_{2,a}^0(D) \| + \| \partial_{t^{l+1}} u(t) | \mathring{H}^m(D) \|$$

$$+ \sum_{k=0}^{l-1} \| (\partial_{t^k} u)(t) | \mathcal{K}_{2,a+2m}^{2m}(D) \|.$$

Integration w.r.t. t, our inductive assumptions, and Theorem 5.2.3 give

$$\| \partial_{t^l} u | L_2([0, T], \mathcal{K}_{2,a+2m}^{2m}(D)) \|$$

$$\lesssim \sum_{k=0}^{l} \| \partial_{t^k} f | L_2([0, T], \mathcal{K}_{2,a}^0(D)) \| + \sum_{k=0}^{l+1} \| \partial_{t^k} f | L_2(D_T) \|.$$

Assume now inductively that our assumption holds for $\eta - 1$ and all derivatives $l \in \mathbb{N}$. We are going to show first that the claim then holds for η and $l = 0$ as well. Looking at

$$Lu = (-1)^m (f - \partial_t u) =: F,$$

we see that $f(t) \in \mathcal{K}_{2,a}^{\eta-2m}(D)$ and by our inductive assumption $\partial_t u(t) \in \mathcal{K}_{2,a+2m}^{\eta-1}(D) \hookrightarrow \mathcal{K}_{2,a}^{\eta-2m}(D)$. Therefore, $F(t) \in \mathcal{K}_{2,a}^{\eta-2m}(D)$ and an application of Lemma 5.2.7 (with $\gamma = m$, $a = -m$, $\gamma' = \eta$, $a' = a$) gives $u(t) \in \mathcal{K}_{2,a+2m}^{\eta}(D)$

with a priori estimate

$$\|u(t)|\mathcal{K}_{2,a+2m}^{\eta}(D)\| \lesssim \|f(t)|\mathcal{K}_{2,a}^{\eta-2m}(D)\| + \|\partial_t u(t)|\mathcal{K}_{2,a}^{\eta-2m}(D)\|$$

$$\lesssim \|f(t)|\mathcal{K}_{2,a}^{\eta-2m}(D)\| + \|\partial_t u(t)|\mathcal{K}_{2,a+2m}^{\eta-1}(D)\|.$$

Integration w.r.t. the parameter t and our inductive assumptions show that the claim is true for η and $l = 0$, i.e.,

$$\|u|L_2([0, T], \mathcal{K}_{2,a+2m}^{\eta}(D))\|$$

$$\lesssim \sum_{k=0}^{1+(\eta-1-2m)} \|\partial_{t^k} f|L_2([0, T], \mathcal{K}_{2,a}^{\eta-2m}(D))\| + \sum_{k=0}^{2+(\eta-1-2m)} \|\partial_{t^k} f|L_2(D_T)\|$$

$$= \sum_{k=0}^{\eta-2m} \|\partial_{t^k} f|L_2([0, T], \mathcal{K}_{2,a}^{\eta-2m}(D))\| + \sum_{k=0}^{1+(\eta-2m)} \|\partial_{t^k} f|L_2(D_T)\|.$$

Suppose now that it is true for η and derivatives $k = 0, \ldots, l - 1$. Differentiating (5.1.1) l-times again gives (5.2.35). From our initial assumptions on f we see that $\partial_{t^l} f(t) \in \mathcal{K}_{2,a}^{\eta-2m}(D)$ and from the inductive assumptions it follows that $\partial_{t^{l+1}} u(t) \in \mathcal{K}_{2,a+2m}^{\eta-1}(D) \hookrightarrow \mathcal{K}_{2,a}^{\eta-2m}(D)$. Moreover, by the inductive assumptions $\partial_{t^k} u(t) \in \mathcal{K}_{2,a+2m}^{\eta}(D)$ for $k = 0, \ldots, l - 1$ and therefore

$$\partial_{t^{l-k}} L(\partial_{t^k} u)(t) \in \mathcal{K}_{2,a}^{\eta-2m}(D).$$

This shows that the right hand side in (5.2.35) satisfies $F(t) \in \mathcal{K}_{2,a}^{\eta-2m}(D)$. Applying Lemma 5.2.7 again (with $\gamma = m$, $a = -m$, $\gamma' = \eta$, $a' = a$) gives $u(t) \in \mathcal{K}_{2,a+2m}^{\eta}(D)$ with a priori estimate

$$\|\partial_{t^l} u(t)|\mathcal{K}_{2,a+2m}^{\eta}(D)\|$$

$$\lesssim \|\partial_{t^l} f(t)|\mathcal{K}_{2,a}^{\eta-2m}(D)\| + \|\partial_{t^{l+1}} u(t)|\mathcal{K}_{2,a}^{\eta-2m}(D)\|$$

$$+ \sum_{k=0}^{l-1} \|\partial_{t^{l-k}} L(\partial_{t^k} u)(t)|\mathcal{K}_{2,a}^{\eta-2m}(D)\|$$

$$\lesssim \|\partial_{t^l} f(t)|\mathcal{K}_{2,a}^{\eta-2m}(D)\| + \|\partial_{t^{l+1}} u(t)|\mathcal{K}_{2,a+2m}^{\eta-1}(D)\|$$

$$+ \sum_{k=0}^{l-1} \|(\partial_{t^k} u)(t)|\mathcal{K}_{2,a+2m}^{\eta}(D)\|.$$

Integration w.r.t. the parameter t together with our inductive assumptions gives

$$\|\partial_{t^l} u | L_2([0, T], \mathcal{K}^\eta_{2,a+2m}(D))\|$$

$$\lesssim \|\partial_{t^l} f(t) | \mathcal{K}^{\eta-2m}_{2,a}(D)\|$$

$$+ \sum_{k=0}^{l+1+(\eta-1-2m)} \|\partial_{t^k} f | L_2([0, T], \mathcal{K}^{\eta-2m}_{2,a}(D))\| + \sum_{k=0}^{l+2+(\eta-1-2m)} \|\partial_{t^k} f | L_2(D_T)\|$$

$$+ \sum_{k=0}^{l-1+(\eta-2m)} \|\partial_{t^k} f | L_2([0, T], \mathcal{K}^{\eta-2m}_{2,a}(D))\| + \sum_{k=0}^{l-1+1+(\eta-2m)} \|\partial_{t^k} f | L_2(D_T)\|$$

$$\lesssim \sum_{k=0}^{l+(\eta-2m)} \|\partial_{t^k} f | L_2([0, T], \mathcal{K}^{\eta-2m}_{2,a}(D))\| + \sum_{k=0}^{l+1+(\eta-2m)} \|\partial_{t^k} f | L_2(D_T)\|,$$

which finishes the proof. $\square$

Remark 5.2.13 In Theorem 5.2.12 compared to Theorem 5.2.9 we only require the parameter a to satisfy $a \in [-m, m]$ and $-\delta^{(k)}_- < a + m < \delta^{(k)}_+$ independent of the regularity parameter η which can be arbitrary high. In particular, for the heat equation on a domain of polyhedral type D (which for simplicity we assume to be a polyhedron with straight edges and faces where θ_k denotes the angle at the edge M_k), we have $\delta^{(k)}_\pm = \frac{\pi}{\theta_k}$, which leads to the restriction

$$-1 \le a < \min\left(1, \frac{\pi}{\theta_k} - 1\right).$$

Therefore, even in the extremal case when $\theta_k = 2\pi$ we can still take $-1 \le a < -\frac{1}{2}$ (resulting in $u \in L_2([0, T], \mathcal{K}^\eta_{a+2}(D))$) being locally integrable since $a + 2 > 0$). Then choosing η arbitrary high, we also cover non-convex polyhedral type domains with our results from Theorem 5.2.12.

5.2.3 Regularity Results in Sobolev and Kondratiev Spaces for Problem II

In this subsection we show that the regularity estimates in Kondratiev and Sobolev spaces as stated in Theorems 5.2.9 and 5.2.3, respectively, carry over to Problem II, provided that ε is sufficiently small. In order to do this we interpret Problem II as a fixed point problem in the following way.

Let D and S be Banach spaces (D and S will be specified in the theorem below) and let $\tilde{L}^{-1} : D \to S$ be the linear operator defined via

$$\tilde{L}u := \frac{\partial}{\partial t}u + (-1)^m Lu. \tag{5.2.36}$$

Problem II is equivalent to

$$\tilde{L}u = f - \varepsilon u^M =: Nu,$$

where $N : S \to D$ is a nonlinear operator. If we can show that N maps S into D, then a solution of Problem II is a fixed point of the problem

$$(\tilde{L}^{-1} \circ N)u = u.$$

Our aim is to apply Banach's fixed point theorem, which will also guarantee uniqueness of the solution, if we can show that $T := (\tilde{L}^{-1} \circ N) : S_0 \to S_0$ is a contraction mapping, i.e., there exists some $q \in [0, 1)$ such that

$$\|T(x) - T(y)|S\| \leq q\|x - y|S\| \quad \text{for all} \quad x, y \in S_0,$$

where the corresponding metric space $S_0 \subset S$ is a small closed ball with center $\tilde{L}^{-1}f$ (the solution of the corresponding linear problem) and suitably chosen radius $R > 0$.

Our main result of this stated in the theorem below.

Theorem 5.2.14 (Nonlinear Sobolev and Kondratiev Regularity) *Let $\tilde{L}$ and N be as described above. Assume the assumptions of Theorem 5.2.9 are satisfied and, additionally, we have $\gamma_m \geq 1$, $m \geq \left[\frac{d}{2}\right] + 1$, and $a \geq \frac{d}{2} - 2$. Let*

$$\mathcal{D}_1 := \bigcap_{k=0}^{\gamma_m} H^k([0, T], \mathcal{K}_{2,a+2m(\gamma_m-k)}^{2m(\gamma_m-k)}(D)), \quad \mathcal{D}_2 := H^{\gamma_m+1}([0, T], L_2(D))$$

and consider the data space

$$\mathcal{D} := \{f \in \mathcal{D}_1 \cap \mathcal{D}_2 : \partial_{t^k} f(0, \cdot) = 0, \quad k = 0, \dots, \gamma_m\}.$$

Moreover, let

$$S_1 := \bigcap_{k=0}^{\gamma_m+1} H^k([0, T], \mathcal{K}_{2,a+2m(\gamma_m-(k-1))}^{2m(\gamma_m-(k-1))}(D)),$$

$$S_2 := H^{m,\gamma_m+2*}(D_T),$$

and consider the solution space $S := S_1 \cap S_2$. *Suppose that* $f \in \mathcal{D}$ *and put* $\eta :=$
$\|f|\mathcal{D}\|$ *and* $r_0 > 1$. *Moreover, we choose* $\varepsilon > 0$ *so small that*

$$\eta^{2(M-1)}\|\tilde{L}^{-1}\|^{2M-1} \le \frac{1}{c\varepsilon M}(r_0 - 1)\left(\frac{1}{r_0}\right)^{2M-1}, \qquad \text{if} \quad r_0\|\tilde{L}^{-1}\|\eta > 1,$$

and

$$\|\tilde{L}^{-1}\| < \frac{r_0 - 1}{r_0}\left(\frac{1}{c\varepsilon M}\right), \qquad \text{if} \quad r_0\|\tilde{L}^{-1}\|\eta < 1,$$

where $c > 0$ *denotes the constant in (5.2.58) resulting from our estimates below.*
Then there exists a unique solution $u \in S_0 \subset S$ *of Problem II, where* S_0 *denotes*
a small ball around $\tilde{L}^{-1}f$ *(the solution of the corresponding linear problem) with*
radius $R = (r_0 - 1)\eta\|\tilde{L}^{-1}\|$.

Proof Let u be the solution of the linear problem $\tilde{L}u = f$. From Theorems 5.2.9
and 5.2.3 we know that

$$\tilde{L}^{-1} : \mathcal{D} \to S$$

is a bounded operator. We need to show

$$u^M \in \mathcal{D} \tag{5.2.37}$$

in order to establish the desired mapping properties of the nonlinear part N, i.e.,

$$Nu = f - \varepsilon u^M \in \mathcal{D}.$$

The fact that $u^M \in \mathcal{D}_1 \cap \mathcal{D}_2$ follows from the estimate (5.2.58): In particular, taking
$v = 0$ in (5.2.58) we get an estimate from above for $\|u^M|\mathcal{D}\|$. The upper bound de-
pends on $\|u|S\|$ and several constants which depend on u but are finite whenever we
have $u \in S$, see also (5.2.48) and (5.2.55). The dependence on R in (5.2.58) comes
from the fact that we choose $u \in B_R(\tilde{L}^{-1}f)$ in S there. However, the same argument
can also be applied to an arbitrary $u \in S$; this would result in a different constant $\tilde{c}$.
In order to have $u^M \in \mathcal{D}$, we still need to show that $\mathrm{Tr}\left(\partial_{t^k}u^M\right) = 0, k = 0, \ldots, \gamma_m$:
Since $u \in S \hookrightarrow H^{\gamma_m+2}([0, T], H^{-m}(D)) \hookrightarrow C^{\gamma_m+1}([0, T], H^{-m}(D))$ we see that
the trace operator $\mathrm{Tr}\left(\partial_{t^k}u\right) := \left(\partial_{t^k}u\right)(0, \cdot)$ is well defined for $k = 0, \ldots, \gamma_m + 1$ (in
particular, since $u \in S \hookrightarrow H^{\gamma_m+1}([0, T], \mathring{H}^m(D)) \hookrightarrow C^{\gamma_m}([0, T], \mathring{H}^m(D))$ the
values of the trace operator $\mathrm{Tr}\left(\partial_{t^k}u\right), k = 0, \ldots, \gamma_m$, belong to H^m). We first show
that $\mathrm{Tr}(\partial_{t^k}u) = 0$ for $k = 0, \ldots, \gamma_m + 1$. For this we use the assumptions on f, i.e.,
$\partial_{t^k}f(0, \cdot) = 0$ for $k = 0, \ldots, \gamma_m$, the initial assumption $u(0, \cdot) = 0$ in Problem II,
and the fact that

$$\mathrm{Tr}(Lu)(t, x) = 0. \tag{5.2.38}$$

Let us briefly sketch the proof of (5.2.38). We show (5.2.38) first for compactly supported functions $u \in C^\infty(D_T)$. Then, the result follows by density arguments (concerning the density of $C^\infty(D_T)$ in S we refer to the explanations given in Remark 3.2.8). For these functions, we get

$$\mathrm{Tr}(Lu)(t, x) = \lim_{t \to 0} Lu(t, x) = L(0, x; D_x) \lim_{t \to 0} u(t, x) = 0,$$

where the second step follows by our smoothness assumptions on the coefficients of L and the fact that $\lim_{t \to 0} D_x^\alpha u(t, x) = D_x^\alpha (\lim_{t \to 0} u(t, x))$ (this is clear for smooth functions $u \in C^\infty(D_T)$, then using density we get the same for $u \in S$). With this we see that

$$(\partial_t u)(0, \cdot) + (-1)^m Lu(0, \cdot) = f(0, \cdot), \qquad \text{i.e.,} \qquad (\partial_t u)(0, \cdot) = 0.$$

Differentiation yields

$$(\partial_{t^2} u)(0, \cdot) + (-1)^m ((\partial_t L)u(0, \cdot) + L(\partial_t u)(0, \cdot)) = \partial_t f(0, \cdot).$$

i.e., $(\partial_{t^2} u)(0, \cdot) = 0$ using a similar argumentation as in (5.2.38). By induction we deduce that $(\partial_{t^k} u)(0, \cdot) = 0$ for all $k = 0, \ldots, \gamma_m + 1$ (in particular, $\| (\partial_{t^k} u)(0, \cdot)|H^m(D)\| = 0$ for $k = 0, \ldots, \gamma_m$). Moreover, since by Theorem 2.2.5 (generalized Sobolev embedding)

$$u^M \in \mathcal{D}_1 \cap \mathcal{D}_2 \hookrightarrow H^{\gamma_m + 1}([0, T], L_2(D)) \hookrightarrow C^{\gamma_m}([0, T], L_2(D)),$$

we see that the trace operator $\mathrm{Tr}\left(\partial_{t^k} u^M\right) := \left(\partial_{t^k} u^M\right)(0, \cdot)$ is well defined for $k = 0, \ldots, \gamma_m$. By (5.2.54) below the term $\| \left(\partial_{t^k} u^M\right)(0, \cdot)|L_2(D)\|$ is estimated from above by powers of $\| \left(\partial_{t^l} u\right)(0, \cdot)|H^m(D)\|$, $l = 0, \ldots, k$. Since all these terms are equal to zero, we obtain $\left(\partial_{t^k} u^M\right)(0, \cdot) = 0$ in the L_2-norm for $k = 0, \ldots, \gamma_m$, which shows that $u^M \in \mathcal{D}$.

Hence, using (5.2.37) we can apply Theorem 5.2.9 now with right hand side Nu. Since

$$(\tilde{L}^{-1} \circ N)(v) - (\tilde{L}^{-1} \circ N)(u) = \tilde{L}^{-1}(f - \varepsilon v^M) - \tilde{L}^{-1}(f - \varepsilon u^M) = \varepsilon \tilde{L}^{-1}(u^M - v^M)$$

one sees that $\tilde{L}^{-1} \circ N$ is a contraction if, and only, if

$$\varepsilon \|\tilde{L}^{-1}(u^M - v^M)|S\| \leq q \|u - v|S\| \qquad \text{for some} \quad q < 1, \qquad (5.2.39)$$

where $u, v \in S_0$ (meaning $u, v \in B_R(\tilde{L}^{-1} f)$ in S). We analyse the resulting condition with the help of the formula $u^M - v^M = (u - v) \sum_{j=0}^{M-1} u^j v^{M-1-j}$.

This together with Theorem 5.2.9 gives

$$\|\tilde{L}^{-1}(u^M - v^M)|S\|$$

$$\leq \|\tilde{L}^{-1}\|\|u^M - v^M|\mathcal{D}\|$$

$$= \|\tilde{L}^{-1}\|\ \left\|u^M - v^M|\mathcal{D}_1 \cap \mathcal{D}_2\right\|$$

$$= \|\tilde{L}^{-1}\|\ \left(\|u^M - v^M|\mathcal{D}_1\| + \|u^M - v^M|\mathcal{D}_2\|\right)$$

$$= \|\tilde{L}^{-1}\|\ \left(\left\|(u - v)\sum_{j=0}^{M-1} u^j v^{M-1-j}|\mathcal{D}_1\right\| + \left\|(u - v)\sum_{j=0}^{M-1} u^j v^{M-1-j}|\mathcal{D}_2\right\|\right)$$

$$= \|\tilde{L}^{-1}\|\ \bigg(\sum_{k=0}^{\gamma_m}\left\|\partial_{t^k}\left[(u-v)\sum_{j=0}^{M-1} u^j v^{M-1-j}\right]|L_2([0,T], \mathcal{K}_{2,a+2m(\gamma_m-k)}^{2m(\gamma_m-k)}(D))\right\|$$

$$+ \sum_{k=0}^{\gamma_m+1}\left\|\partial_{t^k}\left[(u-v)\sum_{j=0}^{M-1} u^j v^{M-1-j}\right]|L_2(D_T)\right\|\bigg)$$

$$\tag{5.2.40}$$

Concerning the derivatives, we use Leibniz's formula twice and we see that

$$\partial_{t^k}(u^M - v^M)$$

$$= \partial_{t^k}\left[(u-v)\sum_{j=0}^{M-1} u^j v^{M-1-j}\right]$$

$$= \sum_{l=0}^{k}\binom{k}{l}\partial_{t^l}(u-v)\cdot\partial_{t^{k-l}}\left(\sum_{j=0}^{M-1} u^j v^{M-1-j}\right)$$

$$= \sum_{l=0}^{k}\binom{k}{l}\partial_{t^l}(u-v)\cdot\left[\left(\sum_{j=0}^{M-1}\sum_{r=0}^{k-l}\binom{k-l}{r}\partial_{t^r}u^j\cdot\partial_{t^{k-l-r}}v^{M-1-j}\right)\right].$$

$$\tag{5.2.41}$$

In order to estimate the terms $\partial_{t^r}u^j$ and $\partial_{t^{k-l-r}}v^{M-1-j}$ we apply Faà di Bruno's formula

$$\partial_{t^r}(f \circ g) = \sum \frac{r!}{k_1!\ldots k_r!}\left(\partial_{t^{k_1+\ldots+k_r}} f \circ g\right)\prod_{i=1}^{r}\left(\frac{\partial_{t^i} g}{i!}\right)^{k_i},\tag{5.2.42}$$

where the sum runs over all r-tuples of nonnegative integers $(k_1, \ldots, k_r)$ satisfying

$$1 \cdot k_1 + 2 \cdot k_2 + \ldots + r \cdot k_r = r. \tag{5.2.43}$$

In particular, from (5.2.43) we see that $k_r \leq 1$, where $r = 1, \ldots, k$. Therefore, the highest derivative $\partial_{t^r} u$ appears at most once. We apply the formula to $g = u$ and $f(x) = x^j$ and make use of the embeddings (2.4.5) and the pointwise multiplier results from Theorem 2.4.9 (i) for $k \leq \gamma_m - 1$. (Note that the restriction '$a > \frac{d}{p}$' in Theorem 2.4.9 (i) is satisfied since in our situation we have $a + 2m \geq m > \frac{d}{2}$ from the assumptions $a \in [-m, m]$ and $m \geq \left[\frac{d}{2}\right] + 1$.) This yields

$$\left\| \partial_{t^r} u^j \mid \mathcal{K}^{2m(\gamma_m - k)}_{2, a + 2m(\gamma_m - k)}(D) \right\|$$

$$\leq c_{r,j} \left\| \sum_{\substack{k_1 + \ldots + k_r \leq j, \\ 1 \cdot k_1 + 2 \cdot k_2 + \ldots + r \cdot k_r = r}} u^{j - (k_1 + \ldots + k_r)} \prod_{i=1}^{r} |\partial_{t^i} u|^{k_i} \mid \mathcal{K}^{2m(\gamma_m - k)}_{2, a + 2m(\gamma_m - k)}(D) \right\|$$

$$\lesssim \sum_{\substack{k_1 + \ldots + k_r \leq j, \\ 1 \cdot k_1 + 2 \cdot k_2 + \ldots + r \cdot k_r = r}} \left\| u \mid \mathcal{K}^{2m(\gamma_m - k)}_{2, a + 2m(\gamma_m - k)}(D) \right\|^{j - (k_1 + \ldots + k_r)}$$

$$\prod_{i=1}^{r} \left\| \partial_{t^i} u \mid \mathcal{K}^{2m(\gamma_m - k)}_{2, a + 2m(\gamma_m - k)}(D) \right\|^{k_i}.$$

$$\tag{5.2.44}$$

For $k = \gamma_m$ we use Theorem 2.4.9(ii). (Note that in Theorem 2.4.9(ii) we require that '$a - 1 \geq \frac{d}{p} - 2$' for the parameter. In our situation below $a - 1$ has to be replaced by a, which leads to our restriction $a \geq \frac{d}{2} - 2$.) Similar as above we obtain

$$\left\| \partial_{t^r} u^j \mid \mathcal{K}^0_{2, a}(D) \right\|$$

$$\leq c_{r,j} \left\| \sum_{\substack{k_1 + \ldots + k_r \leq j, \\ 1 \cdot k_1 + 2 \cdot k_2 + \ldots + r \cdot k_r = r}} u^{j - (k_1 + \ldots + k_r)} \prod_{i=1}^{r} |\partial_{t^i} u|^{k_i} \mid \mathcal{K}^0_{2, a}(D) \right\|$$

$$\lesssim \sum_{\substack{k_1 + \ldots + k_r \leq j, \\ 1 \cdot k_1 + 2 \cdot k_2 + \ldots + r \cdot k_r = r}} \left\| u \mid \mathcal{K}^2_{2, a + 2}(D) \right\|^{j - (k_1 + \ldots + k_r)}$$

$$\left\| \partial_{t^r} u \mid \mathcal{K}^0_{2, a}(D) \right\|^{k_r} \prod_{i=1}^{r-1} \left\| \partial_{t^i} u \mid \mathcal{K}^2_{2, a + 2}(D) \right\|^{k_i}$$

$$\lesssim \sum_{\substack{k_1+\ldots+k_r \le j, \\ 1\cdot k_1+2\cdot k_2+\ldots+r\cdot k_r=r}} \left\| u |\mathcal{K}_{2,a+2m\gamma_m}^{2m\gamma_m}(D) \right\|^{j-(k_1+\ldots+k_r)}$$

$$\left\| \partial_{t^r} u |\mathcal{K}_{2,a+2m(\gamma_m-r)}^{2m(\gamma_m-r)}(D) \right\|^{k_r} \prod_{i=1}^{r-1} \left\| \partial_{t^i} u |\mathcal{K}_{2,a+2m(\gamma_m-i)}^{2m(\gamma_m-i)}(D) \right\|^{k_i}.$$

$$(5.2.45)$$

Note that we require $\gamma_m \ge 1$ in the last step. We proceed similarly for $\partial_{t^{k-l-r}} v^{M-1-j}$. Now (5.2.41) together with (5.2.44) and (5.2.45) inserted in (5.2.40) together with Theorem 2.4.9 give

$$\| \tilde{L}^{-1} \| \| u^M - v^M |\mathcal{D}_1 \|$$

$$\lesssim \| \tilde{L}^{-1} \| \sum_{k=0}^{\gamma_m} \left(\int_0^T \left\| \partial_{t^k} \left[(u-v) \sum_{j=0}^{M-1} u^j v^{M-1-j} \right] |\mathcal{K}_{2,a+2m(\gamma_m-k)}^{2m(\gamma_m-k)}(D) \right\|^2 dt \right)^{1/2}$$

$$\lesssim \| \tilde{L}^{-1} \| \sum_{k=0}^{\gamma_m} \sum_{l=0}^{k} \sum_{j=0}^{M-1} \sum_{r=0}^{k-l} \left(\int_0^T \left\| \partial_{t^l} (u-v) |\mathcal{K}_{2,a+2m(\gamma_m-k)}^{2m(\gamma_m-k)}(D) \right\|^2 \right.$$

$$\left. \left\| \partial_{t^r} u^j |\mathcal{K}_{2,a+2m(\gamma_m-k)}^{2m(\gamma_m-k)}(D) \right\|^2 \left\| \partial_{t^{k-l-r}} v^{M-1-j} |\mathcal{K}_{2,a+2m(\gamma_m-k)}^{2m(\gamma_m-k)}(D) \right\|^2 dt \right)^{1/2}$$

$$(5.2.46)$$

$$\lesssim \| \tilde{L}^{-1} \| \sum_{k=0}^{\gamma_m} \sum_{l=0}^{k} \sum_{j=0}^{M-1} \sum_{r=0}^{k-l} \left(\int_0^T \left\| \partial_{t^l} (u-v) |\mathcal{K}_{2,a+2m(\gamma_m-k)}^{2m(\gamma_m-k)}(D) \right\|^2 \right.$$

$$\sum_{\substack{\kappa_1+\ldots+\kappa_r \le j, \\ \kappa_1+2\kappa_2+\ldots+r\kappa_r=r}} \left\| u |\mathcal{K}_{2,a+2m(\gamma_m-k)}^{2m(\gamma_m-k)}(D) \right\|^{2(j-(\kappa_1+\ldots+\kappa_r))} \prod_{i=0}^{r} \left\| \partial_{t^i} u |\mathcal{K}_{2,a+2m(\gamma_m-i)}^{2m(\gamma_m-i)}(D) \right\|^{2\kappa_i}$$

$$\sum_{\substack{\kappa_1+\ldots+\kappa_{k-l-r} \le M-1-j, \\ \kappa_1+2\kappa_2+\ldots+(k-l-r)\kappa_{k-l-r}=k-l-r}} \left\| v |\mathcal{K}_{2,a+2m(\gamma_m-k)}^{2m(\gamma_m-k)}(D) \right\|^{2(M-1-j-(\kappa_1+\ldots+\kappa_{k-l-r}))}$$

$$\left. \prod_{i=0}^{k-l-r} \left\| \partial_{t^i} v |\mathcal{K}_{2,a+2m(\gamma_m-i)}^{2m(\gamma_m-i)}(D) \right\|^{2\kappa_i} dt \right)^{1/2}$$

$$\lesssim \| \tilde{L}^{-1} \| \sum_{k=0}^{\gamma_m} M \left(\int_0^T \left\| \partial_{t^k} (u-v) |\mathcal{K}_{2,a+2m(\gamma_m-k)}^{2m(\gamma_m-k)}(D) \right\|^2 \right.$$

$$\sum_{\substack{\kappa_1'+\ldots+\kappa_k' \le \min\{M-1,k\}, \\ \kappa_k' \le 1}} \max_{w \in \{u,v\}} \left\| w |\mathcal{K}_{2,a+2m(\gamma_m-k)}^{2m(\gamma_m-k)}(D) \right\|^{2(M-1-(\kappa_1'+\ldots+\kappa_k'))}$$

$$\prod_{i=0}^{k} \max \left\{ \left\| \partial_{t^i} u | \mathcal{K}_{2,a+2m(\gamma_m-i)}^{2m(\gamma_m-i)}(D) \right\|, \left\| \partial_{t^i} v | \mathcal{K}_{2,a+2m(\gamma_m-i)}^{2m(\gamma_m-i)}(D) \right\|, 1 \right\}^{4\kappa_i'} dt \right)^{1/2}$$

$$(5.2.47)$$

$$\lesssim M \|\tilde{L}^{-1}\| \cdot \left\| u - v | \bigcap_{k=0}^{\gamma_m+1} H^k([0,T], \mathcal{K}_{2,a+2m(\gamma_m-(k-1))}^{2m(\gamma_m-(k-1))}(D)) \right\| \cdot$$

$$\max_{w \in \{u,v\}} \max_{l=0,\dots,\gamma_m} \max \left(\left\| \partial_{t^l} w | L_\infty([0,T], \mathcal{K}_{2,a+2m(\gamma_m-l)}^{2m(\gamma_m-l)}(D)) \right\|, 1 \right)^{2(M-1)}.$$

$$(5.2.48)$$

We give some explanations concerning the estimate above. In (5.2.46) the term with $k = \gamma_m$ requires some special care since we have to apply Theorem 2.4.9 (ii). In this case we calculate

$$\left\| \partial_{\gamma_m} \left[(u-v) \left(\sum_{j=0}^{M-1} u^j v^{M-1-j} \right) \right] | \mathcal{K}_{2,a}^0(D) \right\|$$

$$\lesssim \left\| \partial_{\gamma_m}(u-v) | \mathcal{K}_{2,a}^0(D) \right\| \sum_{j=0}^{M-1} \left\| u^j v^{M-1-j} | \mathcal{K}_{2,a+2}^2(D) \right\|$$

$$+ \left\| u - v | \mathcal{K}_{2,a+2}^2(D) \right\| \sum_{j=0}^{M-1} \sum_{r=0}^{\gamma_m} \left\| (\partial_{t^r} u^j)(\partial_{t^{\gamma_m-r}} v^{M-1-j}) | \mathcal{K}_{2,a}^0(D) \right\|$$

$$+ \left\| \sum_{r=1}^{\gamma_m-1} \binom{\gamma_m}{r} \partial_r(u-v) \partial_{\gamma_m-r} \left(\sum_{j=0}^{M-1} \dots \right) | \mathcal{K}_{2,a}^0(D) \right\|.$$

The lower order derivatives in the last line cause no problems since we can (again) apply Theorem 2.4.9(i) as before. The term $\left\| u^j v^{M-1-j} | \mathcal{K}_{2,a+2}^2(D) \right\|$ can now be further estimated with the help of Theorem 2.4.9(i). For the term $\sum_{r=0}^{\gamma_m} \left\| (\partial_{t^r} u^j)(\partial_{t^{\gamma_m-r}} v^{M-1-j}) | \mathcal{K}_{2,a}^0(D) \right\|$ we again use Theorem 2.4.9(ii), then proceed as in (5.2.45) and see that the resulting estimate yields (5.2.46).

Moreover, in (5.2.47) we used the fact that in the step before we have two sums with $\kappa_1 + \dots + \kappa_r \le j$ and $\kappa_1 + \dots + \kappa_{k-l-r} \le M-1-j$, i.e., we have $k-l$ different κ_i's which leads to at most k different κ_i's if $l = 0$. We allow all combinations of κ_i's and redefine the κ_i's in the second sum leading to $\kappa_1', \dots, \kappa_k'$ with $\kappa_1' + \dots + \kappa_k' \le M - 1$ and replace the old conditions $\kappa_1 + 2\kappa_2 + r\kappa_r \le r$ and $\kappa_1 + 2\kappa_2 + (k - l - r)\kappa_{k-l-r} \le k-l-r$ by the weaker ones $\kappa_1' + \dots + \kappa_k' \le k$ and $\kappa_k' \le 1$. This causes no problems since the other terms appearing in this step do not depend on κ_i apart from the product term. There, the fact that some of the old κ_i's from both sums might coincide is reflected in the new exponent $4\kappa_i'$. From Theorem 2.2.5 (Sobolev embedding) we conclude that

$$u, v \in S \hookrightarrow \bigcap_{k=0}^{\gamma_m+1} H^k([0, T], \mathcal{K}^{2m(\gamma_m-(k-1))}_{2,a+2m(\gamma_m-(k-1))}(D))$$

$$\hookrightarrow \bigcap_{k=1}^{\gamma_m+1} C^{k-1,\frac{1}{2}}([0, T], \mathcal{K}^{2m(\gamma_m-(k-1))}_{2,a+2m(\gamma_m-(k-1))}(D))$$

$$\hookrightarrow \bigcap_{k=1}^{\gamma_m+1} C^{k-1}([0, T], \mathcal{K}^{2m(\gamma_m-(k-1))}_{2,a+2m(\gamma_m-(k-1))}(D)) \tag{5.2.49}$$

$$= \bigcap_{l=0}^{\gamma_m} C^l([0, T], \mathcal{K}^{2m(\gamma_m-l)}_{2,a+2m(\gamma_m-l)}(D)), \tag{5.2.50}$$

hence, the term involving the maxima, $\max_{w\in\{u,v\}} \max_{l=0,\dots,\gamma_m} \max(\dots)^{M-1}$ in (5.2.48) is bounded by $\max(R + \|\tilde{L}^{-1} f|S\|, 1)^{M-1}$. Moreover, since u and v are taken from $B_R(\tilde{L}^{-1} f)$ in $S = S_1 \cap S_2$, we obtain from (5.2.48),

$$\|\tilde{L}^{-1}\| \|u^M - v^M|\mathcal{D}_1\|$$

$$\leq c_0 \|\tilde{L}^{-1}\| M \max(R + \|\tilde{L}^{-1} f|S\|, 1)^{2(M-1)} \|u - v|S\|$$

$$\leq c_2 \|\tilde{L}^{-1}\| M \max(R + \|\tilde{L}^{-1}\| \cdot \|f|\mathcal{D}\|, 1)^{2(M-1)} \|u - v|S\|$$

$$= c_2 \|\tilde{L}^{-1}\| M \max(R + \|\tilde{L}^{-1}\|\eta, 1)^{2(M-1)} \|u - v|S\|, \tag{5.2.51}$$

where we put $\eta := \|f|\mathcal{D}\|$ in the last line, c_0 denotes the constant resulting from (5.2.44) and (5.2.48) and $c_2 = c_0 c_1$ with c_1 being the constant from the estimates in Theorem 5.2.9.

We now turn our attention towards the second term $\|\tilde{L}^{-1}\| \|u^M - v^M|\mathcal{D}_2\|$ in (5.2.40) and calculate

$$\|\tilde{L}^{-1}\| \|(u^M - v^M)|\mathcal{D}_2\|$$

$$= \|\tilde{L}^{-1}\| \left\| (u - v) \sum_{j=0}^{M-1} u^j v^{M-1-j} | H^{\gamma_m+1}([0, T], L_2(D)) \right\|$$

$$= \|\tilde{L}^{-1}\| \sum_{k=0}^{\gamma_m+1} \left\| \partial_t^k \left[(u - v) \sum_{j=0}^{M-1} u^j v^{M-1-j} \right] | L_2(D_T) \right\|$$

$$= \|\tilde{L}^{-1}\| \sum_{k=0}^{\gamma_m+1} \left\| \sum_{l=0}^{k} \binom{k}{l} \partial_{t^l}(u-v) \cdot \right.$$

$$\left. \left[\left(\sum_{j=0}^{M-1} \sum_{r=0}^{k-l} \binom{k-l}{r} \partial_{t^r} u^j \cdot \partial_{t^{k-l-r}} v^{M-1-j} \right) \right] |L_2(D_T) \right\|$$

$$\lesssim \|\tilde{L}^{-1}\| \sum_{k=0}^{\gamma_m+1} \left\| \sum_{l=0}^{k} |\partial_{t^l}(u-v)| \cdot \right.$$

$$\left. \left[\left(\sum_{j=0}^{M-1} \sum_{r=0}^{k-l} |\partial_{t^r} u^j \cdot \partial_{t^{k-l-r}} v^{M-1-j}| \right) \right] |L_2(D_T) \right\|,$$

$$(5.2.52)$$

where we used Leibniz's formula twice as in (5.2.41) in the second but last line. Again Faà di Bruno's formula, cf. (5.2.42), is applied in order to estimate the derivatives in (5.2.52). We use a special case of the multiplier result from [RS96, Sect. 4.6.1, Thm. 1(i)], which states that for parameters $s_2 > s_1$, $s_1 + s_2 > d \max\left(0, \frac{2}{p} - 1\right)$, $s_2 > \frac{d}{p}$, and $q \geq \max(q_1, q_2)$, we have for the Triebel-Lizorkin spaces

$$\|uv|F_{p,q_1}^{s_1}\| \lesssim \|u|F_{p,q_2}^{s_2}\| \cdot \|v|F_{p,q_1}^{s_1}\|.$$

In particular, choosing $s_1 = 0$, $s_2 = m > \frac{d}{2}$, $q_1 = q_2 = p = 2$ and using the identity $F_{2,2}^0 = L_2$ and $F_{2,2}^m = H^m$, we obtain

$$\|uv|L_2\| \lesssim \|u|H^m\| \cdot \|v|L_2\|. \qquad (5.2.53)$$

This is exactly the point where our assumption $m \geq \left[\frac{d}{2}\right] + 1$ comes into play, since $s_2 = m > \frac{d}{2}$ is needed. With this we obtain

$$\left\| \partial_{t^r} u^j |L_2(D) \right\|$$

$$\leq c_{r,j} \left\| \sum_{k_1+\ldots+k_r \leq j} u^{j-(k_1+\ldots+k_r)} \prod_{i=1}^{r} |\partial_{t^i} u|^{k_i} |L_2(D) \right\|$$

$$\lesssim \sum_{k_1+\ldots+k_r \leq j} \|u|H^m(D)\|^{j-(k_1+\ldots+k_r)} \prod_{i=1}^{r-1} \|\partial_{t^i} u|H^m(D)\|^{k_i} \|\partial_{t^r} u|L_2(D)\|^{k_r}.$$

$$(5.2.54)$$

Similar for $\partial_{t^{k-l-r}} v^{M-1-j}$. As before, from (5.2.43) we observe $k_r \leq 1$, therefore the highest derivative $u^{(r)}$ appears at most once. Note that since $H^m(D)$ is a multiplication algebra for $m > \frac{d}{2}$, we get (5.2.54) with $L_2(D)$ replaced by $H^m(D)$ as well. Now (5.2.53) and (5.2.54) inserted in (5.2.52) gives

$$\|\tilde{L}^{-1}\| \|u^M - v^M | \mathcal{D}_2\|$$

$$= \|\tilde{L}^{-1}\| \sum_{k=0}^{\gamma_m+1} \left(\int_0^T \left\| \partial_{t^k}(u-v) \sum_{j=0}^{M-1} u^j v^{M-1-j} | L_2(D) \right\|^2 dt \right)^{1/2}$$

$$\lesssim \|\tilde{L}^{-1}\| \sum_{k=0}^{\gamma_m+1} \sum_{l=0}^{k} \left(\int_0^T \left\| \partial_{t^l}(u-v) | H^m(D) \right\|^2 \right.$$

$$\sum_{j=0}^{M-1} \sum_{r=0}^{k-l} \left\| \partial_{t^r} u^j \cdot \partial_{t^{k-l-r}} v^{M-1-j} | L_2(D) \right\|^2 dt \bigg)^{1/2}$$

$$\lesssim \|\tilde{L}^{-1}\| \sum_{k=0}^{\gamma_m+1} \sum_{l=0}^{k} \left(\int_0^T \left\{ \left\| \partial_{t^l}(u-v) | H^m(D) \right\|^2 \right.\right.$$

$$\sum_{j=0}^{M-1} \sum_{\substack{r=0, \\ (k-l-r \neq \gamma_m+1) \wedge (r \neq \gamma_m+1)}}^{k-l} \left\| \partial_{t^r} u^j | H^m(D) \right\|^2 \left\| \partial_{t^{k-l-r}} v^{M-1-j} | H^m(D) \right\|^2$$

$$+ \|u-v| H^m(D)\|^2 \|\partial_{t^{\gamma_m+1}} u^j | L_2(D)\|^2 \|v^{M-1-j} | H^m(D)\|^2$$

$$+ \|u-v| H^m(D)\|^2 \|u^j | H^m(D)\|^2 \|\partial_{t^{\gamma_m+1}} v^{M-1-j} | L_2(D)\|^2 \bigg\} dt \bigg)^{1/2}$$

$$\lesssim \|\tilde{L}^{-1}\| \sum_{k=0}^{\gamma_m+1} \sum_{l=0}^{k} \left(\int_0^T \left\| \partial_{t^l}(u-v) | H^m(D) \right\|^2 \cdot \right.$$

$$\sum_{j=0}^{M-1} \sum_{r=0}^{k-l} \sum_{\substack{\kappa_1+\ldots+\kappa_r \leq j, \\ \kappa_1+2\kappa_2+\ldots+r\kappa_r \leq r}} \left\| u | H^m(D) \right\|^{2(j-(\kappa_1+\ldots+\kappa_r))}$$

$$\begin{cases} \|\partial_{t^r} u | L_2(D)\|^{2\kappa_r} \prod_{i=1}^{r-1} \|\partial_{t^i} u | H^m(D)\|^{2\kappa_i}, & r = \gamma_m + 1, \\ \prod_{i=1}^{r} \|\partial_{t^i} u | H^m(D)\|^{2\kappa_i}, & r \neq \gamma_m + 1 \end{cases}$$

$$\sum_{\substack{\kappa_1+\ldots+\kappa_{k-l-r} \leq M-1-j, \\ \kappa_1+2\kappa_2+\ldots+(k-l-r)\kappa_{k-l-r} \leq k-l-r}} \left\| v | H^m(D) \right\|^{2(M-1-j-(\kappa_1+\ldots+\kappa_{k-l-r}))}$$

$$\begin{cases} \|\partial_{t^r} v | L_2(K)\|^{2\kappa_r} \prod_{i=1}^{k-l-r-1} \|\partial_{t^i} v | H^m(D)\|^{2\kappa_i}, & k-l-r = \gamma_m + 1, \\ \prod_{i=1}^{l-k-r} \|\partial_{t^i} v | H^m(D)\|^{2\kappa_i}, & k-l-r \neq \gamma_m + 1 \end{cases} dt \bigg)^{1/2}$$

$$\lesssim \|\tilde{L}^{-1}\| \sum_{k=0}^{\gamma_m+1} \left(\int_0^T \left\|\partial_{t^k}(u-v)|H^m(D)\right\|^2 \cdot \right.$$

$$M \sum_{\kappa_1'+...+\kappa_k' \le \min\{M-1,k\}} \max_{w \in \{u,v\}} \left\|w|H^m(D)\right\|^{2(M-1-(\kappa_1'+...+\kappa_k'))}$$

$$\left. \begin{cases} \max(\left\|\partial_{t^k} w|L_2(D)\right\|^{4\kappa_k'} \prod_{i=1}^{k-1} \left\|\partial_{t^i} w|H^m(D)\right\|^{4\kappa_i'}, 1), & k = \gamma_m+1, \\ \max(\prod_{i=1}^{k} \left\|\partial_{t^i} w|H^m(D)\right\|^{4\kappa_i'}, 1), & k \ne \gamma_m+1 \end{cases} dt \right)^{1/2}$$

$$\lesssim \|\tilde{L}^{-1}\| M \|u-v|H^{\gamma_m+1}([0,T], H^m(D))\|^2 \max_{w \in \{u,v\}} \max_{i=0,...,\gamma_m} \max$$

$$\left(\left\|\partial_{t^i} w|L_\infty([0,T], H^m(D))\right\|, \left\|\partial_{t^{\gamma_m+1}} w|L_\infty([0,T], L_2(D))\right\|, 1 \right)^{2(M-1)}$$

$$(5.2.55)$$

Similar to (5.2.48) in the calculations above the term $k = \gamma_m + 1$ required some special care. For the redefinition of the κ_i's in the second but last line in (5.2.55) we refer to the explanations given after (5.2.48). From Theorem 2.2.5 we see that

$$u, v \in S \hookrightarrow H^{\gamma_m+1}([0,T], \mathring{H}^m(D)) \cap H^{\gamma_m+2}([0,T], L_2(D))$$

$$\hookrightarrow C^{\gamma_m, \frac{1}{2}}([0,T], \mathring{H}^m(D)) \cap C^{\gamma_m+1, \frac{1}{2}}([0,T], L_2(D))$$

$$\hookrightarrow C^{\gamma_m}([0,T], \mathring{H}^m(D)) \cap C^{\gamma_m+1}([0,T], L_2(D)), \quad (5.2.56)$$

hence the term $\max_{w \in \{u,v\}} \max_{m=0,...,l} \max(...)^{M-1}$ in (5.2.55) is bounded. Moreover, since u and v are taken from $B_R(\tilde{L}^{-1} f)$ in $S_2 = H^{m,\gamma_m+2*}(D_T) = H^{\gamma_m+1}([0,T], \mathring{H}^m(D)) \cap H^{\gamma_m+2}([0,T], H^{-m}(D))$, as in (5.2.51) we obtain from (5.2.55) and (5.2.56),

$$\|\tilde{L}^{-1}\| \|u^M - v^M|\mathcal{D}_2\| \le c_3 \|\tilde{L}^{-1}\| M \max(R + \|\tilde{L}^{-1}\|\eta, 1)^{2(M-1)} \cdot \|u-v|S\|,$$

$$(5.2.57)$$

where we put $\eta := \|f|\mathcal{D}\|$ and c_3 denotes the constant arising from our estimates (5.2.55) and (5.2.56) above. Now (5.2.40) together with (5.2.51) and (5.2.57) yields

$$\|\tilde{L}^{-1}(u^M - v^M)|S\| \le \|\tilde{L}^{-1}\| \|(u^M - v^M)|\mathcal{D}\|$$

$$\le c\|\tilde{L}^{-1}\| M \max(R + \|\tilde{L}^{-1}\|\eta, 1)^{M-1} \|u-v|S\|,$$

$$(5.2.58)$$

where $c = c_2 + c_3$. For $\tilde{L}^{-1} \circ N$ to be a contraction, we therefore require

$$c\varepsilon \|\tilde{L}^{-1}\| M \max(R + \|\tilde{L}^{-1}\|\eta, 1)^{2(M-1)} < 1,$$

cf. (5.2.39). In case of $\max(R + \|\tilde{L}^{-1}\|\eta, 1) = 1$ this leads to

$$\|\tilde{L}^{-1}\| < \frac{1}{c\varepsilon M}. \tag{5.2.59}$$

On the other hand, if $\max(R + \|\tilde{L}^{-1}\|\eta, 1) = R + \|\tilde{L}^{-1}\|\eta$, we choose $R = (r_0 - 1)\eta\|\tilde{L}^{-1}\|$, which gives rise to the condition

$$c\varepsilon\|\tilde{L}^{-1}\|M(r_0\|\tilde{L}^{-1}\|\eta)^{2(M-1)} < 1, \quad \text{i.e.,} \quad \eta^{2(M-1)}\|\tilde{L}^{-1}\|^{2M-1} < \frac{1}{c\varepsilon M}\left(\frac{1}{r_0}\right)^{2(M-1)}. \tag{5.2.60}$$

The next step is to show that $(\tilde{L}^{-1} \circ N)(B_R(\tilde{L}^{-1}f)) \subset B_R(\tilde{L}^{-1}f)$ in S. Since $(\tilde{L}^{-1} \circ N)(0) = \tilde{L}^{-1}(f - \varepsilon 0^M) = \tilde{L}^{-1}f$, we only need to apply the above estimate (5.2.58) with $v = 0$. This gives

$$\varepsilon\|\tilde{L}^{-1}u^M|S\| \leq c\varepsilon\|\tilde{L}^{-1}\|M\max(R + \|\tilde{L}^{-1}\|\eta, 1)^{2(M-1)}(R + \|\tilde{L}^{-1}\|\eta)$$

$$\overset{!}{\leq} R = (r_0 - 1)\eta\|\tilde{L}^{-1}\|,$$

which, in case that $\max(R + \|\tilde{L}^{-1}\|\eta, 1) = 1$, leads to

$$\|\tilde{L}^{-1}\| < \frac{r_0 - 1}{r_0}\left(\frac{1}{c\varepsilon M}\right), \tag{5.2.61}$$

whereas for $\max(R + \|\tilde{L}^{-1}\|\eta, 1) = R + \|\tilde{L}^{-1}\|\eta$ we get

$$\eta^{2(M-1)}\|\tilde{L}^{-1}\|^{2M-1} \leq \frac{1}{c\varepsilon M}(r_0 - 1)\left(\frac{1}{r_0}\right)^{2M-1}. \tag{5.2.62}$$

We see that condition (5.2.61) implies (5.2.59). Furthermore, since

$$(r_0 - 1)\left(\frac{1}{r_0}\right)^{2M-1} = \frac{r_0 - 1}{r_0}\left(\frac{1}{r_0}\right)^{2(M-1)} < \left(\frac{1}{r_0}\right)^{2(M-1)},$$

also condition (5.2.62) implies (5.2.60). Thus, by applying Banach's fixed point theorem in a sufficiently small ball around the solution of the corresponding linear problem, we obtain a unique solution of Problem II. $\qquad\square$

Remark 5.2.15 The restriction $m \geq \left[\frac{d}{2}\right] + 1$ in Theorem 5.2.14 comes from the fact that we require $s_2 = m > \frac{d}{2}$ in (5.2.53). For $d = 3$ this leads to $m \geq 2$. This assumption can probably be weakened, since we expect the solution to satisfy $u \in L_2([0, T], H^s(D))$ for all $s < \frac{3}{2}$, see also Remark 5.4.3 and the explanations given there.

Moreover, the restriction $a \geq \frac{d}{2} - 2$ in Theorem 5.2.14 comes from Theorem 2.4.9(ii) that we applied. Together with the restriction $a \in [-m, m]$ we are looking for $a \in [-\frac{1}{2}, m]$ if the domain is a corner domain, e.g. a smooth cone $K \subset \mathbb{R}^3$ (subject to some truncation). For polyhedral cones with edges M_k, $k = 1, \ldots, l$, we furthermore require $-\delta_-^{(k)} < a + 2m(\gamma_m - i) + m < \delta_+^{(k)}$ for $i = 0, \ldots, \gamma_m$ from Theorem 5.2.9.

5.2.4 Regularity Results in Kondratiev Spaces for Problem III

The Kondratiev spaces on generalized wedges we consider are defined as follows: For a wedge $\mathcal{K} = K \times \mathbb{R}^{d-m} \subset \mathbb{R}^d$ according to Definition 2.1.6 the singular set is $S = \{0\} \times \mathbb{R}^{d-m}$ with dimension $\delta = d - m$. In this case we put $\rho(x) = \min(|x'|, 1)$ where $x' = (x_1, \ldots, x_m) \in \mathbb{R}^m$ and get the corresponding Kondratiev spaces

$$\|u|\mathcal{K}_{p,a}^m(\mathcal{K})\| := \Big(\sum_{|\alpha| \leq m} \int_{\mathcal{K}} |\min(|x'|, 1)^{|\alpha|-a} \partial^\alpha u(x)|^p \, dx \Big)^{1/p}. \qquad (5.2.63)$$

Before formulating the regularity results in these Kondratiev spaces for Problem III we need to introduce the so-called critical exponents. Let Q_{r_0} denote the truncated cylinder

$$Q_{r_0}(0, t_0) := \mathcal{K}_0 \times (t_0 - r_0^2, t_0] \qquad (5.2.64)$$

with constants $r_0 > 0$ and $t_0 \in \mathbb{R}$, where $\mathcal{K}_0$ is the truncated wedge from (2.1.10).

Definition 5.2.16 (Critical Exponent) The *critical exponent* $\lambda_c^+ \equiv \lambda_c^+(\mathcal{K}, \mathcal{L})$ for the operator $\mathcal{L}$ and the generalized wedge $\mathcal{K}$ is defined as the supremum of all λ such that

$$|u(t, x)| \leq C(\lambda, \kappa) \left(\frac{|x'|}{R} \right)^\lambda \sup_{Q_{\kappa R}(0, t_0)} |u| \qquad \text{for} \quad (t, x) \in Q_{R/2}(0, t_0)$$

$$(5.2.65)$$

for a certain $\kappa \in (1/2, 1)$ independent of t_0, R, and u. This inequality must hold for all $t_0 \in \mathbb{R}$, $R > 0$ and $u \in \mathcal{V}_{\text{loc}}(Q_R(0, t_0))$, i.e., the space containing all functions having finite norm

$$\sup_{t \in (t_0 - R'^2, t_0]} \|u(t, \cdot)|L_2(B_{R'}(0))\| + \sum_{i=1}^n \left\| \frac{\partial}{\partial x_i} u|L_2(Q_{R'}(0, t_0)) \right\|$$

for all $R' \in (0, R)$, which additionally satisfy

$$\mathcal{L}u = 0 \quad \text{in} \quad Q_R(0, t_0), \qquad u\big|_{x \in \partial \mathcal{K}} = 0.$$

Furthermore, we define

$$\lambda_c^- \equiv \lambda_c^-(\mathcal{K}, \mathcal{L}) := \lambda_c^+(\mathcal{K}, \hat{\mathcal{L}}),$$

where $\hat{\mathcal{L}}$ is defined as $\mathcal{L}$ in (5.1.8) with $A_{ij}(t)$ replaced by $A_{ij}(-t)$.

Remark 5.2.17 We list some important properties and estimates for the critical exponents $\lambda_c^{\pm}$ for various geometries of $\mathcal{K}$, see [KN14, Sect. 2] for details.

   (i)   The definition (5.2.65) does not depend on $\kappa \in (1/2, 1)$.
   (ii) It can be shown that $\lambda_c^{\pm} > 0$.
   (iii) Let $\mathcal{K}_1 \subset \mathcal{K}_2$. Then $\lambda_c^+(\mathcal{K}_1, \mathcal{L}) \geq \lambda_c^+(\mathcal{K}_2, \mathcal{L})$ and $\lambda_c^-(\mathcal{K}_1, \mathcal{L}) \geq \lambda_c^-(\mathcal{K}_2, \mathcal{L})$.
   (iv) If $A_{ij}(t) = \delta_{ij}$ then

$$\lambda_c^{\pm} \equiv \lambda_1^+ := -\frac{m-2}{2} + \sqrt{\Lambda_1 + \frac{(m-2)^2}{4}}, \tag{5.2.66}$$

   where Λ_1 is the first eigenvalue of the Dirichlet boundary value problem of the Laplace-Beltrami operator on Ω.
   (v)   If K is an acute cone, i.e.,

$$\overline{K} \setminus \{0\} \subset \mathbb{R}_+^m = \{x' \in \mathbb{R}^m : x_1 > 0\},$$

   then $\lambda_c^{\pm} > 1$.
   (vi) If $K \to \mathbb{R}_+^m$, then $\lambda_c^{\pm} \to 1$.
   (vii) $\lambda_c^{\pm} \geq -\frac{m}{2} + \nu\sqrt{\Lambda_1 + \frac{(m-2)^2}{4}}$, where ν is the constant from (5.1.9). In particular, if $\Lambda_1 \to \infty$, then $\lambda_c^{\pm} \to \infty$.

Our aim is to study the regularity of the solution u of Problem III in generalized fractional Sobolev spaces. For this we rely on [KN14, Thm. 1.1], were the authors obtained results concerning the regularity of the solution u in generalized weighted Sobolev spaces. We reformulate their results in terms of our Kondratiev spaces from (5.2.63) as follows.

Theorem 5.2.18 (Kondratiev Regularity) *Let* $2 \leq m \leq d$, $1 < p, q < \infty$, *and let* $\mathcal{K} = K \times \mathbb{R}^{d-m}$ *be a generalized wedge according to Definition 2.1.6. Suppose that*

$$2 - \frac{m}{p} - \lambda_c^+ < \mu < m - \frac{m}{p} + \lambda_c^-,$$

where $\lambda_c^{\pm}$ are the critical exponents from Definition 5.2.16. Furthermore, assume $f \in L_q(\mathbb{R}, \mathcal{K}_{p,-\mu}^0(\mathcal{K}))$. Then there is a solution u of Problem III satisfying

$$\|u|L_q(\mathbb{R}, \mathcal{K}_{p,2-\mu}^2(\mathcal{K})) \cap W_q^1(\mathbb{R}, \mathcal{K}_{p,-\mu}^0(\mathcal{K}))\| \lesssim \|f|L_q(\mathbb{R}, \mathcal{K}_{p,-\mu}^0(\mathcal{K}))\|.$$

5.2.5 Regularity Results in Kondratiev Spaces for Problem IV

In this subsection we consider the fractional Kondratiev spaces $\mathfrak{K}_{p,a}^s(\mathcal{O})$ from Definition 2.4.18, where $\mathcal{O} \subset \mathbb{R}^d$ is a bounded Lipschitz domain. Concerning regularity estimates for Problem IV in these spaces, the following result was proven in [Kim09, Thm. 2.5]. For convenience we adapt the notation from the paper to our needs.

Theorem 5.2.19 (Kondratiev Regularity) *Let $p \in [2, \infty)$, $s \in [0, \infty)$, and Assumption 5.1.2 be satisfied. Then there exists $\beta_0 = \beta_0(p, d, \mathcal{O}) > 0$ such that for*

$$a \in \left(\frac{2-p-\beta_0}{p}, \frac{2-p+\beta_0}{p} \right), \tag{5.2.67}$$

any $f \in L_p([0,T], \mathfrak{K}_{p,a-1}^s(\mathcal{O}))$, and initial data $u_0 \in \mathfrak{K}_{p,a+\frac{p-2}{p}}^{s+2-\frac{2}{p}}(\mathcal{O})$, Problem IV admits a unique solution $u \in L_p([0,T], \mathfrak{K}_{p,a+1}^{s+2}(\mathcal{O}))$ with $\partial_t u \in L_p([0,T], \mathfrak{K}_{p,a-1}^s(\mathcal{O}))$. In particular, we have

$$\sum_{k=0}^{1} \left\| \partial_{t^k} u | L_p([0,T], \mathfrak{K}_{p,a+1-2k}^{s+2-2k}(\mathcal{O})) \right\|$$

$$\leq C \left(\|f|L_p([0,T], \mathfrak{K}_{p,a-1}^s(\mathcal{O}))\| + \|u_0|\mathfrak{K}_{p,a+\frac{p-2}{p}}^{s+2-\frac{2}{p}}(\mathcal{O})\| \right),$$

where $C = C(d, p, s, \theta, \delta_0, K, T, \mathcal{O})$.

Remark 5.2.20 In [KK04] similar results for C^1 domains were established. In particular, the condition on a from (5.2.67) in this case has to be replaced by

$$a \in \left(\frac{1}{p} - 1, \frac{1}{p} \right).$$

5.3   Regularity Results in Fractional Sobolev Spaces

In this section we turn our attention to Problem III. In particular, we investigate the spatial smoothness of the solutions in the fractional Sobolev scale H^s, $s \in \mathbb{R}$. As already outlined in the introduction, the regularity in these spaces is related with the approximation order that can be achieved by numerical schemes based on uniform grid refinements. Our results provide a first attempt to generalize the well-known $H^{3/2}$-Theorem of Jerison and Kenig [JK95] to parabolic PDEs. We are not as general since our setting in Problem III is restricted to wedges according to Definition 2.1.6, whereas the $H^{3/2}$-Theorem holds for general Lipschitz domains. On the other hand we cover more PDEs than just the heat equation.

Our main result is formulated in Theorem 5.3.1 in Sect. 5.3.1. The proof relies on regularity results in Kondratiev spaces from [KN14] and embeddings into the scale of Triebel-Lizorkin spaces derived in [HS18]. Afterwards we discuss our findings in detail, relate our work to corresponding results for elliptic PDEs, and compare it with similar results for the heat equation on polygons presented in [Gri92]. Moreover, in Example 5.3.4 we show that (at least) for the heat equation the upper bounds on the fractional Sobolev smoothness s in Theorem 5.3.1 are optimal.

Subsequently, in Sect. 5.3.2 we focus entirely on the heat equation where the underlying domain is a rotationally invariant cone. For this particular case we are able to give precise upper bounds for the fractional Sobolev smoothness s depending on the opening angle of the cone, cf. Theorem 5.3.6 together with the values provided in Table 5.1.

5.3.1   Fractional Sobolev Regularity of Problem III

Let us briefly explain our strategy. As before when investigating the Besov regularity of solutions to elliptic and parabolic PDEs our central ingredient will be to use regularity results in weighted Sobolev spaces for Problem III as stated in Theorem 5.2.18. We wish to combine these with the embedding from Theorem 2.4.16. One difficulty arises in this context: Theorem 5.2.18 holds for unbounded wedges $\mathcal{K} \subset \mathbb{R}^d$ whereas the embedding result in Theorem 2.4.16 is true for bounded Lipschitz domains $D \subset \mathbb{R}^d$. In order to avoid this problem we consider the truncated wedge $\mathcal{K}_0$ as defined in (2.1.10). Then, the additional difficulty occurs that the Kondratiev norm on the truncated wedge is not just defined by restriction. Instead, the distance to the new corners produced by the truncation from considering $\mathcal{K}_0$ instead of $\mathcal{K}$ have to be taken into account. We solve this problem by multiplying u with a radial cut-off function $\varphi \in C_0^\infty(\mathcal{K}_0)$ satisfying

$$\varphi(x) \equiv \varphi(x', x'') = \begin{cases} 1 & \text{on} \quad (B_{r_0-\varepsilon}^m(0) \cap K) \times B_{r_0-\varepsilon}^{d-m}(0), \\ 0 & \text{on} \quad \mathcal{K}_0 \setminus ((B_{r_0-\frac{\varepsilon}{2}}^m(0) \cap K) \times B_{r_0-\frac{\varepsilon}{2}}^{d-m}(0)), \end{cases} \tag{5.3.1}$$

Fig. 5.2 Illustration of
cut-off function φ when
$m = d = 2$

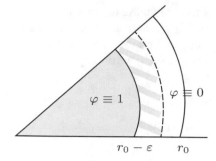

where $x = (x', x'') = (x_1, \ldots, x_d)$ is a point in $\mathbb{R}^d$, $x' = (x_1, \ldots, x_m) \in \mathbb{R}^m$ and
$x'' = (x_{m+1}, \ldots, x_d) \in \mathbb{R}^{d-m}$ (Fig. 5.2).

This truncation process does not induce serious restrictions for when it comes to
practical applications it is clear that only truncated wedges can be considered. Then
the regularity of φu corresponds to the regularity of u as stated in Theorem 5.2.18.
In particular, we obtain

$$\|\varphi u | L_q(\mathbb{R}, \mathcal{K}^2_{2,2-\mu}(\mathcal{K}_0)) \cap W^1_q(\mathbb{R}, \mathcal{K}^0_{2,-\mu}(\mathcal{K}_0))\|$$

$$\lesssim \|\varphi u | L_q(\mathbb{R}, \mathcal{K}^2_{2,2-\mu}(\mathcal{K})) \cap W^1_q(\mathbb{R}, \mathcal{K}^0_{2,-\mu}(\mathcal{K}))\|$$

$$\leq c_\varphi \|u | L_q(\mathbb{R}, \mathcal{K}^2_{2,2-\mu}(\mathcal{K})) \cap W^1_q(\mathbb{R}, \mathcal{K}^0_{2,-\mu}(\mathcal{K}))\|,$$

cf. Remark 2.4.3(vi). This procedure enables us to combine the regularity results in
Kondratiev spaces from Theorem 5.2.18 with the embedding from Theorem 2.4.16,
which shows that Kondratiev spaces (subject to some restrictions on the parameters)
are embedded in the scale of Triebel-Lizorkin spaces. Finally, using the fact that in
some cases Triebel-Lizorkin spaces may coincide with fractional Sobolev spaces we
obtain the following regularity result in the fractional Sobolev scale.

Theorem 5.3.1 (Fractional Sobolev Regularity) *Let* $2 \leq m \leq d$, $1 < q < \infty$,
$1 < p < \infty$, *and let* $\mathcal{K} = K \times \mathbb{R}^{d-m}$ *be a wedge according to Definition 2.1.6 with
singular set of dimension* $\delta = d - m$. *Suppose that*

$$2 - \frac{m}{p} - \lambda_c^+ < \mu < \min\left(m - \frac{m}{p} + \lambda_c^-, 2\right), \tag{5.3.2}$$

where $\lambda_c^\pm$ *are the critical exponents from Definition 5.2.16. Furthermore, assume*
$f \in L_q(\mathbb{R}, \mathcal{K}^0_{p,-\mu}(\mathcal{K}))$ *and let* φ *be the cut-off function from (5.3.1). Then there is
a solution* u *of Problem III satisfying*

$$\varphi u \in L_q(\mathbb{R}, H^s(\mathcal{K}_0)) \quad \text{for any} \quad s < 2 + \frac{d}{2} - d \cdot \max\left(\frac{1}{p}, \frac{\mu}{d-\delta} + \frac{1}{p}, \frac{1}{2}\right).$$

Proof We make use of Theorem 2.4.16. Here we consider the truncated wedge $\mathcal{K}_0$, which is a bounded Lipschitz domain of polyhedral type. By Theorem 5.2.18 and our assumptions we know that $\varphi u(\cdot, t) \in \mathcal{K}^2_{p,2-\mu}(\mathcal{K}_0)$. But then since $2 - \mu > 0$ from (2.4.20) we obtain

$$\mathcal{K}^2_{p,2-\mu}(\mathcal{K}_0) \hookrightarrow F^2_{\tau,2}(\mathcal{K}_0) \tag{5.3.3}$$

if $\tau < p$ and

$$2 - (2 - \mu) < (d - \delta)\left(\frac{1}{\tau} - \frac{1}{p}\right), \quad \text{i.e.,} \quad \frac{1}{\tau} > \frac{\mu}{d - \delta} + \frac{1}{p}.$$

Note that by the support properties of φu we may smoothen the boundary of $\mathcal{K}_0$ at the new corners and edges of the truncated wedge without any problem such that the singular set of the smoothed truncated wedge is the same as the singular set of the unbounded wedge $\mathcal{K}$. In particular, in this case we also have $\delta = d - m$. Now Sobolev's embedding for the Triebel-Lizorkin spaces from Proposition 2.3.4 yields for $s < 2$ and $\tau < 2$ that

$$F^2_{\tau,2}(\mathcal{K}_0) \hookrightarrow F^s_{2,2}(\mathcal{K}_0) = H^s(\mathcal{K}_0), \tag{5.3.4}$$

if $2 - \frac{d}{\tau} \geq s - \frac{d}{2}$, i.e., $s \leq 2 - \frac{d}{\tau} + \frac{d}{2}$. Since $\frac{1}{\tau} > \max\left(\frac{1}{p}, \frac{\mu}{d-\delta} + \frac{1}{p}, \frac{1}{2}\right)$ we obtain as upper bound

$$s < 2 + \frac{d}{2} - d \cdot \max\left(\frac{1}{p}, \frac{\mu}{d - \delta} + \frac{1}{p}, \frac{1}{2}\right), \tag{5.3.5}$$

from which we see that $s < 2$ is always satisfied. Combining the above results (5.3.3) and (5.3.4) yields

$$\varphi u \in L_q(\mathbb{R}, \mathcal{K}^2_{p,2-\mu}(\mathcal{K}_0)) \hookrightarrow L_q(\mathbb{R}, H^s(\mathcal{K}_0))$$

for all s satisfying (5.3.5), which completes the proof. $\square$

Remark 5.3.2 For $\mu \geq 0$ since $f \in L_q(\mathbb{R}, \mathcal{K}^0_{p,-\mu}(\mathcal{K})) \supset L_q(\mathbb{R}, \mathcal{K}^0_{p,0}(\mathcal{K})) = L_q(\mathbb{R}, L_p(\mathcal{K}))$ we may simply assume that the right hand side $f \in L_q(\mathbb{R}, L_p(\mathcal{K}))$. In particular, for $p = 2$ and $\delta = d - m$ the upper bound for s then reads as

$$s < 2 - \frac{d}{m}\mu. \tag{5.3.6}$$

Remark 5.3.3

(i) From Theorem 5.3.1 we obtain (similar as in the elliptic case) a shift by 2 in the fractional Sobolev scale when we consider equations of the form $\mathcal{L}u = f$. In particular, let (5.3.2) be satisfied. Then for right hand side

$f \in L_q(\mathbb{R}, \mathcal{K}^0_{p,-\mu}(\mathcal{K}))$ we have $\varphi f(\cdot, t) \in \mathcal{K}^0_{p,-\mu}(\mathcal{K}_0)$, where φ is the cut-off function defined in (5.3.1). Using (2.4.20) we obtain

$$\mathcal{K}^0_{p,-\mu}(\mathcal{K}_0) \hookrightarrow F^0_{\tau,2}(\mathcal{K}_0) \tag{5.3.7}$$

if $\tau < p$ and

$$\mu < (d - \delta)\left(\frac{1}{\tau} - \frac{1}{p}\right), \quad \text{i.e.,} \quad \frac{1}{\tau} > \frac{\mu}{d - \delta} + \frac{1}{p}.$$

Sobolev's embedding for the Triebel-Lizorkin spaces, cf. Proposition 2.3.4, yields for $s < 0$ and $\tau < 2$ that

$$F^0_{\tau,2}(\mathcal{K}_0) \hookrightarrow F^s_{2,2}(\mathcal{K}_0) = H^s(\mathcal{K}_0), \tag{5.3.8}$$

if $0 - \frac{d}{\tau} \geq s - \frac{d}{2}$, i.e., $s \leq \frac{d}{2} - \frac{d}{\tau}$. Since $\frac{1}{\tau} > \max\left(\frac{1}{p}, \frac{\mu}{d-\delta} + \frac{1}{p}, \frac{1}{2}\right)$ we obtain as upper bound

$$s < \frac{d}{2} - d \cdot \max\left(\frac{1}{p}, \frac{\mu}{d - \delta} + \frac{1}{p}, \frac{1}{2}\right), \tag{5.3.9}$$

and we see that $s < 0$ is always satisfied. Combining the above results (5.3.7) and (5.3.8) yields

$$\varphi f \in L_q(\mathbb{R}, \mathcal{K}^0_{p,-\mu}(\mathcal{K}_0)) \hookrightarrow L_q(\mathbb{R}, H^s(\mathcal{K}_0))$$

for all s satisfying (5.3.9). The results of Theorem 5.3.1 then imply that under these conditions we obtain a solution satisfying $\varphi u \in L_q(\mathbb{R}, H^{s+2}(\mathcal{K}_0))$.

(ii) For the special case when $p = 2$, $d = m = 3$, and $\delta = 0$ (which we may assume by the support of φ since in that case the singular set of $\mathcal{K} = K$ contains only the vertex of the cone), we obtain fractional Sobolev smoothness $s < -\max(\mu, 0)$ for f and $s < 2 - \max(\mu, 0)$ for the solution u. The situation is illustrated in Fig. 5.3. We obtain the maximal Sobolev regularity for the smallest possible value of μ. The condition (5.3.2) in this case reads as $\mu > \frac{1}{2} - \lambda_c^+$ with $\lambda_c^+ > 0$.

(iii) We compare our regularity results with related ones. Concerning the solutions to elliptic problems it is nowadays classical knowledge that their Sobolev regularity depends not only on the properties of the coefficients and the right-hand side, but also on the regularity/roughness of the boundary of the underlying domain. While for smooth coefficients and smooth boundaries we have $u \in H^{s+2}(\Omega)$ for $f \in H^s(\Omega)$, it is well-known that this becomes false for more general domains. In particular, if we only assume Ω to be a Lipschitz domain, then it was shown by Jerison and Kenig [JK95] that in general we only have $u \in H^s$ for all $s \leq 3/2$ for the solution of the Poisson equation, even for

Fig. 5.3 Shift from s to $s + 2$

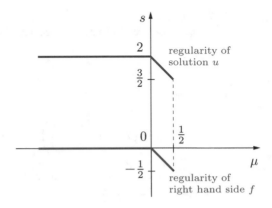

smooth right-hand side f. This behaviour is caused by singularities near the boundary. Moreover, this famous $H^{3/2}$-Theorem implies that the optimal rate of convergence for nonadaptive methods of approximation is just $3/2d$ as long as we do not impose further properties on Ω.

If, additionally, $\Omega \subset \mathbb{R}^2$ is a polygonal domain then it was shown in Grisvard [Gri92, Rem. 2.4.6] that the Poisson equation with Dirichlet boundary conditions, i.e.,

$$\Delta u = f \quad \text{in} \quad \Omega, \qquad u\big|_{\partial\Omega} = 0, \tag{5.3.10}$$

for $f \in L_2(\Omega)$ has a solution $u \in H^s(\Omega)$ for every $s < \frac{\pi}{\max_j \omega_j} + 1$, where ω_j denote the inner angles of the polygon. Furthermore, if we consider bounded polyhedral domains $\Omega \subset \mathbb{R}^3$, the results in Grisvard [Gri92, Sect. 2.6] imply that (5.3.10) has a solution $u \in H^s(\Omega)$ with $s \in (\frac{3}{2}, 2]$, see [Gri92, Cor. 2.6.7]. As in the two dimensional case again the upper bound of s depends on the inner angles of the polyhedron.

In particular, for (5.3.10) considered on corner domains in $\mathbb{R}^3$ in [Dau08] the upper bound is precisely determined by $s < \lambda_c^+ + \frac{3}{2}$.

Turning towards parabolic problems we now see that Theorem 5.3.1 gives corresponding results. Also in this situation μ depends on the inner angle of the cone and the smaller the angles the higher is the regularity. In this context we refer to the results on specific cones derived in Theorem 5.3.6, where this dependence becomes clearer. For the special case that $p = 2$, $d = m = 3$, and $\delta = 0$ already mentioned above, we see from Theorem 5.3.1 that for right hand side $f \in L_2(\mathbb{R}, \mathcal{K}^0_{-\mu,2}(\mathcal{K}))$ Eq. (5.1.7) has a solution

$$\varphi u \in L_2(\mathbb{R}, H^s(\mathcal{K}_0)) \qquad \text{with} \qquad s < 2 - \mu,$$

where $\mu > \frac{1}{2} - \lambda_c^+$ for $\lambda_c^+ \in \left(0, \frac{1}{2}\right]$, i.e., the best we can hope for in this case in general is $s < \min\left(\frac{3}{2} + \lambda_c^+, 2\right)$. We give an example below which demonstrates that these results are optimal, i.e., $s > \frac{3}{2} + \lambda_c^+$ is impossible even for smooth right hand sides f. In this sense, our results can be interpreted as a first step towards the generalization of the $H^{3/2}$-Theorem to parabolic problems. However, we remark that our findings are not as general as the $H^{3/2}$-Theorem from [JK95]. Firstly, the bounded wedges $\mathcal{K}_0$ we consider here are (only) special Lipschitz domains. Secondly, if $\mu < 0$ we have $f \in L_2(\mathbb{R}, \mathcal{K}_{2,-\mu}^0(\mathcal{K})) \subset L_2(\mathbb{R}, \mathcal{K}_{2,0}^0(\mathcal{K})) = L_2(\mathbb{R} \times \mathcal{K})$, so in this case we do not cover all right hand sides $f \in L_2(\mathbb{R} \times \mathcal{K})$.

Our results are also in good agreement with Grisvard [Gri92, Thm. 5.2.1]. There it was shown for the heat equation

$$\frac{\partial u}{\partial t} - \Delta u = f \quad \text{in} \quad \Omega \times (0, \infty), \qquad u\big|_{\partial\Omega \times (0,\infty)} = 0,$$

when $\Omega \subset \mathbb{R}^2$ is a bounded polygonal domain, that for right hand sides $f \in L_2((0, \infty) \times \Omega)$ there is a solution u satisfying

$$u \in L_2((0, \infty), H^s(\Omega)) \quad \text{with} \quad s < \frac{\pi}{\max_j \omega_j} + 1,$$

where ω_j again denote the inner angles of the polygon.

Example 5.3.4 (Optimality of Results for the Heat Equation)   The following problem was treated in [KM87] and indicates that our results are optimal. The authors consider the heat equation on a cone $K \subset \mathbb{R}^m$ as in Definition 2.1.6, where $\Omega = K \cap S^{m-1}$ is of class C^∞ and find representations for the coefficients in the asymptotic expansions of the solutions near the conical point. To be more precise, the following boundary value problem is considered:

$$\frac{\partial u}{\partial t} - \Delta u = 0 \quad \text{in} \quad K \times (0, \infty),$$

$$u\big|_{\partial K \times (0,\infty)} = 0,$$

$$u\big|_{t=0} = \varphi \quad \text{in} \quad K,$$

where $\varphi \in \mathcal{D}(\overline{K} \setminus \{0\})$, satisfying $\varphi(x) = 0$ for $x \in \partial K$. Moreover, let $\{\Lambda_j\}_{j\in\mathbb{N}}$ be the nondecreasing sequence of eigenvalues of the Laplace-Beltrami operator on Ω (with Dirichlet boundary condition) counted with their multiplicities, and let $\{\varphi_j\}_{j\in\mathbb{N}}$ be an orthonormal (in $L_2(\Omega)$) sequence of eigenfunctions corresponding to the eigenvalues Λ_j. Furthermore, by $\lambda_j^{\pm}$ we denote the solutions of the quadratic equation $\lambda(\lambda + d - 2) = \Lambda_j$. In particular, it follows that $\lambda_1^+ \equiv \lambda_c^{\pm}$ from (5.2.66) in this case. Let M be the largest integer such that $\lambda_1^+ \leq \lambda_2^+ \leq \ldots \leq \lambda_M^+ < \frac{1}{2}$. Then the authors show in [KM87] that the solution u is given by

$$u(t, x) = \frac{1}{2} \sum_{j=1}^{M} r^{\lambda_j^+} \varphi_j(\omega) \sum_{k=0}^{m_j} \frac{\left(r^2 \frac{\partial}{\partial t}\right)^k h_j(t)}{k! \Gamma(k + \sigma_j + 1)} + \mathcal{O}(r^{\lambda_{M+1}^+}), \qquad (5.3.11)$$

where $\sigma_j = (\lambda_j^+ - \lambda_j^-)/2, r < 1, m_j = \left[(\lambda_{M+1}^+ - \lambda_j^+)/2\right]$, and

$$h_j(t) = (2\sqrt{t})^{-\lambda_j^+} \int_K \varphi(2\sqrt{t}x) r^{\lambda_j^+} e^{-r^2} \varphi_j(\omega) dx.$$

From the representation (5.3.11) we see that the singularity of the solution u near the vertex behaves like $r^{\lambda_1^+} = r^{\lambda_c^+}$. Therefore, $u(\cdot, t)$ belongs locally to $H^s(K)$ if for some constant $c > 0$ we have

$$\int_0^c r^{2(\lambda_c^+ - s)} r^{m-1} dr \sim \left[r^{2(\lambda_c^+ - s) + m} \right]_{r=0}^c < \infty,$$

which holds for $2(\lambda_c^+ - s) + m > 0$, i.e., $s < \frac{m}{2} + \lambda_c^+$ and shows that Theorem 5.3.1 gives the best result possible in this case. Note that there is the slight discrepancy that the results from [KM87] hold for time axis $t \in (0, \infty)$ whereas our results are established for $t \in \mathbb{R}$. However, this should be immaterial in the context when regarding spacial regularity of the solution u.

Until now we have focused on the spacial regularity of the solution of Problem III. However, as an immediate consequence of Theorem 5.2.18 in combination with Sobolev's embedding theorem we obtain the following Hölder regularity in time.

Corollary 5.3.5 (Hölder Regularity in Time) *Let* $2 \leq m \leq d, 1 < p, q < \infty$, *and let* $\mathcal{K} = K \times \mathbb{R}^{d-m}$ *be a wedge according to Definition 2.1.6. Suppose that*

$$2 - \frac{m}{p} - \lambda_c^+ < \mu < m - \frac{m}{p} + \lambda_c^-,$$

where $\lambda_c^\pm$ *are the critical exponents from Definition 5.2.16. Furthermore, assume* $f \in L_q(\mathbb{R}, \mathcal{K}_{p,-\mu}^0(\mathcal{K}))$. *Then there is a solution* u *of Problem III satisfying*

$$u \in C^{0, 1 - \frac{1}{q}}(\mathbb{R}, \mathcal{K}_{p,-\mu}^0(\mathcal{K})).$$

Proof Theorem 5.2.18 together with Sobolev's embedding theorem for Banach-space valued functions, cf. Theorem 2.2.5, yields

$$u \in L_q(\mathbb{R}, \mathcal{K}^2_{p,2-\mu}(\mathcal{K})) \cap W^1_q(\mathbb{R}, \mathcal{K}^0_{p,-\mu}(\mathcal{K})) \hookrightarrow W^1_q(\mathbb{R}, \mathcal{K}^0_{p,-\mu}(\mathcal{K}))$$

$$\hookrightarrow C^{0,1-\frac{1}{q}}(\mathbb{R}, \mathcal{K}^0_{p,-\mu}(\mathcal{K})),$$

which completes the proof. □

5.3.2 Fractional Sobolev Regularity of the Heat Equation on Specific Smooth Cones

As a special important case of Problem III we now consider the heat equation

$$\frac{\partial}{\partial t}u - \Delta u = f \quad \text{in } K \times \mathbb{R}, \qquad u\big|_{\partial K \times \mathbb{R}} = 0, \tag{5.3.12}$$

on the cone $K \subset \mathbb{R}^3$ (choosing $m = d = 3$ in Definition 2.1.6) and assume that $\Omega = K \cap S^2$ is of class $C^{1,1}$. Using (5.2.66) we see that the critical exponents coincide in this case. Moreover, we have

$$\lambda^{\pm}_c = \lambda^+_1 := -\frac{1}{2} + \sqrt{\Lambda_1 + \frac{1}{4}}, \tag{5.3.13}$$

where Λ_1 is the first eigenvalue of the Dirichlet problem of the Laplace-Beltrami operator in Ω. Regarding the fractional Sobolev regularity of the solution u of (5.3.12) for $p = q = 2$ Theorem 5.3.1 now implies the following.

Theorem 5.3.6 (Fractional Sobolev Regularity of Heat Equation) *Let $K \subset \mathbb{R}^3$ be a smooth cone according to Definition 2.1.6 (choosing $m = d = 3$). Suppose*

$$\max\left(\frac{1}{2} - \lambda^+_1, 0\right) < \mu < \min\left(\frac{3}{2} + \lambda^+_1, 2\right),$$

where $\mu = 0$ is allowed if $\frac{1}{2} - \lambda^+_1 < 0$ and λ^+_1 is defined in (5.3.13). Furthermore, assume $f \in L_2(\mathbb{R} \times K)$ and let φ be the cut-off function from (5.3.1). Then there is a solution u of (5.3.12) satisfying

$$\varphi u \in L_2(\mathbb{R}, H^s(K_0)) \quad \text{for any} \quad s < 2 - \mu.$$

Eigenvalues of the Laplacian on a Spherical Cap

We are interested in exact values of λ_1^+ for particular cones $K \subset \mathbb{R}^3$ in Theorem 5.3.6. In view of (5.3.13), for this we need to determine the first eigenvalue Λ_1 of the Dirichlet problem of the Laplace-Beltrami operator on $\Omega = K \cap S^2$. In the particular case of $\Omega = K \cap S^2$ being a spherical cap precise results are known. Therefore, we now investigate the problem

$$\left\{ \begin{array}{rcl} \Delta_{S^2} w + \Lambda w & = & 0 \quad \text{in } \Omega_{\theta_0} \times \mathbb{R}, \\ w & = & 0 \quad \text{on } \partial\Omega_{\theta_0} \times \mathbb{R}, \end{array} \right\} \tag{5.3.14}$$

where $\Omega_{\theta_0} := K \cap S^2$ is a *spherical cap*, which in polar coordinates is expressed as

$$\left\{ \begin{array}{l} y_1 = \sin\theta \sin\varphi, \\ y_2 = \sin\theta \cos\varphi, \\ y_3 = \cos\theta, \end{array} \right.$$

i.e., $\Omega_{\theta_0} = \{(y_1, y_2, y_3) \in S^2 : \theta \in (0, \theta_0), \varphi \in [0, 2\pi]\}$ (Figs. 5.4 and 5.5). Since Ω_{θ_0} is a compact Riemannian manifold with smooth boundary it follows from [Tay11, Cor. 1.5] that the eigenfunctions w of the Laplace-Beltrami operator in (5.3.14) are smooth.

The operator $\Delta_{S^2} + \lambda$ in this coordinates reads as

$$\Delta_{S^2} w + \Lambda w = \frac{1}{\sin\theta} \frac{\partial}{\partial\theta} \left(\frac{\partial w}{\partial\theta} \sin\theta \right) + \frac{1}{\sin^2\theta} \frac{\partial^2 w}{\partial\varphi^2} + \Lambda w.$$

Solutions to (5.3.14) are expressed by using separation of variables,

$$w(\theta, \varphi) = P(x)\Phi(\varphi) \qquad \text{with} \qquad x = \cos\theta.$$

Fig. 5.4 Spherical cap, angle $\theta_0 < \frac{\pi}{2}$

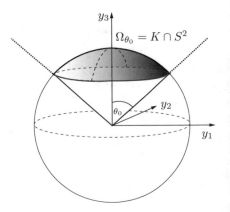

Fig. 5.5 Spherical cap, angle
$\theta_0 > \frac{\pi}{2}$

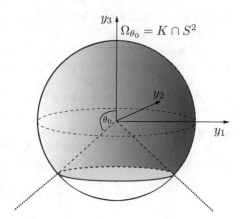

With this we see that

$$0 = \Delta_{S^2} w + \Lambda w$$

$$= \frac{1}{\sin\theta} \frac{\partial}{\partial\theta} \left(\frac{\partial P(\cos\theta)}{\partial\theta} \sin\theta \right) \Phi(\varphi) + \frac{1}{\sin^2\theta} \frac{\partial^2\Phi(\varphi)}{\partial\varphi^2} P(\cos\theta) + \Lambda P(\cos\theta)\Phi(\varphi)$$

$$= \left((1 - x^2) \frac{\partial^2 P}{\partial x^2} - 2x \frac{\partial P}{\partial x} \right) \Phi(\varphi) + \frac{1}{1 - x^2} \frac{\partial^2\Phi(\varphi)}{\partial\varphi^2} P(x) + \Lambda P(x)\Phi(\varphi).$$

$$(5.3.15)$$

Recall that the eigenfunctions $w(\theta, \varphi) = P(x)\Phi(\varphi)$ are smooth (and from [Pee57, Thm. 3] is follows that the eigenfunction corresponding to the smallest positive eigenvalue Λ_1 has no roots). Thus, we can multiply both sides of (5.3.15) by $(1 - x^2)/(P(x)\Phi(\varphi))$ and separating variables yields

$$\frac{1 - x^2}{P(x)} \left((1 - x^2) \frac{\partial^2 P}{\partial x^2} - 2x \frac{\partial P}{\partial x} \right) + \Lambda(1 - x^2) = -\frac{\partial^2\Phi(\varphi)}{\partial\varphi^2} \frac{1}{\Phi(\varphi)} =: m^2.$$

$$(5.3.16)$$

By the regularity of the solutions, $|P(1)| < \infty$, $\Phi(0) = \Phi(2\pi)$, and $\Phi'(0) = \Phi'(2\pi)$ must be satisfied. For convenience below we define $\nu \geq 0$ satisfying $\Lambda := \nu(\nu + 1)$. From (5.3.16) for $P(x)$ and $\Phi(\varphi)$ we see that

$$(1 - x^2) \frac{d^2 P}{dx^2} - 2x \frac{dP}{dx} + \left(\nu(\nu + 1) - \frac{m^2}{1 - x^2} \right) P = 0, \qquad (5.3.17)$$

and

$$\frac{d^2\Phi}{d\varphi^2} + m^2\Phi = 0. \qquad (5.3.18)$$

From the periodicity of $\Phi(\varphi)$ it follows that m is a non-negative integer and any solutions to (5.3.18) are expressed as $\Phi(\varphi) = c_1 \cos(m\varphi) + c_2 \sin(m\varphi)$.

Furthermore, (5.3.17) is known as the *associated Legendre equation*, which has two kinds of (linearly independent) solutions $P = P_\nu^m(x)$ and $Q_\nu^m(x)$ such that $|P_\nu^m(1)| < \infty$ and $|Q_\nu^m(x)| \to \infty$ as $x \to 1$, respectively. From the condition $|P(1)| < \infty$, we only have to treat $P = P_\nu^m(x)$ and, in conclusion, $\Lambda = \nu(\nu+1)$ and $c_1 P_\nu^m(\cos\theta)\cos(m\varphi) + c_2 P_\nu^m(\cos\theta)\sin(m\varphi)$ are eigenvalues and eigenfunctions of $\Delta_{S^2} w + \Lambda w = 0$ on S^2, respectively.

In order to solve the eigenvalue problem (5.3.14), we are required to find solutions to (5.3.17) satisfying the boundary condition

$$P(\cos\theta_0) = 0. \tag{5.3.19}$$

For any fixed $m = 0, 1, 2, \ldots$ there exist infinitely many $\Lambda = \nu(\nu + 1)$ satisfying (5.3.17) and (5.3.19). Since we are interested in the smallest positive eigenvalue Λ_1 of (5.3.14), we have to look for the smallest root $\nu_{\min}$ of the Legendre function P_ν^m such that $P_\nu^m(\cos\theta_0) = 0$. Moreover, from (5.3.13) we see that

$$\lambda_1^+ = -\frac{1}{2} + \sqrt{\nu_{\min}(\nu_{\min} + 1) + \frac{1}{4}} = -\frac{1}{2} + \sqrt{\left(\nu_{\min} + \frac{1}{2}\right)^2} = \nu_{\min},$$

hence, λ_1^+ in this case is determined by the minimal root $\nu_{\min}$ of the Legendre function P_ν^m with respect to the angle θ_0. Table 5.1 below is taken from [Bau86], where the roots of the Legendre functions P_ν^m for different values of θ_0 were computed.

Now we obtain the following regularity result for the heat equation in the special case that K is a convex cone and $\Omega_\theta = K \cap S$ a spherical cap.

Corollary 5.3.7 (Fractional Sobolev Regularity of the Heat Equation on Convex Cones) *Let $K \subset \mathbb{R}^3$ be a convex cone such that $\Omega_{\theta_0} = K \cap S$ is a spherical cap. Assume $f \in L_2(\mathbb{R} \times K)$. Furthermore, let φ be the cut-off function from (5.3.1). Then there is a solution u of (5.3.12) satisfying*

$$\varphi u \in L_2(\mathbb{R}, H^s(K_0)) \quad \text{for any} \quad s < 2.$$

Proof If K is convex, then $\theta_0 \le \frac{\pi}{2}$. But then the values in Table 5.1 imply that $\lambda_1^+ \ge 1$. Therefore, we can choose $\mu = 0$ in Theorem 5.3.6, which completes the proof. $\square$

Remark 5.3.8

(i) A closer look at Table 5.1 reveals that Corollary 5.3.7 remains true for cones K with $\Omega_{\theta_0} = K \cap S$ being a spherical cap and angles $\theta_0 \in [0, 130°)$, since we have $\lambda_1^+ > \frac{1}{2}$ in this case and can therefore choose $\mu = 0$ in Theorem 5.3.6.

(ii) If $0 < \lambda_1^+ \leq 1/2$ Theorem 5.3.6 yields that $\mu > \frac{1}{2} - \lambda_1^+$. Then for $f \in L_2(\mathbb{R}, \mathcal{K}_{2,-\frac{1}{2}+\lambda_1^+}^0(\mathcal{K}))$ and φ being the cut-off function from (5.3.1), the solution u of (5.3.12) satisfies

$$\varphi u \in L_2(\mathbb{R}, H^s(\mathcal{K}_0)) \quad \text{with} \quad s < \frac{3}{2} + \lambda_1^+.$$

(iii) Our results in this section on the fractional Sobolev regularity of Problem III strongly indicate that it might be possible to improve the Besov regularity results for Problems I and II from Sect. 5.4 (although in Problems I and II the time interval is $[0, T]$ for some $T > 0$ whereas for Problem III we consider $\mathbb{R}$) as follows.

The results from Theorem 5.3.6 together with Example 5.4.4 suggest that if $f \in L_2(\mathbb{R}, \mathcal{K}_{2,-\mu}^0(K))$ with $\mu > \max\left(\frac{1}{2} - \lambda_1^+, 0\right)$ and φ being the cut-off function from (5.3.1), then the solution u of the heat equation on a truncated smooth cone $K_0 \subset \mathbb{R}^3$ satisfies

$$\varphi u \in L_2(\mathbb{R}, B_{\tau,\infty}^\gamma(K_0)) \qquad \text{for all} \qquad \frac{1}{2} < \frac{1}{\tau} < \frac{\gamma}{3} + \frac{1}{2}, \qquad (5.3.20)$$

where for the Besov regularity γ we derive the upper bound

$$\gamma < 3(2 - \mu) < \begin{cases} 6 & \text{if } \lambda_1^+ > \frac{1}{2}, \\ \frac{9}{2} + 3\lambda_1^+ & \text{if } 0 < \lambda_1^+ \leq \frac{1}{2}. \end{cases}$$

This implies that on truncated cones K_0 we might always have $\gamma < \frac{9}{2}$ whereas in Example 5.4.4 we only obtain $\gamma < 3$ for non-convex cones.

Moreover, the results from Theorem 5.3.6 can also help to improve the nonlinear Sobolev regularity results from Theorem 5.2.14. There we need the restriction that $m \geq 2$ for the Sobolev regularity of the solution, i.e., $u(\cdot, t) \in H^m(D)$. The reason for this was a multiplier result used in the proof, cf. formula (5.2.53). This multiplier result can now be reformulated with the help of the spaces H^s as long as $s > \frac{3}{2}$.

Finally, let us mention that the obtained results hint, that when it comes to numerical schemes providing constructive approximations of the solutions, adaptive schemes usually would outperform uniform ones. The reason for this lies in the fact that as a role of thumb the convergence order that can be achieved by adaptive algorithms is determined by the regularity γ of the exact solution in the adaptivity scale of Besov spaces (5.3.20), whereas the convergence order for uniform schemes depends on the classical Sobolev smoothness s. From the considerations above we have $\gamma < 3(2 - \mu)$ for the Besov smoothness in contrast to $s < 2 - \mu$ for the smoothness in the fractional Sobolev scale.

5.4   Regularity Results in Besov Spaces

With all preliminary work, in this section we finally come to the presentation of the regularity results in Besov spaces for Problems I, II, and IV, which form the center of our investigations in this chapter. For this purpose, we rely on the results from Sect. 5.2 on regularity in Sobolev and Kondratiev spaces for the respective problems and combine these with the embeddings of Kondratiev spaces into Besov spaces presented in Sect. 2.4.4. It turns out that in all cases studied the Besov regularity is higher than the Sobolev regularity. This indicates that adaptivity pays off when solving these problems numerically. The implications of our results in this respect will be further discussed in Chap. 7.

The Sobolev regularity we are working with (e.g. see Theorem 5.2.2 for Problem I) canonically comes out from the variational formulation of the problem, i.e., we have spacial Sobolev regularity m if the corresponding differential operator is of order $2m$. We give an outlook on how our results could be improved by using regularity results in fractional Sobolev spaces as presented in Sect. 5.3 instead. However, our findings presented in Sect. 5.3 are still at a very early stage: It is planned to do further investigations in this direction in the future.

As an important special case we focus on the heat equation in Example 5.4.4 and give precise upper bounds for the Besov regularity in this setting. Moreover, we again discuss the role of the weight parameter a appearing in our Kondratiev spaces to some extend.

5.4.1   Besov Regularity of Problem I

A combination of Theorem 5.2.9 (Kondratiev regularity A) and the embedding in Theorem 2.4.12 yields the following Besov regularity of Problem I.

Theorem 5.4.1 (Parabolic Besov Regularity A) *Let D be a bounded polyhedral domain in $\mathbb{R}^d$, $d = 3$. Let $\gamma \in \mathbb{N}$ with $\gamma \geq 2m$ and put $\gamma_m := \left\lceil \frac{\gamma-1}{2m} \right\rceil$. Furthermore, let $a \in \mathbb{R}$ with $a \in [-m, m]$. Assume that the right hand side f of Problem I satisfies*

(i) $\partial_{t^k} f \in L_2(D_T) \cap L_2([0, T], \mathcal{K}^{2m(\gamma_m-k)}_{2,a+2m(\gamma_m-k)}(D))$, $k = 0, \ldots, \gamma_m$, *and* $\partial_{t^{\gamma_m+1}} f \in L_2(D_T)$.

(ii) $\partial_{t^k} f(x, 0) = 0$, $k = 0, 1, \ldots, \gamma_m$.

Furthermore, let Assumption 5.2.5 hold for weight parameters $b = a + 2m(\gamma_m - i)$, where $i = 0, \ldots, \gamma_m$, and $b' = -m$. Then for the weak solution $u \in H^{m,\gamma_m+2*}(D_T)$ of Problem I, we have

$$u \in L_2([0, T], B^{\eta}_{\tau,\infty}(D)) \qquad \text{for all} \quad 0 < \eta < \min\left(\gamma, \frac{d}{\delta}m\right), \qquad (5.4.1)$$

where $\frac{1}{2} < \frac{1}{\tau} < \frac{\eta}{d} + \frac{1}{2}$ and δ denotes the dimension of the singular set of D. In particular, for any η satisfying (5.4.1) and τ as above, we have the a priori estimate

$$\|u|L_2([0, T], B^{\eta}_{\tau,\infty}(D))\|$$

$$\lesssim \sum_{k=0}^{\gamma_m} \|\partial_{t^k} f|L_2([0, T], \mathcal{K}^{2m(\gamma_m-k)}_{2,a+2m(\gamma_m-k)}(D))\| + \sum_{k=0}^{\gamma_m+1} \|\partial_{t^k} f|L_2(D_T)\|.$$

Proof According to Theorem 5.2.9 by our assumptions we know $u \in L_2([0, T], \mathcal{K}^{2m(\gamma_m+1)}_{2,a+2m(\gamma_m+1)}(D))$. Together with Theorem 2.4.12 (choosing $k = 0$) we obtain

$$u \in L_2([0, T], \mathcal{K}^{2m(\gamma_m+1)}_{2,a+2m(\gamma_m+1)}(D)) \cap H^{m,\gamma_m+2*}(D_T)$$

$$\hookrightarrow L_2([0, T], \mathcal{K}^{2m(\gamma_m+1)}_{2,a+2m(\gamma_m+1)}(D)) \cap L_2([0, T], H^m(D))$$

$$\hookrightarrow L_2([0, T], \mathcal{K}^{2m(\gamma_m+1)}_{2,a+2m(\gamma_m+1)}(D)) \cap L_2([0, T], B^m_{2,\infty}(D))$$

$$\hookrightarrow L_2([0, T], \mathcal{K}^{\eta}_{2,a+2m(\gamma_m+1)}(D) \cap B^m_{2,\infty}(D))$$

$$\hookrightarrow L_2([0, T], B^{\eta}_{\tau,\infty}(D)),$$

where in the third step we use the fact that $2m(\gamma_m + 1) \geq 2m\left(\frac{\gamma}{2m} - 1 + 1\right) = \gamma$ and choose $\eta \leq \gamma$. Moreover, the condition on a from Theorem 2.4.12 yields

$$m = \min(m, a + 2m(\gamma_m + 1)) > \frac{\delta}{d}\eta.$$

Therefore, the upper bound for η is

$$\eta < \min\left(\gamma, \frac{d}{\delta}m\right).$$

Concerning the restriction on τ, Theorem 2.4.12 with $\tau_0 = 2$ gives

$$\frac{1}{2} < \frac{1}{\tau} < \frac{1}{\tau^*} = \frac{\eta}{d} + \frac{1}{2}.$$

This completes the proof. $\qquad\square$

Remark 5.4.2 (The Parameter a) We discuss the role of the weight parameter in our Kondratiev spaces: Note that on the one hand we require $a + 2m(\gamma_m + 1) > 0$ in order to apply the embedding from Theorem 2.4.12. Since we assume $a \in [-m, m]$ this is always true. On the other hand it should be expected that the derivatives of the solution u have singularities near the boundary of the polyhedral domain. Thus, looking at the highest derivative of $u(t) \in \mathcal{K}^{2m(\gamma_m+1)}_{2,a+2m(\gamma_m+1)}(D)$ we see that we require

$$\sum_{|\alpha|=2m(\gamma_m+1)} \int_D \rho^{-ap}(x)|\partial^\alpha u(t,x)|^p dx < \infty,$$

hence, if $a < 0$ the derivatives of the solution u might be unbounded near the boundary of D. From this it follows that the range

$$-m < a < 0$$

is the most interesting for our considerations.

Remark 5.4.3 The above theorem relies on the fact that Problem I has a weak solution $u \in H^{m,\gamma_m+2*}(D_T) = H^{\gamma_m+1}([0,T], \overset{\circ}{H}{}^m(D)) \cap H^{\gamma_m+2}([0,T], H^{-m}(D)) \hookrightarrow L_2([0,T], H^m(D))$, cf. Theorem 5.2.3. We strongly believe that (in good agreement with the elliptic case) this result can be improved by studying the regularity of Problem I in fractional Sobolev spaces $H^s(D)$. In this case (assuming that the weak solution of Problem I satisfies $u \in L_2([0,T], H^s(D))$ for some $s > 0$) under the assumptions of Theorem 5.4.1, using Theorems 5.2.9 and 2.4.12 (with $k = 0$), we would obtain

$$u \in L_2([0,T], \mathcal{K}^\eta_{2,a'}(D)) \cap L_2([0,T], H^s(D)) \hookrightarrow L_2([0,T], B^\eta_{\tau,\infty}(D)), \tag{5.4.2}$$

where $a' = a + 2m(\gamma_m + 1) \geq a + 2m$ and again $\frac{1}{2} < \frac{1}{\tau} < \frac{\eta}{d} + \frac{1}{2}$ but the restriction on η now reads as

$$\eta < \frac{d}{\delta} \min(s, a'). \tag{5.4.3}$$

For general Lipschitz domains $D \subset \mathbb{R}^3$ we expect that the solution of Problem I (for $m = 1$) is contained in $H^s(D)$ for all $s < \frac{3}{2}$ (as in the elliptic case, Theorem 3.2.21). This would lead to $\eta < \frac{9}{2}$ when $\delta = 1$. For convex domains it probably even holds that $s = 2$ (for the heat equation this was already proven in [Woo07]). First results in this direction can be found in Sect. 5.3, where the underlying domain is a generalized wedge according to Definition 2.1.6. In this context we also refer to the discussion in Remark 5.3.8(iii). More general domains will be studied in the future.

Example 5.4.4 As a parabolic model case for Problem I we consider the heat equation

$$\partial_t u - \Delta u = f \quad \text{on } D_T,$$
$$u\big|_{t=0} = 0 \quad \text{on } D,$$

where $D \subset \mathbb{R}^3$ is a bounded polyhedral domain ($\delta = 1$) and f satisfies the assumptions of Theorem 5.4.1 for some $\gamma \geq 2m$ that we can choose large enough. In particular, since $m = 1$ we see that e.g. for $\gamma = 6$ we have $\gamma_m = \left\lceil \frac{\gamma-1}{2m} \right\rceil = 2$. In this case, the restriction on a in Remark 5.2.11, cf. formula (5.2.32), reads as

$$-1 < a < \min\left(1, \frac{\pi}{\theta_k} - 5\right),$$

where the upper bound is 1 for small angles $\theta_k < \frac{\pi}{6}$. But then $a' = a + 2m(\gamma_m + 1) \geq 5$ since $a \in (-1, 1)$, hence, (5.4.2) holds for

$$\eta < 3\min(s, a') = 3\min(s, 5).$$

If we assume Sobolev regularity $s = m = 1$ of the solution, together what was said above Theorem 5.4.1 yields the maximal Besov regularity

$$\eta < \min(3s, \gamma) = 3, \tag{5.4.4}$$

for the solution u of the heat equation. Furthermore, since the angles $\theta_k < \frac{\pi}{6}$ have to be small, i.e., our polyhedral domain D is convex, we can do even better: In this case we have $s = 2$, cf. [Woo07, Thm. 6.2], thus, the upper bound for the maximal Besov regularity in this case actually is

$$\eta < \min(3s, \gamma) = 3 \cdot 2 = 6.$$

Alternatively, we combine Theorem 5.2.12 (Kondratiev regularity B) and Theorem 2.4.12. This leads to the following regularity result in Besov spaces.

Theorem 5.4.5 (Parabolic Besov Regularity B) *Let D be a bounded polyhedral domain in $\mathbb{R}^d$, $d = 3$. Let $\gamma \in \mathbb{N}$ with $\gamma \geq 2m$. Moreover, let $a \in \mathbb{R}$ with $a \in [-m, m]$. Assume that the right hand side f of Problem I satisfies*

(i) $f \in \bigcap_{l=0}^{\infty} H^l([0, T], L_2(D) \cap \mathcal{K}_{2,a}^{\gamma-2m}(D))$.
(ii) $\partial_{t^l} f(x, 0) = 0, \quad l \in \mathbb{N}_0$.

Furthermore, let Assumption 5.2.5 hold for weight parameters $b = a$ and $b' = -m$. Then for the weak solution $\bigcap_{l=0}^{\infty} u \in H^{m,l+1*}(D_T)$ of Problem I, we have

$$u \in L_2([0, T], B_{\tau,\infty}^{\eta}(D)) \quad \text{for all} \quad 0 < \eta < \min\left(\gamma, \frac{d}{\delta}m\right), \tag{5.4.5}$$

where $\frac{1}{2} < \frac{1}{\tau} < \frac{\eta}{d} + \frac{1}{2}$ *and* δ *denotes the dimension of the singular set of D. In particular, for any* η *satisfying* (5.4.5) *and* τ *as above, we have the a priori estimate*

$$\|u|L_2([0, T], B_{\tau,\infty}^{\eta}(D))\|$$

$$\lesssim \sum_{k=0}^{\gamma-2m} \|\partial_{t^k} f|L_2([0, T], \mathcal{K}_{2,a}^{\gamma-2m}(D))\| + \sum_{k=0}^{(\gamma-2m)+1} \|\partial_{t^k} f|L_2(D_T)\|.$$

Proof According to Theorem 5.2.12 by our assumptions we know $u \in L_2([0, T], \mathcal{K}_{2,a+2m}^{\gamma}(D))$. Together with Theorem 2.4.12 (choosing $k = 0$) we obtain

$$u \in L_2([0, T], \mathcal{K}_{2,a+2m}^{\gamma}(D)) \cap H^{m,1*}(D_T)$$

$$\hookrightarrow L_2([0, T], \mathcal{K}_{2,a+2m}^{\gamma}(D)) \cap L_2([0, T], H^m(D))$$

$$\hookrightarrow L_2([0, T], \mathcal{K}_{2,a+2m}^{\gamma}(D)) \cap L_2([0, T], B_{2,\infty}^m(D))$$

$$\hookrightarrow L_2([0, T], \mathcal{K}_{2,a+2m}^{\eta}(D) \cap B_{2,\infty}^m(D))$$

$$\hookrightarrow L_2([0, T], B_{\tau,\infty}^{\eta}(D)),$$

where $\eta \leq \gamma$ in the second but last line. Moreover, the condition on the parameter 'a' from Theorem 2.4.12 yields

$$m = \min(m, a + 2m) > \frac{\delta}{d}\eta.$$

Therefore, the upper bound for η is

$$\eta < \min\left(\gamma, \frac{d}{\delta}m\right).$$

Concerning the restriction on τ, Theorem 2.4.12 with $\tau_0 = 2$ gives

$$\frac{1}{2} < \frac{1}{\tau} < \frac{1}{\tau^*} = \frac{\eta}{d} + \frac{1}{2}.$$

This finishes the proof.   □

5.4.2   Besov Regularity of Problem II

Concerning the Besov regularity of Problem II, we proceed in the same way as before for Problem I: Combining Theorem 5.2.14 (Nonlinear Sobolev and

Kondratiev regularity) with the embeddings from Theorem 2.4.12 we derive the following result.

Theorem 5.4.6 (Nonlinear Besov Regularity) *Let the assumptions of Theorems 5.2.14 and 5.2.9 be satisfied. In particular, as in Theorem 5.2.14 for* $\eta := \|f|\mathcal{D}\|$ *and* $r_0 > 1$, *we choose* $\varepsilon > 0$ *so small that*

$$\eta^{2(M-1)}\|\tilde{L}^{-1}\|^{2M-1} \le \frac{1}{c\varepsilon M}(r_0 - 1)\left(\frac{1}{r_0}\right)^{2M-1}, \qquad \text{if} \quad r_0\|\tilde{L}^{-1}\|\eta > 1,$$

(5.4.6)

and

$$\|\tilde{L}^{-1}\| < \frac{r_0 - 1}{r_0}\left(\frac{1}{c\varepsilon M}\right), \qquad \text{if} \quad r_0\|\tilde{L}^{-1}\|\eta < 1. \tag{5.4.7}$$

Then there exists a solution u of Problem II, which satisfies $u \in B_0 \subset B$,

$$B := L_2([0, T], B^{\alpha}_{\tau,\infty}(D)),$$

for all $0 < \alpha < \min\left(\frac{d}{\delta}m, \gamma\right)$, *where* δ *denotes the dimension of the singular set of D,* $\frac{1}{2} < \frac{1}{\tau} < \frac{\alpha}{d} + \frac{1}{2}$, *and* B_0 *is a small ball around* $\tilde{L}^{-1}f$ *(the solution of the corresponding linear problem) with radius* $R = C\tilde{C}(r_0 - 1)\eta\|\tilde{L}^{-1}\|$, *see Fig. 5.6.*

Proof This is a consequence of the regularity results in Kondratiev and Sobolev spaces from Theorem 5.2.14. To be more precise, Theorem 5.2.14 establishes the

Fig. 5.6 Nonlinear solution in B_0

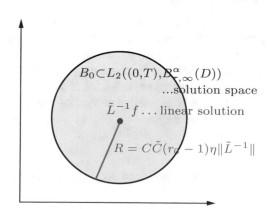

$B_0 \subset L_2((0,T), B^{\alpha}_{\tau,\infty}(D))$
...solution space

$\tilde{L}^{-1}f$ ... linear solution

$R = C\tilde{C}(r_0 - 1)\eta\|\tilde{L}^{-1}\|$

existence of a fixed point u in

$$S_0 \subset S := \bigcap_{k=0}^{\gamma_m+1} H^k([0, T], \mathcal{K}_{2,a+2m(\gamma_m-(k-1))}^{2m(\gamma_m-(k-1))}(D)) \cap H^{m,\gamma_m+2*}(D_T)$$

$$\hookrightarrow \bigcap_{k=0}^{\gamma_m+1} H^k([0, T], \mathcal{K}_{2,a+2m(\gamma_m-(k-1))}^{2m(\gamma_m-(k-1))}(D))$$

$$\cap H^{\gamma_m+1}([0, T], H^m(D)) \cap H^{\gamma_m+2}([0, T], H^{-m}(D))$$

$$\hookrightarrow L_2([0, T], \mathcal{K}_{2,a+2m(\gamma_m+1)}^{2m(\gamma_m+1)}(K) \cap H^m(D)) =: S'.$$

This together with the embedding results for Besov spaces from Theorem 2.4.12 (choosing $k = 0$) completes the proof, in particular, we calculate for the solution (cf. the proof of Theorem 5.4.1)

$$\|u - \tilde{L}^{-1}f | L_2([0, T], B_{\tau,\infty}^\alpha(D))\|$$
$$\leq C \|u - \tilde{L}^{-1}f | L_2([0, T], \mathcal{K}_{2,a+2m(\gamma_m+1)}^{2m(\gamma_m+1)}(D) \cap H^m(D))\|$$
$$= C \|u - \tilde{L}^{-1}f | S'\| \leq C\tilde{C} \|u - \tilde{L}^{-1}f | S\|$$
$$\leq C\tilde{C}(r_0 - 1)\eta \|\tilde{L}^{-1}\|. \tag{5.4.8}$$

Furthermore, it can be seen from (5.4.8) that new constants C and $\tilde{C}$ appear when considering the radius R around the linear solution where the problem can be solved compared to Theorem 5.2.14. □

Remark 5.4.7 A few words concerning the parameters appearing in Theorem 5.4.6 (and also Theorem 5.2.14) seem to be in order. Usually, the operator norm $\|\tilde{L}^{-1}\|$ as well as ε are fixed; but we can change η and r_0 according to our needs. From this we deduce that by choosing η small enough the condition (5.4.7) can always be satisfied. Moreover, it is easy to see that the smaller the nonlinear perturbation $\varepsilon > 0$ is, the larger we can choose the radius R of the ball B_0 where the solution of Problem II is unique.

5.4.3 Hölder-Besov Regularity of Problem I

So far we have not exploited the fact that Theorem 5.2.9 (Kondratiev regularity A) not only provides regularity properties of the solution u of Problem I but also of its partial derivatives $\partial_{t^k} u$. In this subsection we use this fact in combination with Theorem 2.2.5 (Generalized Sobolev's embedding theorem) in order to obtain some mixed Hölder-Besov regularity results on the whole space-time cylinder D_T.

For parabolic SPDEs, results in this direction have been obtained in [CKLL13]. However, for SPDEs, the time regularity is limited in nature. This is caused by the non-smooth character of the driving processes. Typically, Hölder regularity $C^{0,\beta}$ can be obtained, but not more. In contrast to this, it is well-known that deterministic parabolic PDEs are smoothing in time. Therefore, in the deterministic case considered here, higher regularity results in time can be obtained compared to the probabilistic setting.

Theorem 5.4.8 (Hölder-Besov Regularity) *Let D be a bounded polyhedral domain in $\mathbb{R}^d$, $d = 3$. Moreover, let $\gamma \in \mathbb{N}$ with $\gamma \geq 4m + 1$ and put $\gamma_m := \left[\frac{\gamma-1}{2m}\right]$. Furthermore, let $a \in \mathbb{R}$ with $a \in [-m, m]$. Assume that the right hand side f of Problem I satisfies*

(i) $\partial_{t^k} f \in L_2(D_T) \cap L_2([0, T], \mathcal{K}_{2,a+2m(\gamma_m-k)}^{2m(\gamma_m-k)}(D))$, $k = 0, \ldots, \gamma_m$, and
  $\partial_{t^{\gamma_m+1}} f \in L_2(D_T)$,
(ii) $\partial_{t^k} f(x, 0) = 0$, $k = 0, 1, \ldots, \gamma_m$.

Let Assumption 5.2.5 hold for weight parameters $b = a + 2m(\gamma_m - i)$, where $i = 0, \ldots, \gamma_m$ and $b' = -m$. Then for the solution $u \in H_2^{m,\gamma_m+2*}(D_T)$ of Problem I, we have

$$u \in C^{\gamma_m-2,\frac{1}{2}}([0, T], B_{\tau,\infty}^{\eta}(D)) \quad \text{for all} \quad 0 < \eta < \min\left(\frac{d}{\delta}, 4\right)m,$$

where $\frac{1}{2} < \frac{1}{\tau} < \frac{\eta}{d} + \frac{1}{2}$ and δ denotes the dimension of the singular set of D. In particular, we have the a priori estimate

$$\|u|C^{\gamma_m-2,\frac{1}{2}}([0, T], B_{\tau,\infty}^{\eta}(D))\|$$

$$\lesssim \sum_{k=0}^{\gamma_m} \|\partial_{t^k} f|L_2([0, T], \mathcal{K}_{2,a+2m(\gamma_m-k)}^{2m(\gamma_m-k)}(D))\| + \sum_{k=0}^{\gamma_m+1} \|\partial_{t^k} f|L_2(D_T)\|,$$

where the constant is independent of u and f.

Proof Theorems 5.2.9 and 5.2.3 show together with Theorems 2.4.12 and 2.2.5, that under the given assumptions on the initial data f, we have for $k \leq \gamma_m - 2$,

$$u \in H^{k+1}([0, T], \mathcal{K}_{2,a+2m(\gamma_m-k)}^{2m(\gamma_m-k)}(D)) \cap H^{\gamma_m+1}([0, T], H^m(D))$$

$$\hookrightarrow H^{k+1}([0, T], \mathcal{K}_{2,a+2m(\gamma_m-k)}^{2m(\gamma_m-k)}(D) \cap H^m(D))$$

$$\hookrightarrow C^{k,\frac{1}{2}}([0, T], \mathcal{K}_{2,a+2m(\gamma_m-k)}^{2m(\gamma_m-k)}(D) \cap H^m(D))$$

$$\hookrightarrow C^{k,\frac{1}{2}}([0, T], \mathcal{K}_{2,a+2m(\gamma_m-k)}^{\eta}(D) \cap H^m(D))$$

$$\hookrightarrow C^{k,\frac{1}{2}}([0, T], B_{\tau,\infty}^{\eta}(D)),$$

where in the third step we require $\eta \leq 2m(\gamma_m - k)$ and by Theorem 2.4.12 we get the additional restriction

$$m = \min(m, a + 2m(\gamma_m - k)) \geq \frac{\delta}{d}\eta, \quad \text{i.e.,} \quad \eta < \frac{d}{\delta}m.$$

Therefore, the upper bound on η reads as $\eta < \min\left(\frac{d}{\delta}m, 2m(\gamma_m - k)\right)$ since $k \leq \gamma_m - 2$, which for $k = \gamma_m - 2$ yields $\eta < \min\left(\frac{d}{\delta}, 4\right)m$.   □

Remark 5.4.9

(i) For $\gamma \geq 2m + 1$ and $k = \gamma_m - 1$ we have $\eta \leq \min\left(\frac{d}{\delta}, 2\right)m$ in the theorem above. For $\gamma \geq 2m$ and $k = \gamma_m$ we get $\eta = 0$.
(ii) From the proof of Theorem 5.4.8 above it can be seen that the solution satisfies

$$u \in C^{k,\frac{1}{2}}([0, T], \mathcal{K}^{2m(\gamma_m-k)}_{2,a+2m(\gamma_m-k)}(D)),$$

implying that for high regularity in time, which is displayed by the parameter k, we have less spacial regularity in terms of $2m(\gamma_m - k)$.

5.4.4 Besov Regularity of Problem IV

We turn our attention towards Besov regularity results for parabolic PDEs on general Lipschitz domains using the regularity results in Kondratiev spaces from [Kim09] together with embeddings of Kondratiev into Besov spaces from [Cio13]. For general Lipschitz domains the definition of the Kondratiev spaces from Definition 2.4.18 is different when compared to polyhedral domains, since in this case the whole boundary $\partial \mathcal{O}$ might coincide with the singular set. Therefore, the singularities induced by the boundary have a much stronger influence. Consequently, the results on Besov regularity for Problem I on domains of polyhedral type in Sect. 5.4.1 turn out to be much stronger compared to the Lipschitz case in Theorem 5.4.10. We refer to the discussion in Remark 5.4.11 below.

On the other hand, a bit suprising, it turns out that the obtained spatial Besov regularity results in the deterministic setting of Problem IV are more or less the same as for the case of SPDEs that was already studied in [Cio13] based on [Kim04, Kim12]. However, for the time regularity we nevertheless expect a significant difference.

Concerning Besov regularity for the solutions of Problem IV a combination of the Kondratiev regularity results from Theorem 5.2.19 together with the embedding from Theorem 2.4.20 gives the following.

Theorem 5.4.10 (Besov Regularity) *Let $p \in [2, \infty)$, $s \in [0, \infty)$, and Assumption 5.1.2 be satisfied. Then there exists $\beta_0 = \beta_0(p, d, \mathcal{O}) > 0$ such that for*

$$a \in \left(\frac{2 - p - \beta_0}{p}, \frac{2 - p + \beta_0}{p} \right),$$

any $f \in L_p([0, T], \mathfrak{K}^s_{p,a-1}(\mathcal{O}))$, and initial data $u_0 \in \mathfrak{K}^{s+2-\frac{2}{p}}_{p,a+\frac{p-2}{p}}(\mathcal{O})$, Problem IV admits a solution

$$u \in L_p([0, T], B^\alpha_{\tau,\tau}(\mathcal{O})), \qquad \frac{1}{\tau} = \frac{\alpha}{d} + \frac{1}{p}, \qquad 0 < \alpha < \min\left\{ s + 2, (a+1)\frac{d}{d-1} \right\}.$$

In particular, we have

$$\left\| u | L_p([0, T], B^\alpha_{\tau,\tau}(\mathcal{O})) \right\| \leq C \left(\|f|L_p([0, T], \mathfrak{K}^s_{p,a-1}(\mathcal{O}))\| + \left\| u_0 | \mathfrak{K}^{s+2-\frac{2}{p}}_{p,a+\frac{p-2}{p}}(\mathcal{O}) \right\| \right),$$

where $C = C(d, p, s, \theta, \delta_0, K, T, \mathcal{O})$.

Remark 5.4.11 We calculate for $a + 1$ appearing in the upper bound for α that

- $a + 1 \in \left(\dfrac{2 - \beta_0}{p}, \dfrac{2 + \beta_0}{p} \right)$ for Lipschitz domains $\mathcal{O}$,

- $a + 1 \in \left(\dfrac{1}{p}, 1 + \dfrac{1}{p} \right)$ for C^1 domains $\mathcal{O}$,

cf. Remark 5.2.20 concerning C^1 domains. In particular, for $p = 2$ and $d = 3$ as an upper bound for α we get

$$\alpha < \begin{cases} \min\left\{ s + 2, \left(1 + \frac{\beta_0}{2} \right) \frac{3}{2} \right\} & \text{for Lipschitz domains} \\ \min\left\{ s + 2, \frac{9}{4} \right\} & \text{for } C^1 \text{ domains} \end{cases} \Bigg\} \leq \frac{9}{4}$$

(every C^1 domain is also a Lipschitz domain), whereas Theorem 5.4.1 yields $\alpha < 3$ (since we have $m = 1$ in (5.1.10)).

Example 5.4.12 (Heat Equation) In [Kim09, Lem. 3.10] it is shown that the heat equation

$$\partial_t u = \Delta u \quad \text{on } \mathcal{O}_T, \qquad u(\cdot, 0) = 0,$$

has a solution $u \in L_p([0, T], \mathcal{K}^2_{p, \frac{2}{p}}(\mathcal{O}))$. Using Theorem 5.4.10 we obtain for the Besov regularity of the solution (now $\gamma \gg 0$) that

$$u \in L_p\left([0, T], B^\alpha_{\tau,\tau}(\mathcal{O}) \right) \quad \text{with} \quad \frac{1}{\tau} = \frac{\alpha}{d} + \frac{1}{p} \quad \text{and} \quad 0 < \alpha < \frac{2d}{p(d-1)}.$$

For $p = 2$ and $d = 3$ this even yields $\alpha < \frac{3}{2}$. However, comparing this with our considerations in Example 5.4.4, we see that for a domain of polyhedral type $D \subset \mathbb{R}^3$ our Besov regularity results are better than what can be expected for the heat equation on arbitrary Lipschitz domains. We also refer to [AG12] in this context, where the investigations (subject to some restrictions) lead to $\alpha < \frac{3}{2}s$, where s stands for the (fractional) Sobolev regularity of the solution.

Remark 5.4.13 Since β_0 depends also on p, Theorem 5.2.19 is not applicable in general if $p > 2$ as there might be a problem to fulfill the assumptions on a. In this case Theorem 5.2.19 does not yield the existence of a solution $u \in L_p(\mathcal{O}_T)$ even if the data of the equation is assumed to be arbitrarily smooth. As a deterministic counterexample the heat equation is discussed after [Cio13, Thm. 3.13]. Thus, also in the deterministic case, one should distinguish between $p \in [2, \infty)$ and $p \in [2, p_0)$ as was done in [Cio13, Thms. 3.13, 5.2].

Chapter 6
Regularity Theory for Hyperbolic PDEs

In this chapter we study linear hyperbolic equations (6.1.1) of second order on special Lipschitz domains according to Definition 2.1.8. For these kinds of equations regularity estimates in Kondratiev spaces were derived in [LT15], which enable us to treat these equations in a similar way as the parabolic problems in Chap. 5.

We proceed as follows: We exploit the regularity estimates in Kondratiev spaces from [LT15] and obtain Besov smoothness of the solutions to our hyperbolic problem with the help of the results from Theorem 2.4.14. There we established embeddings between Kondratiev spaces and Besov spaces on special Lipschitz domains by using the wavelet expansions of the Besov spaces and estimating the corresponding wavelet coefficients.

6.1 The Fundamental Hyperbolic Problem

Throughout this chapter let $\Omega_T := (0, T] \times \Omega$. We consider the following hyperbolic initial-boundary value problem of second order

$$
\begin{cases}
\frac{\partial^2}{\partial t^2} u + L(t, x, D_x)u = f(t, x) & \text{in } \Omega_T, \\
u\big|_{[0,T] \times \partial\Omega} = 0, \\
u(0, x) = \frac{\partial}{\partial t} u(0, x) = 0 & \text{in } \Omega.
\end{cases}
\tag{6.1.1}
$$

Here Ω is the special Lipschitz domain from Definition 2.1.8 and L is a linear differential operator of second order on Ω_T of the following form

$$
L(t, x, D_x)u = -\sum_{i,j=1}^{d} \frac{\partial}{\partial x_j}\left(a_{ij}(t, x)\frac{\partial u}{\partial x_i}\right) + \sum_{i=1}^{d} b_i(t, x)\frac{\partial u}{\partial x_i} + c(t, x)u,
$$

© The Author(s), under exclusive license to Springer Nature Switzerland AG 2021
C. Schneider, *Beyond Sobolev and Besov*, Lecture Notes in Mathematics 2291,
https://doi.org/10.1007/978-3-030-75139-5_6

where the coefficients are real-valued functions on Ω_T satisfying

$$a_{ij}(t, x), \ b_i(t, x), \ c(t, x) \in C^{m+1}(\Omega_T), \quad m \in \mathbb{N}_0. \tag{6.1.2}$$

Moreover, assume that the coefficients of L and their derivatives are bounded on Ω_T. Suppose that $a_{ij} = a_{ji}$ $(i, j = 1, \ldots, n)$ are continuous in $x \in \overline{\Omega}$ and uniformly elliptic with respect to $t \in [0, T]$, i.e., there exists a constant $\mu_0 > 0$ such that

$$\sum_{i,j=1}^{d} a_{ij}(t, x)\xi_i\xi_j \geq \mu_0|\xi|^2 \qquad \text{for all} \quad \xi \in \mathbb{R}^d \setminus \{0\}, \quad (t, x) \in \Omega_T.$$

It is possible to reduce the operator L with coefficients at $P \in l_0$, $t \in [0, T]$, to its canonical form

$$L_0^{(2)} := -\sum_{i,j=1}^{2} a_{ij}(t, P)\frac{\partial^2}{\partial x_i \partial x_j},$$

cf. [LT15, p. 460] and the references given there. Via this reduction it can be realized, that after a linear transformation of coordinates the half-spaces $T_1(P)$ and $T_2(P)$ (see Definition 2.1.8 for details) go over into hyperplanes T_1' and T_2', respectively. Furthermore, the angle β at (t, P) is transformed to

$$\omega(t, P) = \arctan \frac{\left[a_{11}(t, P)a_{22}(t, P) - a_{12}^2(t, P)\right]^{1/2}}{a_{22}(t, P)\cot\beta - a_{12}(t, P)}. \tag{6.1.3}$$

The value $\omega(t, P)$ does not depend on the method by which $L_0^{(2)}$ is reduced to its canonical form. Moreover, the function $\omega(t, P)$ is infinitely differentiable and $\omega(t, P) > 0$. Since Ω is bounded it follows that the manifold l_0 is compact and we put

$$\omega := \max_{P \in l_0, t \in [0,T]} \omega(t, P). \tag{6.1.4}$$

Remark 6.1.1 (Existence and Uniqueness of Weak Solution) Note that our assumptions on the coefficients a_{ij}, b_i, and c of L given above are slightly weaker compared to Theorem 3.4.4 (choose $m = 0$ in (6.1.2)). However, the proof of Theorem 3.4.4 carries over to this setting. In particular, it was observed in [LT15, Thm. 2.1], that under the above assumptions on the coefficients, for $f \in C([0, T], L_2(\Omega))$ problem (6.1.1) has a unique weak solution u which satisfies

$$\|u(t)|H^1(\Omega)\|^2 + \|\partial_t u(t)|L_2(\Omega)\|^2 \leq C\int_0^t \|f(s)|L_2(\Omega)\|^2 ds \qquad \text{for all} \qquad t \in (0, T].$$

6.2   Regularity Results in Kondratiev Spaces

For the special Lipschitz domains Ω from Definition 2.1.8 the corresponding Kondratiev spaces we consider are defined as in (2.4.1) with singular set $S = l_0$. Thus, the space $\mathcal{K}_{p,a}^m(\Omega)$ contains all functions $u(x)$ such that

$$\|u|\mathcal{K}_{p,a}^m(\Omega)\| := \left(\sum_{|\alpha| \le m} \int_\Omega |\rho(x)|^{p(|\alpha|-a)} |D_x^\alpha u(x)|^p dx \right)^{1/p} < \infty, \qquad (6.2.1)$$

where the weight function $\rho : \Omega \to [0, 1]$ now is the smooth distance to l_0, i.e., ρ is a smooth function and in the vicinity of l_0 it is equivalent to $\mathrm{dist}(x, l_0)$. The spaces $C([0, T], \mathcal{K}_{p,a}^m(\Omega))$ are defined in an obvious way, cf. (2.4.4).

Concerning the regularity of the solution of (6.1.1) in these spaces, a reformulation of the results in [LT15, Thm. 2.1, 2.2] gives the following.

Lemma 6.2.1 (Kondratiev Regularity) *Let* $\Omega \subset \mathbb{R}^d$ *be a special Lipschitz domain according to Definition 2.1.8. Furthermore, let* $m \in \mathbb{N}$, $m \ge 2$, $m + 1 - a < \frac{\pi}{\omega}$ *with* $a \in [0, 1]$, *and assume* f *satisfies*

(i) $\partial_{t^j} f \in C([0, T], \mathcal{K}_{2,m-j-a}^{m-j}(\Omega))$, $0 \le j \le m - 1$,
(ii) $\partial_{t^j} f(x, 0) = 0$, $0 \le j \le m - 2$.

Moreover, let the coefficients of L *satisfy the assumptions stated in Sect. 6.1, in particular,* (6.1.2). *Then for the weak solution* u *of problem* (6.1.1) *we have*

$$\partial_{t^j} u \in C([0, T], \overset{\circ}{H}^1(\Omega)) \cap C([0, T], \mathcal{K}_{2,m-j-a}^{m-j}(\Omega)), \qquad 0 \le j \le m,$$

and the following a priori estimate holds

$$\sum_{j=0}^m \|\partial_{t^j} u | C([0, T], \mathcal{K}_{2,m-j-a}^{m-j}(\Omega))\| \lesssim \sum_{j=0}^{m-1} \left\| \partial_{t^j} f | C([0, T], \mathcal{K}_{2,m-1-j-a}^{m-1-j}(\Omega)) \right\|.$$

Remark 6.2.2 For the restriction on a in Lemma 6.2.1 to make sense we require

$$m < \frac{\pi}{\omega}, \quad \text{i.e.,} \quad \omega < \frac{\pi}{m}.$$

Unfortunately, this causes a restriction on the angle β of our domain Ω, i.e., β has to be small, cf. (6.1.3) and (6.1.4).

6.3   Regularity Results in Besov Spaces

Now Lemma 6.2.1 in combination with Theorem 2.4.15 gives the following result concerning the Besov regularity of the solution of (6.1.1).

Theorem 6.3.1 (Hyperbolic Besov Regularity) *Let $\Omega \subset \mathbb{R}^d$ be a special Lipschitz domain according to Definition 2.1.8. Furthermore, let $m \in \mathbb{N}$, $m \geq 2$, $m + 1 - a < \frac{\pi}{\omega}$ with $a \in [0, 1]$, and assume f satisfies*

(i) $\partial_{t^j} f \in C([0, T], \mathcal{K}^{m-j}_{2, m-j-a}(\Omega))$, $0 \leq j \leq m - 1$,
(ii) $\partial_{t^j} f(x, 0) = 0$, $0 \leq j \leq m - 2$.

Moreover, let the coefficients of L the assumptions stated in Sect. 6.1, in particular, (6.1.2). Then for the weak solution u of problem (6.1.1) we have

$$u \in C([0, T], B^r_{\tau, \tau}(\Omega)), \qquad \frac{1}{\tau} = \frac{r}{d} + \frac{1}{p}, \qquad 0 \leq r < \min\left\{m, \frac{d}{d-1}\right\},$$

and the following a priori estimate holds

$$\|u|C([0, T], B^r_{\tau, \tau}(\Omega))\| \lesssim \sum_{j=0}^{m-1} \|\partial_{t^j} f | C([0, T], \mathcal{K}^{m-1-j}_{2, m-1-j-a}(\Omega))\|.$$

Proof According to Lemma 6.2.1 we have $u \in C([0, T], H^1(\Omega)) \cap C([0, T], \mathcal{K}^m_{2, m-a}(\Omega))$ for $m + 1 - a < \frac{\pi}{\omega}$. Now using Theorem 2.4.15 with $s = 1$ and $\frac{d-2}{d} r < m - a$ yields the desired embedding result. As for the restriction on a, using $m \geq 2$ and $d \geq 3$, we further observe that

$$\frac{d-2}{d} r < \frac{d-2}{d} \min\left(m, \frac{d}{d-1}\right) = \frac{d-2}{d} \cdot \frac{d}{d-1} \leq 1,$$

therefore, $a < m - \frac{d-2}{d} r$ is always satisfied since by our assumptions $a \leq 1$. $\square$

Remark 6.3.2 There are more results in [LT15] than we used in this chapter. In particular, in [LT15, Thm. 3.5] also weighted Sobolev regularity of nonlinear hyperbolic problems was investigated. Therefore, using a fixed point theorem as in Sect. 5.4.2, it should be possible to study Besov regularity of nonlinear hyperbolic problems as well. But this is out of our scope for now and will possibly be treated in a forthcoming paper.

Chapter 7
Applications to Adaptive Approximation Schemes

In the introduction we already sketched why we expect that the results proved in Chaps. 4–6 will have some impact concerning the theoretical foundation of adaptive algorithms. In this chapter, we want to return to these relationships in more detail.

Therefore, we apply our previous Besov regularity results in order to obtain convergence rates for adaptive algorithms, in particular, either adaptive wavelet algorithms or adaptive finite element algorithms. For both approaches, in the elliptic case algorithms are known, cf. [CDD03], which provably perform at the optimal convergence rate in the following sense: If the solution u belongs to a related Approximation class $\mathcal{A}^\alpha$ (to be specified below), i.e., the error for the optimal approximation is proportional to $N^{-\alpha}$, then the algorithm indeed produces an approximation with error proportional to $N^{-\alpha}$. Therein N corresponds to the number of degrees of freedom used in the construction of the approximation, and it also corresponds to the computational cost of the algorithm. Thus, in order to analyze the potential performance of adaptive solvers for the problem at hand, we have to study relations between the Approximation classes and regularity classes in which solutions exist.

In Sects. 7.1 and 7.2 we treat in detail the semilinear elliptic problem (4.1.1) from Chap. 4. Afterwards, Sect. 7.3 gives an outlook for parabolic and hyperbolic problems.

7.1 N-Term Approximation and Adaptive Wavelet Algorithms

It is nowadays well-known that certain Besov spaces are closely related to approximation spaces for N-term wavelet approximation. Let us start with adaptive wavelet algorithms as e.g. discussed in [CDD01, Ste09]. To describe related results, let $\Psi = \{\psi_\lambda : \lambda \in \Lambda\}$ be a wavelet system with sufficiently high differentiability

© The Author(s), under exclusive license to Springer Nature Switzerland AG 2021
C. Schneider, *Beyond Sobolev and Besov*, Lecture Notes in Mathematics 2291,
https://doi.org/10.1007/978-3-030-75139-5_7

and vanishing moments, such that all relevant (unweighted) Sobolev and Besov spaces can be characterized in terms of expansion coefficients w.r.t. Ψ, see Sect. 2.3. Then, the best thing we can expect from an adaptive numerical algorithm based on this wavelet basis is that it realizes the convergence order of best N-term wavelet approximation schemes. In this sense, best N-term wavelet approximation serves as the benchmark for the performance of adaptive algorithms.

Let X be some Banach space. The *error of best N-term approximation* is defined as

$$\sigma_N(u; X) = \inf_{\Gamma \subset \Lambda: \#\Gamma \le N} \inf_{c_\lambda} \left\| u - \sum_{\lambda \in \Gamma} c_\lambda \psi_\lambda \Big| X \right\|,$$

i.e., as the name suggests we consider the best approximation by linear combinations of the basis functions consisting of at most N terms. Of course, in the context of the numerical approximation of the solutions to operator equations, such an approximation scheme would never be implementable because this would require the knowledge of all wavelet coefficients, i.e., of the solution itself. Nevertheless, it has been shown that the recently developed adaptive wavelet algorithms indeed asymptotically realize the same order of approximation [CDD01, Ste09]! To quantify the approximation rate, we introduce the *approximation classes* $\mathcal{A}_q^\alpha(X)$, $\alpha > 0, 0 < q \le \infty$ by requiring

$$\|u|\mathcal{A}_q^\alpha(X)\| = \left(\sum_{N=0}^{\infty} \left((N+1)^\alpha \sigma_N(u; X) \right)^q \frac{1}{N+1} \right)^{1/q} < \infty, \qquad (7.1.1)$$

if $0 < q < \infty$ as well as

$$\|u|\mathcal{A}_\infty^\alpha(X)\| = \sup_{N \ge 0} (N+1)^\alpha \sigma_N(u; X) < \infty.$$

Then a famous result of DeVore et al. [DJP92] may be formulated as

$$\mathcal{A}_\tau^{m/d}(L_p(\mathbb{R}^d)) = B_{\tau,\tau}^m(\mathbb{R}^d), \qquad \frac{1}{\tau} = \frac{m}{d} + \frac{1}{p}. \qquad (7.1.2)$$

For our purposes we shall consider a result from [DNS06, Thm. 11, p. 586], which reads as

$$B_{\tau,q}^{m+s}(D) \hookrightarrow \mathcal{A}_\infty^{m/d}(H_p^s(D)), \qquad \frac{1}{\tau} < \frac{m}{d} + \frac{1}{p}, \qquad (7.1.3)$$

where $s \in \mathbb{R}$, $\tau < p$ and $m > 0$, independent of the parameter $0 < q \le \infty$. Consequently, as we see from (7.1.2) and (7.1.3), the optimal approximation rate that can be achieved by adaptive wavelet schemes depends on the Besov smoothness of the unknown solution in the adaptivity scale (1.0.3). In contrast to this, the convergence order of classical nonadaptive (uniform) schemes depends on the Sobolev smoothness of the solution, see e.g. [Hac92, DDD97] for details.

Looking at semilinear elliptic problems (4.1.1), the results presented in Chap. 4 imply that the Besov regularity of the unknown solutions is much higher than the Sobolev regularity, which justifies the use of spatial adaptive wavelet algorithms. In particular, Theorem 4.3.1, together with (2.3.5) and (7.1.3), applied with $s = 1$ and $p = 2$, gives the following result.

Theorem 7.1.1 (Approximation in H^1) *Let $D \subset \mathbb{R}^d$, $d = 2, 3$, be a bounded Lipschitz domain of polyhedral type with singular set S of dimension l. Furthermore, let $\bar{a}, m, g, f, \eta, C$ be as in Theorem 4.2.18 and assume a satisfies*

$$\frac{d}{2} - 1 \leq a < \bar{a} \qquad \text{and} \qquad \frac{ml}{d} < a \,.$$

Then there exists a solution $u \in H^1(D)$ of problem (4.1.1) which belongs to the approximation class $\mathcal{A}_\infty^{m/d}(H^1(D))$, i.e., it satisfies the estimate

$$\sigma_N(u; H^1(D)) \lesssim N^{-m/d} \| f | \mathcal{K}_{2,a-1}^{m-1}(D) \|$$

for N-term wavelet approximation.

Proof Observe that our assumption $\frac{ml}{d} < a$ is equivalent to

$$\frac{m-a}{d-l} + \frac{1}{2} < \frac{m}{d} + \frac{1}{2} = \frac{2m+d}{2d} \,.$$

Hence, there exist parameters τ fulfilling (7.1.3) as well as the assumptions of Theorem 4.3.1. We conclude that we have a solution

$$u \in F_{\tau,2}^{m+1}(D) \hookrightarrow B_{\tau,\infty}^{m+1}(D) \hookrightarrow \mathcal{A}_\infty^{m/d}(H^1(D))$$

of problem (4.1.1). A reformulation of this inclusion gives the claimed approximation result.   □

Remark 7.1.2

(i) The reader should observe that in the case $d = 2$ the lower bound for a reads as $a > 0$ (since $l = 0$). Therefore, Theorem 7.1.1 implies that, by increasing the Kondratiev regularity m of the right–hand side f and of the coefficients $a_{i,j}$, $1 \leq i, j \leq 2$, solutions with arbitrarily high Kondratiev regularity exist. This means that, in principle, these solutions can be approximated by best N-term wavelet approximation up to any order! But here we have to mention that the condition (4.2.18)(c), which has to be satisfied for $\eta = \| f | \mathcal{K}_{2,a-1}^{m-1}(D) \|$ and $\vartheta = C$ from (4.2.33) implies that $\| f | \mathcal{K}_{2,a-1}^{m-1}(D) \|$ has to be sufficiently small which is a serious restriction. On the other hand, high orders of best N-term wavelet approximation are hard to realize in practice since one has to work

with wavelet bases that characterize the corresponding approximation classes,
e.g. the Besov spaces.

(ii) Of course nonlinearities of the form $g(x, \xi) = \xi^n$, $n \in \mathbb{N}$, $n \geq 2$, are admissible in Theorem 7.1.1. In such a situation we may replace Theorem 4.2.18 by Corollary 4.2.23. As a consequence, we may change the phrase *'Then there exists a solution $u \in H^1(D)$ of problem (4.1.1)'* into *'Then there is a unique solution $u \in H^1(D)$ of problem (4.1.1) in the small closed ball K_0'*.
Also from the practical point of view we consider Theorem 7.1.1 as important. Indeed, semilinear problems with nonlinear polynomial part are the standard test cases for adaptive algorithms, see, e.g. [CDD03].

(iii) It is of course not surprising that for $d = 3$ our results are more restrictive. Due to the upper bound $\overline{a}$, we cannot choose m arbitrarily high except the case that $l = 0$ (and $\overline{a} > 1/2$). One particular case, namely problems on smooth cones, has been studied before. The 3D-results of Theorem 7.1.1 can be improved in this situation. Indeed, it has been shown in [KMR97, Thm. 6.1.1], that in this case the upper bound $\overline{a}$ can be avoided. Instead, there is a countable set of parameters a which is excluded.

Concerning approximation in $L_2(D)$ we obtain the following result from (7.1.3).

Theorem 7.1.3 (Approximation in L_2) *Let $D \subset \mathbb{R}^d$, $d = 2, 3$, be a bounded Lipschitz domain of polyhedral type with singular set S of dimension l. Furthermore, let $\overline{a}, m, g, f, \eta, C$ be as in Theorem 4.2.18 and assume a satisfies*

$$\frac{d}{2} - 1 \leq a < \overline{a} \qquad and \qquad \frac{(m+1)l}{d} - 1 < a \,.$$

Then there exists a solution $u \in L_2(D)$ of problem (4.1.1) which belongs to the approximation class $\mathcal{A}_{\infty}^{(m+1)/d}(L_2(D))$, i.e., it satisfies the estimate

$$\sigma_N(u; L_2(D)) \lesssim N^{-(m+1)/d} \| f \, | \mathcal{K}_{2,a-1}^{m-1}(D) \|$$

for N-term wavelet approximation.

Proof By Theorem 4.3.1 we have a solution $u \in F_{\tau,2}^{m+1}(D)$ of problem (4.1.1) for some $0 < \tau < 2$, where this time the interval

$$\frac{m-a}{d-l} + \frac{1}{2} < \frac{1}{\tau} \leq \frac{2m+2+d}{2d} \tag{7.1.4}$$

gives the admissible τ there (the upper bound guarantees that $F_{\tau,2}^{m+1}(D) \hookrightarrow L_2(D)$). Moreover, we need

$$F_{\tau,2}^{m+1}(D) \hookrightarrow B_{\tau,\infty}^{m+1}(D) \hookrightarrow \mathcal{A}_{\infty}^{(m+1)/d}(L_2(D)),$$

Fig. 7.1 Approximation lines in H^1 and L_2

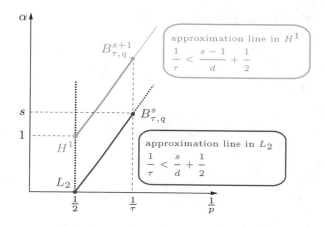

which holds for $1/\tau < (m+1)/d + 1/2$, see (7.1.3). We choose τ such that $1/\tau$ is close to $(m+1)/d + 1/2$. Then (7.1.4) turns into

$$\frac{m-a}{d-l} < \frac{m+1}{d} \qquad \text{if, and only, if} \qquad \frac{(m+1)l}{d} - 1 < a \,.$$

$\square$

Figure 7.1 illustrates the approximation lines in H^1 and L_2 as provided by (7.1.3).

7.2 Finite Element Approximation

Recent results by Gaspoz and Morin [GM14, GM09] show that the basic relationships outlined above also carry over to discretization schemes based on finite elements with adaptive h-refinement.

The starting point is an initial triangulation $\mathcal{T}_0$ of the polyhedral domain D. Furthermore, $\mathbb{T}$ denotes the family of all conforming, shape-regular partitions $\mathcal{T}$ of D obtained from $\mathcal{T}_0$ by refinement using bisection rules (Fig. 7.2). Moreover, $V_\mathcal{T}$ denotes the *finite element space* of continuous piecewise polynomials of degree at most r, i.e.,

$$V_\mathcal{T} = \left\{ v \in C(\overline{D}) : v|_T \in \mathcal{P}_r \text{ for all } T \in \mathcal{T} \right\}.$$

In this setting the counterpart to the quantity $\sigma_N(u; X)$ is given by

$$\sigma_N^{FE}(u; X) = \min_{\substack{\mathcal{T} \in \mathbb{T}: \\ \#\mathcal{T} - \#\mathcal{T}_0 \leq N}} \inf_{v \in V_\mathcal{T}} \|u - v|X\|.$$

Fig. 7.2 Triangulation of domain D

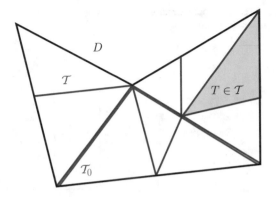

Then [GM14, Thm. 2.2] gives direct estimates,

$$\sigma_N^{FE}\left(u;\, L_p(D)\right) \leq C\, N^{-s/d}\|f\,|B_{\tau,\tau}^s(D)\|\,,$$

as well as

$$\sigma_N^{FE}\left(u;\, B_{p,p}^\alpha(D)\right) \leq C\, N^{-s/d}\|f\,|B_{\tau,\tau}^{s+\alpha}(D)\|\,,$$

where $1 < p < \infty$, $0 < \alpha < r+1$, $0 < s + \alpha \leq r + \frac{1}{\tau_*}$, $\tau_* = \min(1, \tau)$, and $\frac{1}{\tau} < \frac{s}{d} + \frac{1}{p}$. In [Han14] it was shown that this extends to embeddings

$$B_{\tau,\infty}^s(D) \hookrightarrow \mathcal{A}_{\infty,FE}^{s/d}(L_p(D)) \quad \text{and} \quad B_{\tau,\infty}^{s+\alpha}(D) \hookrightarrow \mathcal{A}_{\infty,FE}^{s/d}(B_{p,p}^\alpha(D)),$$
$$(7.2.1)$$

where the approximation class $\mathcal{A}_{\infty,FE}^{s/d}(X)$ is defined as in (7.1.1) with σ_N being replaced by σ_N^{FE}. The embeddings (7.2.1) are the immediate counterparts of (7.1.3). We obtain the following counterpart of Theorem 7.1.1.

Theorem 7.2.1 (Finite Element Approximation in H^1) *In the setting of Theorem 7.1.1 the function u belongs to the approximation space $\mathcal{A}_{\infty,FE}^{m/d}(H^1(D))$, i.e., we have the estimate*

$$\sigma_N^{FE}(u;\, H^1(D)) \lesssim N^{-m/d}\|f\,|\mathcal{K}_{2,a-1}^{m-1}(D)\|$$

for finite element approximation on shape-regular conforming triangulations with adaptive h-refinement.

Proof The solution u from Theorem 7.1.1 satisfies $u \in F_{\tau,2}^{m+1}(D) \hookrightarrow B_{\tau,2}^{m+1}(D) \hookrightarrow B_{\tau,\infty}^{m+1}(D)$, where we used (2.3.5) together the elementary embeddings for Besov spaces from Proposition 2.3.4(i). Together with the second embedding in (7.2.1) this finishes the proof.   □

7.3 Time-Marching Adaptive Algorithms

The regularity results presented in Chaps. 5 and 6 for parabolic and hyperbolic problems, respectively, indicate that for each $t \in (0, T)$ the Besov regularity of the unknown solutions (of the problems studied there) is much higher than the corresponding Sobolev regularity. Therefore, the use of spatial adaptive wavelet algorithms is also justified for these kinds of problems (which is in good agreement with the elliptic problem from Chap. 4 discussed in Sect. 7.1). This corresponds to the classical time-marching schemes such as the Rothe method. We refer e.g. to the monographs [Lan01, Tho06] for a detailed discussion.

Of course, it would be tempting to employ adaptive wavelet strategies in the whole space-time cylinder. First results in this direction for parabolic evolution problems have been reported in [SS09]. To justify these schemes also for the parabolic and hyperbolic problems that we studied, Besov regularity in the whole space-time cylinder has to be established. This is out of the scope of this manuscript but will be subject to future investigations.

Moreover, by what was outlined in Sect. 7.2, our regularity results presented for parabolic and hyperbolic problems also indicate that the use of adaptive time-marching schemes based on finite elements is justified.

Part II
Traces in Function Spaces

Chapter 8
Traces on Lipschitz Domains

In this chapter we investigate traces of functions $f \in \mathbf{B}_{p,q}^s(\Omega)$ on the boundary Γ of Lipschitz domains Ω. Our presentation is organized as follows:

In Sect. 8.1 we show that the trace operator from the source space $\mathbf{B}_{p,q}^{s+\frac{1}{p}}(\Omega)$ into the target space $\mathbf{B}_{p,q}^s(\Gamma)$ is bounded where $0 < s < 1$. The newly developed non-smooth atomic decompositions for the spaces $\mathbf{B}_{p,q}^s$ from Sect. 2.3.5 are well suited in order to proof this result: The additional freedom we gain in the choice of non-smooth atoms enables us to show that the restriction of a smooth atom to the boundary Γ of a Lipschitz domain is a suitable non-smooth boundary atom.

Afterwards, in Sect. 8.2, we consider the 'converse' problem and establish the existence of a (nonlinear) extension operator from $\mathbf{B}_{p,q}^s(\Gamma)$ to $\mathbf{B}_{p,q}^{s+\frac{1}{p}}(\Omega)$. Here we use Whitney's decomposition in order to show that we can extend non-smooth atoms on the boundary in an adequate way to obtain suitable atoms for decompositions in the spaces $\mathbf{B}_{p,q}^{s+\frac{1}{p}}(\Omega)$.

Section 8.3 combines the previous results and contains the main trace theorems on Lipschitz domains. The limiting case $s = 0$ is also considered. Moreover, as a by-product we obtain corresponding trace results on Lipschitz domains for Triebel-Lizorkin spaces $\mathbf{F}_{p,q}^s(\Omega)$, defined via atomic decompositions. The papers [Sch10] and [Sch11b], dealing with traces on hyperplanes and smooth domains, respectively, might be considered as forerunners of the trace results established in this chapter. Nevertheless, the methods we use now are completely different.

Finally, Sect. 8.4, contains yet another application of the non-smooth atomic decompositions. There we deal with pointwise multipliers $M(\mathbf{B}_{p,q}^s(\mathbb{R}^d))$ in the respective function spaces and discuss under which circumstances the characteristic function χ_Ω of a bounded domain $\Omega \subset \mathbb{R}^d$ is a pointwise multiplier in $\mathbf{B}_{p,q}^s(\mathbb{R}^d)$, this way establishing a connection between pointwise multipliers and certain fundamental notion of fractal geometry, so-called h-sets, cf. Definition 8.4.6.

Before we start our investigations let us briefly explain our understanding of the trace operator in this context, since when dealing with L_p functions the pointwise

© The Author(s), under exclusive license to Springer Nature Switzerland AG 2021
C. Schneider, *Beyond Sobolev and Besov*, Lecture Notes in Mathematics 2291,
https://doi.org/10.1007/978-3-030-75139-5_8

trace has no obvious meaning: Let $Y(\Gamma)$ denote one of the spaces $\mathbf{B}^{\sigma}_{u,v}(\Gamma)$ or $L_u(\Gamma)$. Since $\mathcal{S}(\Omega)$ is dense in $\mathbf{B}^s_{p,q}(\Omega)$ for $0 < p, q < \infty$ (both spaces can be interpreted as restrictions of their counterparts defined on $\mathbb{R}^d$), one asks first whether there is a constant $c > 0$ such that

$$\|\mathrm{Tr}\, \varphi | Y(\Gamma)\| \leq c \|\varphi | \mathbf{B}^s_{p,q}(\Omega)\| \quad \text{for all } \varphi \in \mathcal{S}(\Omega), \tag{8.0.1}$$

where $\mathcal{S}(\Omega)$ stands for the restriction of the Schwartz space $\mathcal{S}(\mathbb{R}^d)$ to a domain Ω. If this is the case, then one defines $\mathrm{Tr}\, f \in Y(\Gamma)$ for $f \in \mathbf{B}^s_{p,q}(\Omega)$ by completion and obtains

$$\|\mathrm{Tr}\, f | Y(\Gamma)\| \leq c \|f | \mathbf{B}^s_{p,q}(\Omega)\|, \quad f \in \mathbf{B}^s_{p,q}(\Omega),$$

for the linear and bounded trace operator

$$\mathrm{Tr}\, : \mathbf{B}^s_{p,q}(\Omega) \hookrightarrow Y(\Gamma).$$

Remark 8.0.1 We can extend (8.0.1) to spaces $\mathbf{B}^s_{p,q}(\Omega)$ with $p = \infty$ and/or $q = \infty$ by using embeddings for B- and F-spaces from [HS09, Sch09a]. The results stated there can be generalized to domains Ω, since the spaces $\mathbf{B}^s_{p,q}(\Omega)$ are defined by restriction of the corresponding spaces on $\mathbb{R}^d$, cf. Remark 2.3.12.

If $p = \infty$, we have that $\mathbf{B}^s_{\infty,q}(\Omega)$ with $s > 0$ is embedded in the space of continuous functions and $\mathrm{Tr}\,$ makes sense pointwise. If $q = \infty$,

$$\mathbf{B}^s_{p,\infty}(\Omega) \hookrightarrow \mathbf{B}^{s-\varepsilon}_{p,1}(\Omega) \quad \text{for any } \varepsilon > 0.$$

Let $s > \frac{1}{p}$ and $\varepsilon > 0$ be small enough such that one has

$$s > s - \varepsilon > \frac{1}{p}.$$

Since by [Tri08b, Rem. 13] traces are independent of the source spaces and of the target spaces one can now define $\mathrm{Tr}\,$ for $\mathbf{B}^s_{p,\infty}(\Omega)$ by restriction of $\mathrm{Tr}\,$ for $\mathbf{B}^{s-\varepsilon}_{p,1}(\Omega)$ to $\mathbf{B}^s_{p,\infty}(\Omega)$. Hence (8.0.1) is always meaningful.

8.1   Boundedness of the Trace Operator

In this section we prove our first main theorem which states that the trace operator $\mathrm{Tr}\,$ from above is linear and bounded from the Besov space $\mathbf{B}^{s+\frac{1}{p}}_{p,q}(\Omega)$ into $\mathbf{B}^s_{p,q}(\Gamma)$. In order to proof this result we rely on the non-smooth atomic decompositions from Sect. 2.3.5 for the spaces $\mathbf{B}^s_{p,q}(\Gamma)$. To be precise, we take an optimal atomic decomposition of $f \in \mathbf{B}_{p,q}^{s+\frac{1}{p}}(\Omega)$ and show that the restriction of an atom to the boundary Γ of a Lipschitz domain is a suitable Lip^{Γ}-atom according to Definition 2.3.29. From this we deduce that $\mathrm{Tr}\, f = f|_{\Gamma}$ results in a non-smooth atomic decomposition in $\mathbf{B}_{p,q}^{s}(\Gamma)$.

Theorem 8.1.1 (Boundedness of Trace Operator) *Let $d \geq 2$, $0 < p, q \leq \infty$, $0 < s < 1$, and let $\Omega \subset \mathbb{R}^d$ be a bounded Lipschitz domain with boundary Γ. Then the operator*

$$\mathrm{Tr} : \mathbf{B}_{p,q}^{s+\frac{1}{p}}(\Omega) \longrightarrow \mathbf{B}_{p,q}^{s}(\Gamma) \qquad (8.1.1)$$

is linear and bounded.

Proof The linearity of the operator follows directly from its definition as discussed above. To prove the boundedness, we take an optimal representation of a smooth function $f \in \mathbf{B}_{p,q}^{s+\frac{1}{p}}(\Omega)$ as described in Theorem 2.3.16, formula (2.3.24), i.e.,

$$f = \sum_{j=0}^{\infty} \sum_{m \in \mathbb{Z}^d} {}^{j,\Omega}\lambda_{j,m} a_{j,m} \quad \text{with} \quad \|f|\mathbf{B}_{p,q}^{s+\frac{1}{p}}(\Omega)\| \sim \|\lambda|b_{p,q}^{s+\frac{1}{p}}(\Omega)\|. \qquad (8.1.2)$$

We put

$$\mathrm{Tr}\, f := \left(\sum_{j,m} {}^{j,\Omega}\lambda_{j,m} a_{j,m} \right)\Big|_{\Gamma} = \sum_{j,m} {}^{j,\Gamma}\lambda_{j,m} a_{j,m}\Big|_{\Gamma} = \sum_{j,m} {}^{j,\Gamma}\lambda_{j,m} a_{j,m}^{\Gamma}. \qquad (8.1.3)$$

The proof follows by Theorem 2.3.31 (Non-smooth atomic decomposition) and the following four facts:

(i) $a_{j,m}^{\Gamma}$ are Lip^{Γ}-atoms.

(ii) $\|\lambda|b_{p,q}^{s}(\Gamma)\| \lesssim \|\lambda|b_{p,q}^{s+\frac{1}{p}}(\Omega)\|$.

(iii) The decomposition (8.1.3) converges in $L_p(\Gamma)$.

(iv) The trace operator Tr coincides with the trace operator discussed above, cf. (8.0.1).

To prove the first point, we observe that

$$\mathrm{supp}\, a_{j,m}^{\Gamma} \subset \mathrm{supp}\, a_{j,m} \cap \Gamma \subset Q_{j,m}^{\Gamma}.$$

Furthermore, we have $\|a_{j,m}^{\Gamma}|L_{\infty}(\Gamma)\| \leq \|a_{j,m}|L_{\infty}(\tilde{d} Q_{j,m})\| \leq c$ and

$$\sup_{\substack{x,y \in Q_{j,m}^{\Gamma} \\ x \neq y}} \frac{a_{j,m}^{\Gamma}(x) - a_{j,m}^{\Gamma}(y)}{|x - y|} \leq \sup_{\substack{x,y \in \tilde{d} Q_{j,m} \\ x \neq y}} \frac{a_{j,m}(x) - a_{j,m}(y)}{|x - y|} \lesssim 2^{j}.$$

The proof of the second point follows directly by

$$
\|\lambda|b_{p,q}^{s}(\Gamma)\| = \left(\sum_{j} 2^{j(s-\frac{d-1}{p})q} \left(\sum_{m}^{j,\Gamma} |\lambda_{j,m}|^{p} \right)^{q/p} \right)^{1/p}
$$

$$
\leq \left(\sum_{j} 2^{j\left[(s+\frac{1}{p})-\frac{d}{p} \right]q} \left(\sum_{m}^{j,\Omega} |\lambda_{j,m}|^{p} \right)^{q/p} \right)^{1/p} = \|\lambda|b_{p,q}^{s+\frac{1}{p}}(\Omega)\|.
$$

The proof of the third point follows in the same way as the proof in Step 3 of Theorem 2.3.31.

The proof of (iv) is based on the fact that for $f \in \mathcal{S}(\Omega)$ there is an optimal atomic decomposition (8.1.2) which converges also pointwise. This may be observed by a detailed inspection of [HN07]. Therefore also the series (8.1.3) converges pointwise and the trace operator Tr may be understood in the pointwise sense for smooth f.   □

8.2 Construction of an Extension Operator

In order to compute the exact trace space we still need to construct an extension operator

$$
Ext : \mathbf{B}_{p,q}^{s}(\Gamma) \longrightarrow \mathbf{B}_{p,q}^{s+\frac{1}{p}}(\Omega)
$$

and prove its boundedness (after having established the boundedness of the trace operator in Theorem 8.1.1). Our main result is stated in Theorem 8.2.7. A big problem in this context is to show that we can extend the Lip^{Γ}-atoms from the source spaces in a nice way to obtain suitable atoms for the target spaces. This will be done in Lemma 8.2.6.

8.2.1 Extension of Atoms

Our approach is based on the classical Whitney decomposition of $\mathbb{R}^{d} \setminus \Gamma$ and the corresponding decomposition of unity. We summarise the most important properties of this method in the next Lemma and refer to [Ste70, pp. 167–170] and [JW84, pp. 21–26] for details and proofs.

Lemma 8.2.1 (Whitney Decomposition)

1. Let $\Gamma \subset \mathbb{R}^d$ be a closed set. Then there exists a collection of cubes $\{Q_i\}_{i \in \mathbb{N}}$, such that

   (i) $\mathbb{R}^d \setminus \Gamma = \bigcup_i Q_i$.
   (ii) The interiors of the cubes are mutually disjoint.
   (iii) The inequality

$$\operatorname{diam} Q_i \leq \operatorname{dist}(Q_i, \Gamma) \leq 4 \operatorname{diam} Q_i$$

   holds for every cube Q_i. Here $\operatorname{diam} Q_i$ is the diameter of Q_i and $\operatorname{dist}(Q_i, \Gamma)$ is its distance from Γ.
   (iv) Each point of $\mathbb{R}^d \setminus \Gamma$ is contained in at most N_0 cubes $6/5 \cdot Q_i$, where N_0 depends only on d.
   (v) If Γ is the boundary of a Lipschitz domain then there is a number $\gamma > 0$, which depends only on d, such that $\sigma(\gamma Q_i \cap \Gamma) > 0$ for all $i \in \mathbb{N}$.

2. The are C^∞ functions $\{\psi_i\}_{i \in \mathbb{N}}$ such that

   (i) $\sum_i \psi_i(x) = 1$ for every $x \in \mathbb{R}^d \setminus \Gamma$.
   (ii) $\operatorname{supp} \psi_i \subset 6/5 \cdot Q_i$.
   (iii) For every $\alpha \in \mathbb{N}_0^d$ there is a constant A_α such that $|D^\alpha \psi_i(x)| \leq A_\alpha (\operatorname{diam} Q_i)^{-|\alpha|}$ holds for all $i \in \mathbb{N}$ and all $x \in \mathbb{R}^d$.

If a is a Lipschitz function on the Lipschitz boundary Γ of Ω, then the Whitney extension operator Ext is defined by

$$\operatorname{Ext} a(x) = \begin{cases} a(x), & x \in \Gamma, \\ \sum_i \mu_i \psi_i(x), & x \in \Omega, \end{cases} \tag{8.2.1}$$

where we use the notation of Lemma 8.2.1 and $\mu_i := \frac{1}{\sigma(\gamma Q_i \cap \Gamma)} \int_{\gamma Q_i \cap \Gamma} a(y) d\sigma(y)$ with the number $\gamma > 0$ as described in Lemma 8.2.1. It satisfies $\operatorname{Tr} \circ \operatorname{Ext} a = a$ for a Lipschitz continuous on Γ. This follows directly from the celebrated Whitney's extension theorem (cf. [JW84, p. 23]) as Γ is a closed set if Ω is a bounded Lipschitz domain.

Lemma 8.2.2 Let a be a Lipschitz function on the Lipschitz boundary Γ of Ω. Then $\operatorname{Ext} a \in C^\infty(\Omega)$ and

$$\max_{|\alpha|=k} |D^\alpha \operatorname{Ext} a(x)| \leq c_k \delta(x)^{1-k} \cdot \|a|\operatorname{Lip}(\Gamma)\|, \quad k \in \mathbb{N}, \quad x \in \Omega. \tag{8.2.2}$$

Here, $\delta(x)$ is the distance of x to Γ and c_k depends only on k and Ω.

Proof First, let us note that

$$D^\alpha \operatorname{Ext} a(x) = \sum_i \mu_i D^\alpha \psi_i(x), \qquad x \in \Omega, \quad \alpha \in \mathbb{N}_0^d, \quad |\alpha| = k.$$

By Lemma 8.2.1 we have for every $x \in \Omega$

$$|D^\alpha \psi_i(x)| \le c_k \delta(x)^{-k}, \qquad |\alpha| = k,$$

and

$$\sum_i D^\alpha \psi_i(x) = D^\alpha \sum_i \psi_i(x) = 0.$$

Furthermore, the Lipschitz continuity of a implies

$$|\mu_i - \mu_j| \lesssim \delta(x) \cdot \|a|\operatorname{Lip}(\Gamma)\| \tag{8.2.3}$$

for $x \in \operatorname{supp} \psi_i \cap \operatorname{supp} \psi_j$. To justify (8.2.3), we consider natural numbers i and j with $x \in \operatorname{supp} \psi_i \cap \operatorname{supp} \psi_j$, chose any $x_i \in \gamma Q_i \cap \Gamma$ and $x_j \in \gamma Q_j \cap \Gamma$ and calculate

$$|\mu_i - \mu_j| \le \left| \frac{1}{\sigma(\gamma Q_i \cap \Gamma)} \int_{\gamma Q_i \cap \Gamma} a(x) d\sigma(x) - a(x_i) \right| + |a(x_i) - a(x_j)|$$

$$+ \left| a(x_j) - \frac{1}{\sigma(\gamma Q_j \cap \Gamma)} \int_{\gamma Q_j \cap \Gamma} a(x) d\sigma(x) \right|$$

$$\le \|a|\operatorname{Lip}(\Gamma)\| \cdot \big\{ \operatorname{diam}(\gamma Q_i \cap \Gamma) + |x_i - x_j| + \operatorname{diam}(\gamma Q_j \cap \Gamma) \big\}$$

$$\lesssim \|a|\operatorname{Lip}(\Gamma)\| \cdot \big\{ \operatorname{diam}(Q_i) + |x_i - x| + |x - x_j| + \operatorname{diam}(Q_j) \big\}$$

$$\lesssim \delta(x) \cdot \|a|\operatorname{Lip}(\Gamma)\|.$$

Let us now fix $x \in \Omega$ and denote by $\{i_1, \ldots, i_N\}$, $N \le N_0$, the indices for which x lies in the support of ψ_i. Then we write

$$\left| \sum_{j=1}^N \mu_{i_j} D^\alpha \psi_{i_j}(x) \right| \le \left| \sum_{j=1}^N (\mu_{i_j} - \mu_{i_1}) D^\alpha \psi_{i_j}(x) \right| + \left| \sum_{j=1}^N \mu_{i_1} D^\alpha \psi_{i_j}(x) \right|$$

$$\le \sum_{j=1}^N |\mu_{i_j} - \mu_{i_1}| \cdot |D^\alpha \psi_{i_j}(x)| \lesssim \delta(x)^{1-k} \cdot \|a|\operatorname{Lip}(\Gamma)\|.$$

$\square$

Remark 8.2.3 Let a be a function defined on Γ as in Lemma 8.2.2 with $\operatorname{diam}(\operatorname{supp} a) \le 1$. Then the extension operator from Lemma 8.2.2 may be combined with a multiplication with a smooth cut-off function. This ensures, that (8.2.2) still holds and, in addition, diam (supp $\operatorname{Ext} a$) $\lesssim 1$.

The following lemma describes a certain geometrical property of Lipschitz domains, which shall be useful later on. It resembles very much the notion of Minkowski content, cf. [Fal90].

Lemma 8.2.4 *Let* $\Omega \subset \mathbb{R}^d$ *be a bounded Lipschitz domain and let* $k \in \mathbb{N}$. *Let* $h \in \mathbb{R}^d$ *with* $0 < |h| \le 1$ *and put* $\Omega^h = \{x \in \Omega : [x, x + kh] \subset \Omega\}$. *Furthermore, for* $j \in \mathbb{N}_0$ *we define* $\Omega_j^h = \{x \in \Omega^h : 2^{-j} \le \min_{y \in [x, x+kh]} \delta(y) \le 2^{-j+1}\}$, *where* $\delta(y) = \operatorname{dist}(y, \Gamma)$. *Then*

$$|\Omega_j^h| \lesssim 2^{-j} \tag{8.2.4}$$

with a constant independent of j *and* h.

Proof To simplify the notation, we shall assume that Ω is a simple Lipschitz domain, cf. Remark 2.1.2(ii), of the type $\Omega = \{(x', x_d) = (x_1, \ldots, x_{d-1}, x_d) \in \mathbb{R}^d : x_d > \psi(x'), |x'| < 1\}$, where ψ is a Lipschitz function, and we identify Γ with $\{(x', x_d) : x_d = \psi(x'), |x'| < 1\}$.

Step 1. First, let us observe that

$$\operatorname{dist}(x, \Gamma) \approx (x_d - \psi(x')) \quad \text{for} \quad x = (x', x_d) \in \Omega \tag{8.2.5}$$

and the constants in this equivalence depend only on the Lipschitz constant of ψ. The simple proof of this fact is based on the inner cone property of Lipschitz domains. We refer to [Ste70, Ch. VI, Sect. 3.2, Lem. 2] for details.

Step 2. Let $j \in \mathbb{N}_0$ and $0 < |h| \le 1$ be fixed and let

$$y = (y', y_d) \in \Omega_j^h$$

and let also

$$\tilde{y} = (y', \tilde{y}_d) \in \Omega_j^h$$

with $\tilde{y}_d > y_d$ (Fig. 8.1).

As $\tilde{y} \in \Omega_j^h$, there is a $t_0 \in [0, k]$ such that $\operatorname{dist}(\tilde{y} + t_0 h, \Gamma) \le 2^{-j+1}$.

Then we use $\psi(y' + t_0 h) < t_0 h_d + y_d$ (which follows from $y \in \Omega^h$ and $y + t_0 h \in \Omega$) and (8.2.5) to get

$$\tilde{y}_d - y_d = [\tilde{y}_d + t_0 h_d - \psi(y' + t_0 h')] + [\psi(y' + t_0 h') - t_0 h_d - y_d]$$

$$\lesssim \operatorname{dist}(\tilde{y} + t_0 h, \Gamma) \lesssim 2^{-j}. \tag{8.2.6}$$

Fig. 8.1 Grey strip with measure 2^{-j} belongs to Ω_j^h

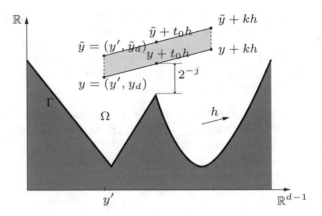

Step 3.   Using (8.2.6), we observe that the set $\Omega(x') = \{x_d \in \mathbb{R} : (x', x_d) \in \Omega_j^h\}$ has for every $|x'| < 1$ length smaller then $c\,2^{-j}$. From this the inequality (8.2.4) quickly follows.   □

We shall use this geometrical observation together with the extension operator (8.2.1) to prove the following.

Lemma 8.2.5 *Let $\Omega \subset \mathbb{R}^d$ be a bounded Lipschitz domain and let Γ be its boundary. Let a be a Lipschitz function on Γ. Let $0 < p \leq \infty$, $0 < s < \infty$, and $k \in \mathbb{N}$ with $0 < s < k < 1/p + 1$. Then the extension operator defined by (8.2.1) satisfies*

$$\|\mathrm{Ext}\,a|\mathbf{B}_{p,p}^s(\Omega)\| \lesssim \|a|\mathrm{Lip}(\Gamma)\| \tag{8.2.7}$$

with the constant independent of $a \in \mathrm{Lip}(\Gamma)$.

Proof Using the definition of Besov spaces via differences, we obtain

$$\|\mathrm{Ext}\,a|\mathbf{B}_{p,p}^s(\Omega)\| \lesssim \|\mathrm{Ext}\,a|\mathbf{B}_{p,\infty}^{s'}(\Omega)\|$$
$$\lesssim \|\mathrm{Ext}\,a|L_p(\Omega)\| + \sup_{0<|h|\leq 1} |h|^{-s'} \|\Delta_h^k \mathrm{Ext}\,a(\cdot, \Omega)|L_p(\Omega)\|,$$

for $s' > 0$ with $s < s' < k$. Furthermore, we observe that one may modify the definition of $\Delta_h^r f(x, \Omega)$ given in (2.3.19) to be zero also if the whole segment $[x, x + kh]$ is not a subset of Ω. This follows by a detailed inspection of [Tri83, Sect. 2.5.12] as well as [Dis03] and [DS93], which are all based on the integration in cones.

Using the definition of μ_i, cf. (8.2.1) and below, the first term may be estimated easily as

$$\|\mathrm{Ext}\,a|L_p(\Omega)\| \lesssim \|\mathrm{Ext}\,a|L_\infty(\Omega)\| \leq \|a|L_\infty(\Gamma)\|.$$

To estimate the second term, we shall need the following relationship between differences and derivatives. If $f \in C^k(\mathbb{R}^d)$ and $x, h \in \mathbb{R}^d$, we put $g(t) = f(x + th)$ for $t \in \mathbb{R}$ and obtain

$$\Delta_h^k f(x) = \Delta_1^k g(0) = \int_0^k g^{(k)}(t) B_k(t) dt, \qquad (8.2.8)$$

where B_k is the standard B-spline of order k, i.e., the k-fold convolution of $\chi_{[0,1]}$ given by $B_k = \chi_{[0,1]} * \cdots * \chi_{[0,1]}$. Although (8.2.8) is a classical result of approximation theory (cf. [DL93, Sect. 4.7]), let us give a short proof using Fubini's Theorem and induction over k:

$$\Delta_1^{k+1} g(0) = \Delta_1^k g(1) - \Delta_1^k g(0) = \int_0^k (g^{(k)}(t+1) - g^{(k)}(t)) B_k(t) dt$$

$$= \int_0^k B_k(t) \int_t^{t+1} g^{(k+1)}(u) du \, dt = \int_0^{k+1} g^{(k+1)}(u) \int_{u-1}^u B_k(t) dt \, du$$

$$= \int_0^{k+1} g^{(k+1)}(u) B_{k+1}(u) du.$$

Hence if $[x, x + kh] \subset \Omega$ for some $x \in \Omega$, we obtain

$$|\Delta_h^k \operatorname{Ext} a(x, \Omega)| \lesssim |h|^k \int_0^k \max_{|\alpha|=k} |D^\alpha \operatorname{Ext} a(x + th)| \cdot B_k(t) dt$$

$$\lesssim |h|^k \cdot \|a|\operatorname{Lip}(\Gamma)\| \cdot \int_0^k \delta(x + th)^{1-k} \cdot B_k(t) dt.$$

Let us fix $h \in \mathbb{R}^d$ with $0 < |h| \le 1$ and put $\Omega^h = \{x \in \Omega : [x, x + kh] \subset \Omega\}$ as in Lemma 8.2.4. We obtain

$$|h|^{-s'} \|\Delta_h^k \operatorname{Ext} a(\cdot, \Omega)|L_p(\Omega)\| \lesssim |h|^{k-s'} \|a|\operatorname{Lip}(\Gamma)\|$$

$$\times \left(\int_{\Omega^h} \left(\int_0^k \delta(x + th)^{1-k} \cdot B_k(t) dt \right)^p dx \right)^{1/p}$$

$$\lesssim \|a|\operatorname{Lip}(\Gamma)\| \left(\int_{\Omega^h} \max_{y \in [x, x+kh]} \delta(y)^{(1-k)p} dx \right)^{1/p}$$

$$\lesssim \|a|\operatorname{Lip}(\Gamma)\| \left(\sum_{j=0}^\infty 2^{-j(1-k)p} |\Omega_j^h| \right)^{1/p}.$$

This, together with Lemma 8.2.4 and $k < 1/p + 1$ finishes the proof. $\qquad \square$

Lemma 8.2.6 (Extension of Atoms) *Let* $0 < s' < 1$ *be fixed. There is a nonlinear extension operator (denoted by* **Ext***), which extends* Lip^Γ*-atoms* $a_{j,m}$ *to* $(s' + 1/p, p)$*-atoms on* $\mathbb{R}^d$.

Proof As the definition of Lip^Γ-atoms as well as the definition of $(s' + 1/p, p)$-atoms works with $a_j(2^{-j}\cdot)$, by homogeneity arguments it is enough to prove

$$\|\mathbf{Ext}\, a_{0,m}|\mathbf{B}_{p,p}^{s'+1/p}(\mathbb{R}^d)\| \lesssim \|a_{0,m}|\mathrm{Lip}(\Gamma)\| \tag{8.2.9}$$

for Lip^Γ-atoms $a_{j,m}$ with $j = 0$. First we show that

$$\|\mathrm{Ext}\, a_{0,m}|\mathbf{B}_{p,p}^{s'+1/p}(\Omega)\| \lesssim \|a_{0,m}|\mathrm{Lip}(\Gamma)\| \tag{8.2.10}$$

for the extension operator constructed in (8.2.1). Let $0 < s' < 1$ and $0 < p \leq \infty$. We observe, that Lemma 8.2.5 implies (8.2.10) for all $0 < s' < 1$ for which there is a $k \in \mathbb{N}_0$ with

$$s' + 1/p < k < 1 + 1/p.$$

In Fig. 8.2 these points correspond to all $(s', \frac{1}{p})$ in the gray-shaded triangles.

Then Lemma 2.3.34 yields (8.2.10) for all $0 < s' < 1$ and $0 < p \leq \infty$ with $s_0 = s_1 = s'$ and $p_0 < p < p_1$ chosen in an appropriate way, see Fig. 8.2.

Finally, by Remark 2.3.12, we know that there is a function (denoted by **Ext** $a_{0,m}$), such that

$$\|\mathbf{Ext}\, a_{0,m}|\mathbf{B}_{p,p}^{s'+1/p}(\mathbb{R}^d)\| \lesssim \|\mathrm{Ext}\, a_{0,m}|\mathbf{B}_{p,p}^{s'+1/p}(\Omega)\|.$$

This together with (8.2.10) finishes the proof of (8.2.9).   □

Fig. 8.2 Parameters $\left(s, \frac{1}{p}\right)$ allowing extension of atoms

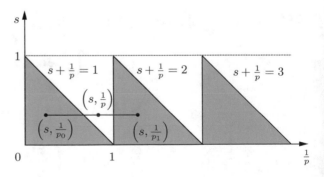

8.2.2   Boundedness of the Extension Operator

With the preliminary work on the extensions of atoms from the last subsection we are now able to establish the existence of an extension operator from $\mathbf{B}_{p,q}^s(\Gamma)$ to $\mathbf{B}_{p,q}^{s+\frac{1}{p}}(\Omega)$.

Theorem 8.2.7 *Let $d \geq 2$ and $\Omega \subset \mathbb{R}^d$ be a bounded Lipschitz domain with boundary Γ. Then for $0 < s < 1$ and $0 < p, q \leq \infty$ there is a bounded nonlinear extension operator*

$$Ext : \mathbf{B}_{p,q}^s(\Gamma) \longrightarrow \mathbf{B}_{p,q}^{s+\frac{1}{p}}(\Omega). \qquad (8.2.11)$$

Proof Let $f \in \mathbf{B}_{p,q}^s(\Gamma)$ with optimal decomposition in the sense of Theorem 2.3.31, i.e.,

$$f(x) = \sum_{j=0}^{\infty} \sum_{m \in \mathbb{Z}^n} \lambda_{j,m} a_{j,m}^{\Gamma}(x), \qquad (8.2.12)$$

where $a_{j,m}^{\Gamma}$ are Lip$^{\Gamma}$-atoms, (8.2.12) converges in $L_p(\Gamma)$, and $\|f|\mathbf{B}_{p,q}^s(\Gamma)\| \sim \|\lambda|b_{p,q}^s(\Gamma)\|$.

We use the extension operator constructed in Lemma 8.2.6 and define by

$$Ext\, f := \sum_{j=0}^{\infty} \sum_{m \in \mathbb{Z}^n} \lambda_{j,m} (\mathbf{Ext}\, a_{j,m}^{\Gamma})|_{\Omega} \qquad (8.2.13)$$

an atomic decomposition of f in the space $\mathbf{B}_{p,q}^{s+1/p}(\Omega)$ with non-smooth $(s' + 1/p, p)$-atoms $\mathbf{Ext}\, a_{j,m}^{\Gamma}$, where $s < s' < 1$. The convergence of (8.2.13) in $L_p(\Omega)$ follows in the same way as in the proof of Step 3 of Theorem 2.3.31.

Together with $\|\lambda|b_{p,q}^s(\Gamma)\| \sim \|\lambda|b_{p,q}^{s+1/p}(\Omega)\|$, this shows that

$$\|Ext\, f|\mathbf{B}_{p,q}^{s+1/p}(\Omega)\| \lesssim \|\lambda|b_{p,q}^{s+1/p}(\Omega)\| \sim \|\lambda|b_{p,q}^s(\Gamma)\| < \infty$$

is bounded. $\square$

8.3   Trace Results for Spaces $\mathbf{B}_{p,q}^s(\Omega)$, $\mathbf{F}_{p,q}^s(\Omega)$

After having established the boundedness of the trace operator and the existence of an extension operator in Sects. 8.1 and 8.2, respectively, in this section we now formulate the main trace theorems. In particular, we present the general trace theorem for Besov spaces on Lipschitz domains with parameter $0 < s < 1$ and show

how the results can be used to derive a corresponding trace theorem for Triebel-Lizorkin spaces as well, using that the trace is independent of q in this case. Finally, we also have a look at the limiting case when $s = 0$, where we rely on previous trace results from [Sch11a] on $\eth$-sets and observe that the boundary of a Lipschitz domain is just a special case of a $\eth$-set with $\eth = d - 1$.

8.3.1   The General Case $0 < s < 1$

Theorems 8.1.1 and 8.2.7 together now allow us to state the general result for traces on Lipschitz domains without any restrictions on the parameters s, p and q.

Theorem 8.3.1 (Trace Theorem) *Let $d \geq 2$ and $\Omega \subset \mathbb{R}^d$ be a bounded Lipschitz domain with boundary Γ. Then for $0 < s < 1$ and $0 < p, q \leq \infty$,*

$$\operatorname{Tr} \mathbf{B}_{p,q}^{s+\frac{1}{p}}(\Omega) = \mathbf{B}_{p,q}^{s}(\Gamma). \tag{8.3.1}$$

Theorem 8.3.1 extends the trace results for B-spaces obtained in [Sch11b, Thm. 2.4] from C^k domains with $k > s$ to Lipschitz domains.

Moreover, the corresponding results for traces of F-spaces carry over as well from smooth domains to Lipschitz domains: The main ingredient in [Sch11b, Thm. 2.6] was to show the independence of the trace on q (for F-spaces defined on smooth domains) and using the coincidence $\mathbf{F}_{p,p}^s(\Omega) = \mathbf{B}_{p,p}^s(\Omega)$, cf. Remark 2.3.39, and the results for B-spaces. Since the spaces $\mathbf{F}_{p,q}^s(\Omega)$ on Lipschitz domains are defined via atomic decompositions, cf. Definition 2.3.38, in order to show their independence on q it suffices to know that the corresponding sequence spaces $f_{p,q}^s(\Gamma)$ are independent of q, i.e.,

$$f_{p,q}^s(\Gamma) = b_{p,p}^s(\Gamma), \tag{8.3.2}$$

when Γ is the boundary of a Lipschitz domain Ω. This is proven in [Tri06, Prop. 9.22, p. 394] for Γ being a compact porous set in $\mathbb{R}^d$ with [FJ90] as an important forerunner. Since in [Tri08a, Prop. 3.6] it is shown that the boundaries $\partial \Omega = \Gamma$ of (ε, δ)-domains Ω are porous and bounded Lipschitz domains are special (ε, δ)-domains, cf. the explanations and references in Remark 2.3.12, the result follows immediately.

Therefore, from the above considerations we obtain the following trace result for F-spaces.

Corollary 8.3.2 (Trace Theorem) *Let $d \geq 2$ and $\Omega \subset \mathbb{R}^d$ be a bounded Lipschitz domain with boundary Γ. Then for $0 < s < 1$, $0 < p < \infty$, and $0 < q \leq \infty$,*

$$\operatorname{Tr} \mathbf{F}_{p,q}^{s+\frac{1}{p}}(\Omega) = \mathbf{B}_{p,p}^{s}(\Gamma). \tag{8.3.3}$$

8.3.2 The Limiting Case s = 0

We briefly discuss what happens in the limiting case $s = 0$. In [Sch11a, Thm. 2.7] traces for Besov and Triebel-Lizorkin spaces on $\eth$-sets Γ, $0 < \eth < d$, were studied. In particular, it was shown that for $0 < p < \infty$ and $0 < q \leq \infty$, one has

$$\operatorname{Tr} \mathbf{B}_{p,q}^{\frac{d-\eth}{p}} (\mathbb{R}^d) = L_p(\Gamma), \qquad 0 < q \leq \min(1, p), \tag{8.3.4}$$

and

$$\operatorname{Tr} \mathbf{F}_{p,q}^{\frac{d-\eth}{p}} (\mathbb{R}^d) = L_p(\Gamma), \qquad 0 < p \leq 1. \tag{8.3.5}$$

Since the boundary Γ of a Lipschitz domain Ω is a $\eth$-set with $\eth = d - 1$ and B- and F-spaces on domains Ω are defined as restrictions of the corresponding spaces on $\mathbb{R}^d$, cf. Remarks 2.3.12 and 2.3.39(i), we obtain the following result in the limiting case.

Corollary 8.3.3 (Traces in the Limiting Case) *Let $d \geq 2$ and $\Omega \subset \mathbb{R}^d$ be a bounded Lipschitz domain with boundary Γ. Furthermore, let $0 < p < \infty$ and $0 < q \leq \infty$.*

(i) Then

$$\operatorname{Tr} \mathbf{B}_{p,q}^{\frac{1}{p}} (\Omega) = L_p(\Gamma), \qquad 0 < q \leq \min(1, p). \tag{8.3.6}$$

(ii) Furthermore,

$$\operatorname{Tr} \mathbf{F}_{p,q}^{\frac{1}{p}} (\Omega) = L_p(\Gamma), \qquad 0 < p \leq 1. \tag{8.3.7}$$

8.4 Application: Pointwise Multipliers in Function Spaces

The non-smooth atomic decompositions for the spaces $\mathbf{B}_{p,q}^s(\mathbb{R}^d)$ from Theorem 2.3.23 were the main tool for establishing trace theorems on Lipschitz domains in this chapter. As another application of these decompositions, in this section we now deal with pointwise multipliers in the respective function spaces.

A function m in $L_{\min(1,p)}^{\text{loc}}(\mathbb{R}^d)$ is called a *pointwise multiplier* for $\mathbf{B}_{p,q}^s(\mathbb{R}^d)$ if

$$f \mapsto mf$$

generates a bounded map in $\mathbf{B}_{p,q}^s(\mathbb{R}^d)$. The collection of all multipliers for $\mathbf{B}_{p,q}^s(\mathbb{R}^d)$ is denoted by $M(\mathbf{B}_{p,q}^s(\mathbb{R}^d))$. In the following, let ψ stand for a non-negative C^∞

function with

$$\text{supp}\,\psi \subset \{y \in \mathbb{R}^d : |y| \leq \sqrt{d}\} \tag{8.4.1}$$

and

$$\sum_{l \in \mathbb{Z}^d} \psi(x - l) = 1, \qquad x \in \mathbb{R}^d. \tag{8.4.2}$$

Definition 8.4.1 Let $s > 0$ and $0 < p, q \leq \infty$. We define the space $\mathbf{B}^s_{p,q,\text{selfs}}(\mathbb{R}^d)$ to be the set of all $f \in L^{\text{loc}}_{\min(1,p)}(\mathbb{R}^d)$ such that

$$\|f|\mathbf{B}^s_{p,q,\text{selfs}}(\mathbb{R}^d)\| := \sup_{j \in \mathbb{N}_0, l \in \mathbb{Z}^d} \|\psi(\cdot - l)f(2^{-j}\cdot)|\mathbf{B}^s_{p,q}(\mathbb{R}^d)\| \tag{8.4.3}$$

is finite.

Remark 8.4.2 The study of pointwise multipliers is one of the key problems of the theory of function spaces. As far as classical Besov spaces and (fractional) Sobolev spaces with $p > 1$ are concerned we refer to [Maz85], [MS85], and [MS09]. Pointwise multipliers in general spaces $B^s_{p,q}(\mathbb{R}^d)$ and $F^s_{p,q}(\mathbb{R}^d)$ have been studied in great detail in [RS96, Ch. 4].

Selfsimilar spaces were first introduced in [Tri03] and then considered in [Tri06, Sect. 2.3]. Corresponding results for anisotropic function spaces may be found in [MPP07]. We also mention their forerunners, the uniform spaces $\mathbf{B}^s_{p,q,\text{unif}}(\mathbb{R}^d)$ containing all $f \in L^{\text{loc}}_{\min(1,p)}(\mathbb{R}^d)$ such that

$$\|f|\mathbf{B}^s_{p,q,\text{unif}}(\mathbb{R}^d)\| := \sup_{l \in \mathbb{Z}^d} \|\psi(\cdot - l)f|\mathbf{B}^s_{p,q}(\mathbb{R}^d)\| < \infty,$$

studied in detail in [RS96, Sect. 4.9]. As stated in [KS02], for these spaces it is known that

$$M(\mathbf{B}^s_{p,q}(\mathbb{R}^d)) = \mathbf{B}^s_{p,q,\text{unif}}(\mathbb{R}^d), \qquad 1 \leq p \leq q \leq \infty, \quad s > \frac{d}{p},$$

cf. [SS99] concerning the proof. Selfsimilar spaces are also closely connected with pointwise multipliers. We shall use the abbreviation

$$\mathbf{B}^s_{p,\text{selfs}}(\mathbb{R}^d) := \mathbf{B}^s_{p,p,\text{selfs}}(\mathbb{R}^d).$$

One can easily show

$$\mathbf{B}^s_{p,q,\text{selfs}}(\mathbb{R}^d) \hookrightarrow L_\infty(\mathbb{R}^d). \tag{8.4.4}$$

To see this applying homogeneity gives

$$\|\psi(\cdot - l)f(2^{-j}\cdot)|\mathbf{B}^s_{p,q}(\mathbb{R}^d)\| \sim 2^{j\frac{d}{p}}\|\psi(2^j \cdot -l)f|L_p(\mathbb{R}^d)\|$$

$$+ 2^{-j(s-\frac{d}{p})}\left(\int_0^1 t^{-sq}\omega_r(\psi(2^j \cdot -l)f,t)_p^q \frac{dt}{t}\right)^{1/q}$$

uniformly for all $j \in \mathbb{N}_0$ and $l \in \mathbb{Z}^d$. Consequently,

$$2^{jd}\int_{\mathbb{R}^d}|\psi(2^jy-l)|^p|f(y)|^pdy \le c\|f|\mathbf{B}^s_{p,q,\text{selfs}}(\mathbb{R}^d)\|^p. \qquad (8.4.5)$$

Thus, the right-hand side of (8.4.5) is just a uniform bound for $|f(\cdot)|^p$ at its Lebesque points, cf. [Ste93, Cor. on p. 13], which proves the desired embedding (8.4.4).

Definition 8.4.3 Let $s > 0$ and $0 < p, q \le \infty$. We define

$$\mathbf{B}^{s+}_{p,q,\text{selfs}}(\mathbb{R}^d) := \bigcup_{\sigma > s}\mathbf{B}^\sigma_{p,q,\text{selfs}}(\mathbb{R}^d).$$

We have the following relation between pointwise multipliers and selfsimilar spaces.

Theorem 8.4.4 Let $s > 0$ and $0 < p, q \le \infty$. Then

(i) $\mathbf{B}^{s+}_{p,q,\text{selfs}}(\mathbb{R}^d) \subset M(\mathbf{B}^s_{p,q}(\mathbb{R}^d)) \hookrightarrow \mathbf{B}^s_{p,q,\text{selfs}}(\mathbb{R}^d)$.
(ii) Additionally, if $0 < p \le 1$,

$$M(\mathbf{B}^s_p(\mathbb{R}^d)) = \mathbf{B}^s_{p,\text{selfs}}(\mathbb{R}^d).$$

Proof We first prove the right-hand side embedding in (i). Let $m \in M(\mathbf{B}^s_{p,q}(\mathbb{R}^d))$. An application of the homogeneity property from Theorem 2.3.10 yields

$$\|\psi(\cdot - l)m(2^{-j}\cdot)|\mathbf{B}^s_{p,q}(\mathbb{R}^d)\| \sim 2^{-j(s-\frac{d}{p})}\|\psi(2^j \cdot -l)m|\mathbf{B}^s_{p,q}(\mathbb{R}^d)\|$$

$$\lesssim 2^{-j(s-\frac{d}{p})}\|m|M(\mathbf{B}^s_{p,q}(\mathbb{R}^d))\|$$

$$\cdot\|\psi(2^j \cdot -l)|\mathbf{B}^s_{p,q}(\mathbb{R}^d)\|$$

$$= 2^{-j(s-\frac{d}{p})}\|m|M(\mathbf{B}^s_{p,q}(\mathbb{R}^d))\| \cdot \|\psi(2^j\cdot)|\mathbf{B}^s_{p,q}(\mathbb{R}^d)\|$$

$$\sim \|m|M(\mathbf{B}^s_{p,q}(\mathbb{R}^d))\|\|\psi|\mathbf{B}^s_{p,q}(\mathbb{R}^d)\|$$

$$\lesssim \|m|M(\mathbf{B}^s_{p,q}(\mathbb{R}^d))\|$$

for all $l \in \mathbb{Z}^d$, $j \in \mathbb{N}_0$, and hence,

$$\|m|\mathbf{B}^s_{p,q,\text{selfs}}(\mathbb{R}^d)\| = \sup_{j \in \mathbb{N}_0, l \in \mathbb{Z}^d} \|\psi(\cdot - l)m(2^{-j})|\mathbf{B}^s_{p,q}(\mathbb{R}^d)\|$$

$$\lesssim \|m|M(\mathbf{B}^s_{p,q}(\mathbb{R}^d))\|.$$

We make use of the non-smooth atomic decompositions for $\mathbf{B}^s_{p,q}(\mathbb{R}^d)$ from Theorem 2.3.23 in order to prove the first inclusion in (i). Let $m \in \mathbf{B}^\sigma_{p,q,\text{selfs}}$ with $\sigma > s$. Let $f \in \mathbf{B}^s_{p,q}(\mathbb{R}^d)$ with optimal smooth atomic decomposition

$$f = \sum_{j=0}^{\infty} \sum_{l \in \mathbb{Z}^d} \lambda_{j,l} a_{j,l} \quad \text{with} \quad \|f|\mathbf{B}^s_{p,q}(\mathbb{R}^d)\| \sim \|\lambda|b^s_{p,q}\|, \tag{8.4.6}$$

where $a_{j,m}$ are K-atoms with $K > \sigma$. Then

$$mf = \sum_{j=0}^{\infty} \sum_{l \in \mathbb{Z}^d} \lambda_{j,l} (ma_{j,l}), \tag{8.4.7}$$

and we wish to prove that, up to normalizing constants, the $ma_{j,l}$ are (σ, p)-atoms. The support condition is obvious:

$$\text{supp } ma_{j,l} \subset \text{supp } a_{j,l} \subset \tilde{d}Q_{j,l}, \qquad j \in \mathbb{N}_0, l \in \mathbb{Z}^d.$$

If $l = 0$ we put $a_j = a_{j,l}$. Note that

$$\text{supp } a_j(2^{-j}) \subset \{y : |y_i| \le \frac{\tilde{d}}{2}\}$$

and we can assume that

$$\psi(y) > 0 \quad \text{if} \quad y \in \{x : |x_i| \le \tilde{d}\}.$$

Then—using multiplier assertions from [Sch11b, Prop. 2.15(ii)]—we have for any $g \in \mathbf{B}^\sigma_{p,q}(\mathbb{R}^d)$,

$$\|a_j(2^{-j})\psi^{-1} g|\mathbf{B}^\sigma_{p,q}(\mathbb{R}^d)\| \lesssim \|a_j(2^{-j})\psi^{-1}|C^K(\mathbb{R}^d)\| \|g|\mathbf{B}^\sigma_{p,q}(\mathbb{R}^d)\|$$

$$\lesssim \|g|\mathbf{B}^\sigma_{p,q}(\mathbb{R}^d)\|$$

and hence

$$\|a_j(2^{-j})\psi^{-1}|M(\mathbf{B}_{p,q}^{\sigma}(\mathbb{R}^d))\| \lesssim 1, \qquad j \in \mathbb{N}_0. \tag{8.4.8}$$

By (8.4.8) and the homogeneity property we then get, for any $\sigma > \sigma' > s$ and $j \in \mathbb{N}_0$,

$$\|(ma_j)(2^{-j}\cdot)|\mathbf{B}_p^{\sigma'}(\mathbb{R}^d)\| \lesssim \|m(2^{-j}\cdot)a_j(2^{-j}\cdot)|\mathbf{B}_{p,q}^{\sigma}(\mathbb{R}^d)\|$$

$$\lesssim \|a_j(2^{-j}\cdot)\psi^{-1}|M(\mathbf{B}_{p,q}^{\sigma}(\mathbb{R}^d))\|\|m(2^{-j}\cdot)\psi|\mathbf{B}_{p,q}^{\sigma}(\mathbb{R}^d)\|$$

$$\lesssim \|m(2^{-j}\cdot)\psi|\mathbf{B}_{p,q}^{\sigma}(\mathbb{R}^d)\|. \tag{8.4.9}$$

In the case of $a_{j,l}$ with $l \in \mathbb{Z}^d$ one arrives at (8.4.9) with $a_{j,l}$ and $\psi(\cdot - l)$ in place of a_j and ψ, respectively. Hence,

$$\|ma_{j,l}(2^{-j}\cdot)|\mathbf{B}_p^{\sigma'}(\mathbb{R}^d)\| \lesssim \sup_{j,l} \|m(2^{-j}\cdot)\psi(\cdot - l)|\mathbf{B}_{p,q}^{\sigma}(\mathbb{R}^d)\|$$

$$= \|m|\mathbf{B}_{p,q,\mathrm{selfs}}^{\sigma}(\mathbb{R}^d)\|, \qquad j \in \mathbb{N}_0, \ l \in \mathbb{Z}^d, \tag{8.4.10}$$

and therefore, $ma_{j,l}$ is a (σ', p)-atom where $\sigma' > s$. By Theorem 2.3.23, in view of (8.4.7), $mf \in \mathbf{B}_{p,q}^s(\mathbb{R}^d)$ and

$$\|mf|\mathbf{B}_{p,q}^s(\mathbb{R}^d)\| \leq \|\lambda|b_{p,q}^s\|\|m|\mathbf{B}_{p,q,\mathrm{selfs}}^{\sigma}(\mathbb{R}^d)\| \sim \|f|\mathbf{B}_{p,q}^s\|\|m|\mathbf{B}_{p,q,\mathrm{selfs}}^{\sigma}(\mathbb{R}^d)\|,$$

which completes the proof of (i).

We now prove (ii): Restricting ourselves to $p = q$, let now $m \in \mathbf{B}_{p,\mathrm{selfs}}^s(\mathbb{R}^d)$. We can modify (8.4.9) by choosing $\sigma' = \sigma = s$,

$$\|(ma_j)(2^{-j}\cdot)|\mathbf{B}_p^s(\mathbb{R}^d)\| = \|m(2^{-j}\cdot)a_j(2^{-j}\cdot)|\mathbf{B}_p^s(\mathbb{R}^d)\|$$

$$\lesssim \|a_j(2^{-j}\cdot)\psi^{-1}|M(\mathbf{B}_p^s(\mathbb{R}^d))\|\|m(2^{-j}\cdot)\psi|\mathbf{B}_p^s(\mathbb{R}^d)\|$$

$$\lesssim \|m(2^{-j}\cdot)\psi|\mathbf{B}_p^s(\mathbb{R}^d)\|, \tag{8.4.11}$$

yielding for general atoms $a_{j,l}$,

$$\|ma_{j,l}(2^{-j}\cdot)|\mathbf{B}_{p,}^s(\mathbb{R}^d)\| \lesssim \sup_{j,l} \|m(2^{-j}\cdot)\psi(\cdot - l)|\mathbf{B}_p^s(\mathbb{R}^d)\|$$

$$= \|m|\mathbf{B}_{p,\mathrm{selfs}}^s(\mathbb{R}^d)\|, \qquad j \in \mathbb{N}_0, \ l \in \mathbb{Z}^d. \tag{8.4.12}$$

Since $p \leq 1$, we have that $\mathbf{B}_p^s(\mathbb{R}^d)$ is a p-Banach space. From (8.4.6), using (8.4.7) and (8.4.12), we obtain

$$\|mf|\mathbf{B}_p^s(\mathbb{R}^d)\|^p \leq \sum_{j=0}^{\infty} \sum_{l \in \mathbb{Z}^d} |\lambda_{j,l}|^p 2^{j(s-\frac{d}{p})p} 2^{-j(s-\frac{d}{p})p} \|ma_{j,l}|\mathbf{B}_p^s(\mathbb{R}^d)\|^p$$

$$\sim \|\lambda|b_{p,p}^s\|^p \|(ma_{j,l})(2^{-j} \cdot)|\mathbf{B}_p^s(\mathbb{R}^d)\|^p$$

$$\lesssim \|\lambda|b_{p,p}^s\|^p \|m|\mathbf{B}_{p,\mathrm{selfs}}^s(\mathbb{R}^d)\|^p. \tag{8.4.13}$$

Hence $m \in M(\mathbf{B}_p^s(\mathbb{R}^d))$ and, moreover, $\mathbf{B}_{p,\mathrm{selfs}}^s(\mathbb{R}^d) \hookrightarrow M(\mathbf{B}_p^s(\mathbb{R}^d))$. The other embedding follows from part (i). $\qquad\square$

Remark 8.4.5 It remains open whether it is possible or not to generalize Theorem 8.4.4(ii) to the case when $p \neq q$. The problem in the proof given above is the estimate (8.4.13), which only holds if $p = q$.

Characteristic Functions as Multipliers The final part of this section is devoted to the question in which function spaces the characteristic function χ_Ω of a domain $\Omega \subset \mathbb{R}^d$ is a pointwise multiplier. We contribute to this question mainly as an application of Theorem 8.4.4. The results shed some light on a relationship between some fundamental notion of fractal geometry and pointwise multipliers in function spaces. For complementary remarks and studies in this direction we refer to [Tri03].

There are further considerations of a similar kind in the literature, asking for geometric conditions on the domain Ω such that the corresponding characteristic function χ_Ω provides multiplier properties, cf. [Gul84, Gul85], [FJ90], and [RS96, Sect. 4.6.3].

Definition 8.4.6 Let Γ be a non-empty compact set in $\mathbb{R}^d$. Let h be a positive non-decreasing function on the interval $(0, 1]$. Then Γ is called a *h-set*, if there is a finite Radon measure $\mu \in \mathbb{R}^d$ with

$$\operatorname{supp} \mu = \Gamma \quad \text{and} \quad \mu(B_r(\gamma)) \sim h(r), \qquad \gamma \in \Gamma, \ 0 < r \leq 1. \tag{8.4.14}$$

Remark 8.4.7 A measure μ with (8.4.14) satisfies the so-called *doubling condition*, meaning there is a constant $c > 0$ such that

$$\mu(B_{2r}(\gamma)) \leq c\mu(B_r(\gamma)), \qquad \gamma \in \Gamma, \ 0 < r < 1. \tag{8.4.15}$$

We refer to [Tri03, p. 476] for further explanations.

Theorem 8.4.8 *Let Ω be a bounded domain in $\mathbb{R}^d$. Moreover, let $\sigma > 0, 0 < p < \infty, 0 < q \leq \infty$, and let $\Gamma = \partial\Omega$ be an h-set with*

$$\sup_{j \in \mathbb{N}_0} \sum_{k=0}^{\infty} 2^{k\sigma q} \left(\frac{h(2^{-j})}{h(2^{-j-k})} 2^{-kd} \right)^{q/p} < \infty, \tag{8.4.16}$$

(with the usual modifications if $q = \infty$). Let $\mathbf{B}^\sigma_{p,q,\text{selfs}}(\mathbb{R}^d)$ be the spaces defined in (8.4.3). Then

$$\chi_\Omega \in \mathbf{B}^\sigma_{p,q,\text{selfs}}(\mathbb{R}^d).$$

Proof It simplifies the argument, and causes no loss of generality, to assume diam $\Omega < 1$. We define

$$\Omega^k = \{x \in \Omega : 2^{-k-2} \leq \text{dist}(x, \Gamma) \leq 2^{-k}\}, \qquad k \in \mathbb{N}_0.$$

Moreover, let

$$\{\varphi_l^k : k \in \mathbb{N}_0, \ l = 1, \ldots, M_k\} \subset C_0^\infty(\Omega)$$

be a resolution of unity,

$$\sum_{k \in \mathbb{N}_0} \sum_{l=1}^{M_k} \varphi_l^k(x) = 1 \qquad \text{if } x \in \Omega, \tag{8.4.17}$$

with

$$\text{supp}\,\varphi_l^k \subset \{x : |x - x_l^k| \leq 2^{-k}\} \subset \Omega^k$$

and

$$|D^\alpha \varphi_l^k(x)| \lesssim 2^{|\alpha|k}, \qquad |\alpha| \leq K,$$

where $K \in \mathbb{N}$ with $K > \sigma$. It is well known that resolutions of unity with the required properties exist. We now estimate the number M_k in (8.4.17). Combining the fact that the measure μ satisfies the doubling condition (8.4.15) together with (8.4.14) we arrive at

$$M_k h(2^{-k}) \lesssim 1, \qquad k \in \mathbb{N}_0. \tag{8.4.18}$$

Since the φ_l^k in (8.4.17) are K-atoms according to Definition 2.3.15, we obtain

$$\|\chi_\Omega|\mathbf{B}^\sigma_{p,q}(\mathbb{R}^d)\|^q \leq \sum_{k=0}^\infty 2^{k(\sigma - d/p)q} M_k^{q/p} \lesssim \sum_{k=0}^\infty 2^{k\sigma q}\left(\frac{2^{-kd}}{h(2^{-k})}\right)^{q/p} < \infty. \tag{8.4.19}$$

This shows that $\chi_\Omega \in \mathbf{B}^\sigma_{p,q}(\mathbb{R}^d)$. We now prove that $\chi_\Omega \in \mathbf{B}^\sigma_{p,q,\text{selfs}}(\mathbb{R}^d)$. We consider the non-negative function $\psi \in C^\infty(\mathbb{R}^d)$ satisfying (8.4.1) and (8.4.2). By the definition of selfsimilar spaces, it suffices to consider

$$\chi_\Omega(2^{-j}\cdot)\psi,$$

assuming in addition that $0 \in 2^j\Gamma = \{2^j\gamma = (2^j\gamma_1,\ldots,2^j\gamma_d) : \gamma \in \Gamma\}$, $j \in \mathbb{N}$. Let μ^j be the image measure of μ with respect to the dilations $y \mapsto 2^j y$. Then we obtain

$$\mu^j(B_{\sqrt{d}}(0) \cap 2^j\Gamma) \sim h(2^{-j}), \qquad j \in \mathbb{N}_0.$$

We apply the same argument as above to $B_{\sqrt{d}}(0) \cap 2^j\Omega$ and $B_{\sqrt{d}}(0) \cap 2^j\Gamma$ in place of Ω and Γ, respectively. Let M_k^j be the counterpart of the above number M_k. Then

$$M_k^j h(2^{-j-k}) \lesssim h(2^{-j}), \qquad j \in \mathbb{N}_0, \; k \in \mathbb{N}_0,$$

is the generalization of (8.4.18) we are looking for, which completes the proof.   □

In view of Theorem 8.4.4 we have the following result.

Corollary 8.4.9 *Let Ω be a bounded domain in $\mathbb{R}^d$. Moreover, let $\sigma > 0$, $0 < p < \infty$, $0 < q \le \infty$, and let $\Gamma = \partial\Omega$ be a h-set satisfying (8.4.16). Then*

$$\chi_\Omega \in M(\mathbf{B}_{p,q}^s(\mathbb{R}^d)) \qquad for \quad 1 < p < \infty, \quad 0 < s < \sigma,$$

and

$$\chi_\Omega \in M(\mathbf{B}_p^\sigma(\mathbb{R}^d)) \qquad for \quad 0 < p \le 1.$$

Remark 8.4.10 As for the assertion (8.4.16) we mention that

$$\sup_{j\in\mathbb{N}_0,k\in\mathbb{N}_0} 2^{k\sigma}\left(\frac{h(2^{-j})}{h(2^{-j-k})}2^{-kd}\right)^{1/p} < \infty$$

is the adequate counterpart for $\mathbf{B}_{p,\infty}^\sigma(\mathbb{R}^d)$. In the special case of ∂-sets, which corresponds to $h(t) \sim t^\partial$, the condition (8.4.16) therefore corresponds to

$$\sigma < \frac{d-\partial}{p} \qquad or \qquad \sigma = \frac{d-\partial}{p} \quad and \quad q = \infty.$$

For bounded Lipschitz domains Ω, i.e., $\partial = d - 1$, Theorem 8.4.8 therefore yields $\chi_\Omega \in \mathbf{B}_{p,q,\text{selfs}}^\sigma(\mathbb{R}^d)$ if

$$\sigma < \frac{1}{p} \qquad or \qquad \sigma = \frac{1}{p} \quad and \quad q = \infty. \qquad (8.4.20)$$

These results are sharp since there exists a Lipschitz domain Ω in $\mathbb{R}^d$ such that

$$\chi_\Omega \in \mathbf{B}_{p,\infty,\text{selfs}}^{\frac{1}{p}}(\mathbb{R}^d) \qquad \text{and} \qquad \chi_\Omega \notin \mathbf{B}_{p,q}^{\frac{1}{p}}(\mathbb{R}^d) \quad \text{if} \quad 0 < q < \infty.$$

In order to see this let $\Omega = \left[-\frac{1}{2}, \frac{1}{2}\right]^d$. Observing that

$$\omega_r(\chi_\Omega, t)_p \lesssim t^{\frac{1}{p}}$$

one calculates

$$\left(\int_0^1 t^{-\sigma q} \omega_r(\chi_\Omega, t)_p^q \frac{dt}{t}\right)^{1/q} \lesssim \left(\int_0^1 t^{(\frac{1}{p}-\sigma)q} \frac{dt}{t}\right)^{1/q}$$

which is finite if, and only if, σ satisfies (8.4.20). Therefore, in view of Theorem 8.4.4, concerning Lipschitz domains there is an

alternative s.t. either the trace of $\mathbf{B}_{p,q}^\sigma(\mathbb{R}^d)$ on Γ

exists or χ_Ω is a pointwisemultiplier for $\mathbf{B}_{p,q}^\sigma(\mathbb{R}^d)$,

as was conjected for F-spaces in [Tri02, p. 36]: For smoothness $\sigma > \frac{1}{p}$ we have traces according to Theorem 8.3.1 whereas for $\sigma < \frac{1}{p}$ we know that χ_Ω is a pointwise multiplier for $\mathbf{B}_{p,q}^\sigma(\mathbb{R}^d)$. The limiting case $\sigma = \frac{1}{p}$ needs to be discussed separately: according to Corollary 8.3.3 we have traces for B-spaces with $q \leq \min(1, p)$, but χ_Ω is (possibly) only a multiplier for $\mathbf{B}_{p,\infty}^{1/p}(\mathbb{R}^d)$. There remains a 'gap' for spaces

$$\mathbf{B}_{p,q}^{1/p}(\mathbb{R}^d) \qquad \text{when} \qquad \min(1, p) < q < \infty.$$

Chapter 9
Traces of Generalized Smoothness Morrey Spaces on Domains

In this chapter we deal with traces of functions in generalized smoothness Morrey spaces on the boundary of C^k domains Ω. Our results remain valid for the usual Besov-Morrey spaces $\mathcal{N}^s_{u,p,q}(\Omega)$, Triebel-Lizorkin-Morrey spaces $\mathcal{E}^s_{u,p,q}(\Omega)$, and Triebel-Lizorkin type spaces $F^{s,\tau}_{p,q}(\Omega)$, which are all included in our scales as special cases. Moreover, we also show how the trace results can be transferred to Besov-type spaces $B^{s,\tau}_{p,q}(\Omega)$, which are not covered by this approach (for an overview of the spaces and their relations we refer to the mindmap on page viii).

We start in Sect. 9.1 with ideas from [FR95] and present a lift operator within the scale of smoothness Morrey spaces which, in particular, has some 'good' mapping properties on $\mathbb{R}^d_+$ according to Corollary 9.1.4. Afterwards, this lift operator is used in Sect. 9.2 to construct an extension operator for the function spaces from $\mathbb{R}^d_+$ to $\mathbb{R}^d$. For the proof we rely on quarkonial decompositions for the spaces from Theorem 2.5.12.

In Sect. 9.3 we then continue by the following standard method: we use a resolution of unity of the domain together with multiplier and diffeomorphism properties of the smoothness Morrey spaces and the equivalent quasi-norm from Proposition 2.5.20. This reduces the trace problem on domains to the study of traces on the upper half plane $\mathbb{R}^d_+$. The extension operator from Sect. 9.2 then transfers the problem from $\mathbb{R}^d_+$ to the hyperplane $\mathbb{R}^{d-1}$. Finally, we take advantage of already existing trace results on hyperplanes, cf. Theorem 9.3.1, in order to complete our task. Afterwards, in Sect. 9.3.2, we investigate the corresponding trace problem for Besov-type spaces $B^{s,\tau}_{p,q}(\Omega)$. We proceed in the same way as described before, looking for suitable substitutes of the above listed items within this scale of spaces. The results we obtain for $B^{s,\tau}_{p,q}(\Omega)$ are in good agreement with the corresponding trace results for the spaces $F^{s,\tau}_{p,q}(\Omega)$, which are covered by the results from Theorem 9.3.2 for the smoothness Morrey spaces.

Finally, in Sect. 9.4 we apply the trace results for the Besov-type spaces in order to obtain some *a priori* estimates for solutions of elliptic boundary value problems, which extend the results from [ElB05].

© The Author(s), under exclusive license to Springer Nature Switzerland AG 2021
C. Schneider, *Beyond Sobolev and Besov*, Lecture Notes in Mathematics 2291,
https://doi.org/10.1007/978-3-030-75139-5_9

Our understanding of the trace operator in the present chapter slightly differs from the explanations given in Chap. 8, formula (8.0.1) and below: as was outlined there, one usually uses the fact that $S(\mathbb{R}^d)$ is dense in some function space $X(\mathbb{R}^d)$, since then the trace is completely defined by density arguments. However, this is not available when dealing with generalized smoothness Morrey spaces. Therefore, we have proceed in a different way and rely on atomic decomposition techniques. This approach was already used in [NNS16, Saw10], where the authors dealt with traces on hyperplanes in the setting of smoothness Morrey spaces. To be precise, assume that $d \geq 2$. For $x = (x_1, \cdots, x_d) \in \mathbb{R}^d$, we put $x' = (x_1, \cdots, x_{d-1}) \in \mathbb{R}^{d-1}$ and in this case we might also write $x = (x', x_d)$. Our understanding of the trace operator on the hyperplane $\mathbb{R}^{d-1} = \{(x', 0) : x' \in \mathbb{R}^{d-1}\}$ of $\mathbb{R}^d$ in the context of generalized smoothness Morrey spaces is as follows: If f is a smooth function, e.g. $f \in S(\mathbb{R}^d)$, it makes sense to consider the restriction of f pointwise on the hyperplane and define the trace operator $\text{Tr}_{\mathbb{R}^{d-1}}$ by

$$(\text{Tr}_{\mathbb{R}^{d-1}} f)(x') := f(x', 0), \quad x' \in \mathbb{R}^{d-1}. \tag{9.0.1}$$

Let $X(\mathbb{R}^d) := \mathcal{A}^s_{\mathcal{M}^\varphi_{p,q}}(\mathbb{R}^d)$ denote a generalized smoothness Morrey space according to Definition 2.5.3. In order to define the trace for $f \in X(\mathbb{R}^d)$ we use the atomic decomposition from Theorem 2.5.9, i.e.,

$$f = \sum_{j=0}^{\infty} \sum_{m \in \mathbb{Z}^n} \lambda_{j,m} a_{j,m} \tag{9.0.2}$$

and define $\text{Tr}_{\mathbb{R}^{d-1}} f$ by

$$\text{Tr}_{\mathbb{R}^{d-1}} f := \sum_{j=0}^{\infty} \left(\sum_{m \in \mathbb{Z}^n} \lambda_{j,m} a_{j,m}(\cdot', 0) \right). \tag{9.0.3}$$

This definition makes sense since the proof of the atomic decomposition from Theorem 2.5.9 reveals that $\lambda_{j,m} a_{j,m}$ is obtained canonically from $f \in X(\mathbb{R}^d)$, meaning there exists a continuous linear operator $I_{j,m}$ from $X(\mathbb{R}^d)$ into the space of L_∞ functions with compact support such that $I_{j,m}(f) := \lambda_{j,m} a_{j,m}$.

If $f \in S(\mathbb{R}^d)$ then (9.0.3) actually coincides with (9.0.1). This can be seen as follows: the limit in (9.0.2) takes place in $C(\overline{\mathbb{R}^d})$ because $f \in B^\varepsilon_{\infty,\infty}(\mathbb{R}^d) \hookrightarrow C(\overline{\mathbb{R}^d})$ for all $\varepsilon \in (0, 1)$. But this in turn implies pointwise convergence since

$$\lim_{J \to \infty} \left(\sup_{x \in \mathbb{R}^d} \left| f(x) - \sum_{j=0}^{J} \sum_{m \in \mathbb{Z}^n} \lambda_{j,m} a_{j,m}(x) \right| \right) = 0,$$

demonstrating that (9.0.3) is well-defined.

9.1 A Lift Operator

In this section we use the family of lift operators $\{J_\sigma\}_{\sigma\in\mathbb{R}}$ due to Franke and Runst [FR95], which goes back to Triebel [Tri78], and extend it to our setting such that it has the following properties:

- J_σ is an isomorphism from $\mathcal{A}^s_{\mathcal{M}^\varphi_{p,q}}(\mathbb{R}^d)$ to $\mathcal{A}^{s-\sigma}_{\mathcal{M}^\varphi_{p,q}}(\mathbb{R}^d)$.
- J_σ and $J_{-\sigma}$ are inverse to each other.
- If $f \in \mathcal{S}'(\mathbb{R}^d)$ is supported on $\mathbb{R}^{d-1} \times (-\infty, 0]$, so is $J_\sigma f$.

Note that by [NNS16, Prop. 3.2] the lift operators $(1 - \Delta)^{\sigma/2}$ also satisfy the above conditions except the third one, which is crucial for us.

We start with the construction of a function on which our family of operators is build on. Let $\eta \in \mathcal{S}(\mathbb{R})$ be a positive real-valued function with $\operatorname{supp}\eta \subset [-2, -1]$ and $\int_\mathbb{R} \eta(x)dx = 2$. For any $0 \leq \varepsilon \ll 1$, we define a holomorphic function ψ_ε on $\mathbb{C}$ by

$$\psi_\varepsilon(z) := \int_{-\infty}^0 \eta(t)e^{-i\varepsilon tz}dt - iz.$$

Let $\mathbb{H} := \{z \in \mathbb{C} : \operatorname{Im}(z) > 0\}$ and $\overline{\mathbb{H}} := \{z \in \mathbb{C} : \operatorname{Im}(z) \geq 0\}$. Furthermore, we consider the domain $\Omega = \{z \in \mathbb{C} : |z| > 4, \operatorname{Re}(z) > 0\}$. If $z \in \mathbb{C}$ satisfies $|z| > 4$ and $\operatorname{Im}(z) > 0$, it follows that $-iz \in \Omega$. Hence, we see

$$\operatorname{dist}(\psi_\varepsilon(z), \Omega) \leq |\psi_\varepsilon(z) + iz| = \left|\int_\infty^0 \eta(t)e^{-i\varepsilon tz}dt\right| < 2. \tag{9.1.1}$$

If $z \in \mathbb{C}$ satisfies $|z| \leq 4$ and $\operatorname{Im}(z) \geq 0$, then we have $\operatorname{Re}(\psi_0(z)) = 2 + \operatorname{Im}(z)$. Thus, for any $0 < \varepsilon \ll 1$, we obtain

$$\operatorname{Re}(\psi_\varepsilon(z)) = \int_{-\infty}^0 \eta(t)e^{\varepsilon t\operatorname{Im}(z)}\cos(\varepsilon t\operatorname{Re}(z))dt + \operatorname{Im}(z) \geq \frac{3}{2},$$

since the integrand is continuous. If $\varepsilon > 0$ is a sufficiently small number, as illustrated in Fig. 9.1 we see that ψ_ε maps $\overline{\mathbb{H}}$ to

$$\Omega_0 := \{z \in \mathbb{C} : \operatorname{Re}(z) > 1\} \cup \{z \in \mathbb{C} : |\operatorname{Im}(z)| > 1\}.$$

Below fix a small $\varepsilon > 0$. We select a branch-cut of $\log$ on $\mathbb{C} \setminus (-\infty, 0]$ such that $\log 1 = 0$. Then we define $z^a = \exp(a \log z)$ for $z \in \mathbb{C} \setminus (-\infty, 0]$. For any $\sigma \in \mathbb{R}$, we define the function $\Phi^{(\sigma)} : \mathbb{R}^{d-1} \times \overline{\mathbb{H}} \to \mathbb{C}$ by

$$\Phi^{(\sigma)}(x', z_n) := \left(\langle x'\rangle \psi_\varepsilon\left(\frac{z_n}{\langle x'\rangle}\right)\right)^\sigma, \quad z_n \in \overline{\mathbb{H}},$$

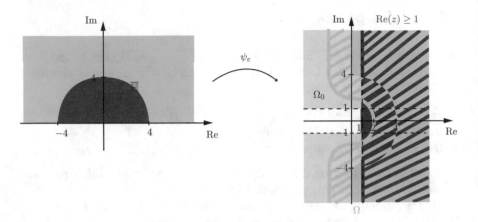

Fig. 9.1 Function ψ_ε and its mapping properties

which is well-defined for $\sigma \in \mathbb{R}$. Here we put $\langle x' \rangle = \sqrt{1 + x_1^2 + \ldots + x_{d-1}^2}$. Setting $\Phi^{(1)} = \Phi$ we have the following lemma clarifying the behaviour of Φ with respect to differentiation, cf. [Saw10, Lem. 4.3].

Lemma 9.1.1 (Behaviour of Φ) *For any multi-index* $\alpha \in \mathbb{N}_0^d$, *there exists a constant* $c_\alpha > 0$ *such that*

$$|\partial^\alpha \Phi(x', z_d)| \le c_\alpha \left(\langle x' \rangle + |z_d| \right)^{1 - |\alpha|} \tag{9.1.2}$$

for all $(x', z_d) \in \mathbb{R}^{d-1} \times \overline{\mathbb{H}}$. *Furthermore, we can even arrange that* c_0 *satisfies*

$$c_0^{-1} \left(\langle x' \rangle + |z_d| \right) \le |\Phi(x', z_d)| \le c_0 \left(\langle x' \rangle + |z_d| \right) \tag{9.1.3}$$

for all $(x', z_d) \in \mathbb{R}^{d-1} \times \overline{\mathbb{H}}$.

We will also use the same symbol $\Phi^{(\sigma)}$ for $\Phi^{(\sigma)}|_{\mathbb{R}^{d-1} \times \mathbb{R}}$. Then by Theorem 2.5.16 and Lemma 9.1.1, we derive the following proposition.

Proposition 9.1.2 (Properties of Lift Operator) *Let* $0 < p < \infty$, $0 < q \le \infty$, *and* $\varphi \in \mathcal{G}_p$. *Additionally assume that* φ *satisfies* (2.5.2) *when* $q < \infty$ *and* $\mathcal{A} = \mathcal{E}$. *Then for any* $\sigma \in \mathbb{R}$ *we have the following properties:*

(i) $J_\sigma := \mathcal{F}^{-1}[\Phi^{(\sigma)} \mathcal{F}]$ *is a linear isomorphism between* $\mathcal{A}^s_{\mathcal{M}^\varphi_{p,q}}(\mathbb{R}^d)$ *and* $\mathcal{A}^{s-\sigma}_{\mathcal{M}^\varphi_{p,q}}(\mathbb{R}^d)$.

(ii) $J_{-\sigma}$ *is the inverse operator of* J_σ.

(iii) *For any* $f \in \mathcal{A}^s_{\mathcal{M}^\varphi_{p,q}}(\mathbb{R}^d)$, *we have* $\| J_\sigma f | \mathcal{A}^{s-\sigma}_{\mathcal{M}^\varphi_{p,q}}(\mathbb{R}^d) \| \sim \| f | \mathcal{A}^s_{\mathcal{M}^\varphi_{p,q}}(\mathbb{R}^d) \|$.

Proof Let μ_0, μ be functions as in Definition 2.5.3, and let $R \in \mathbb{N}$ be such that

$$\operatorname{supp} \mu_0 \subset \{x \in \mathbb{R}^d : |x| \leq 2^R\} \quad \text{and} \quad \operatorname{supp} \mu \subset \{x \in \mathbb{R}^d : 2^{-R} \leq |x| \leq 2^R\}.$$

We consider the case $\mathcal{A}^s_{\mathcal{M}^\varphi_{p,q}}(\mathbb{R}^d) = \mathcal{E}^s_{\mathcal{M}^\varphi_{p,q}}(\mathbb{R}^d)$ as the other case follows in an analogous way. We have

$$\left\| J_\sigma f \mid \mathcal{E}^{s-\sigma}_{\mathcal{M}^\varphi_{p,q}}(\mathbb{R}^d) \right\| = \left\| \left(\sum_{j=0}^{\infty} 2^{j(s-\sigma)q} |\mathcal{F}^{-1}(\mu_j \Phi^{(\sigma)} \mathcal{F} f)(\cdot)|^q \right)^{1/q} \mid \mathcal{M}^\varphi_p(\mathbb{R}^d) \right\|.$$

Let $\phi \in \mathcal{S}(\mathbb{R}^d)$ be such that

$$\phi(x) := 1 \quad \text{if} \quad 2^{-R} \leq |x| \leq 2^R \quad \text{and} \quad \operatorname{supp} \phi \subset \{x \in \mathbb{R}^d : 2^{-R-1} \leq |x| \leq 2^{R+1}\}.$$

Then

$$\mathcal{F}^{-1}(\mu_j \Phi^{(\sigma)} \mathcal{F} f) = \mathcal{F}^{-1}(\Phi^{(\sigma)} \phi(2^{-j}\cdot) \mu_j \mathcal{F} f), \quad j \in \mathbb{N}.$$

Applying Theorem 2.5.16 with $\nu \in \mathbb{N}$ such that $\nu > \frac{d}{\min(1,p,q)} + \frac{d}{2}$ and

$$H_j(x) := 2^{-\sigma j} \Phi^{(\sigma)}(x) \phi(2^{-j} x), \quad j \in \mathbb{N},$$

we obtain

$$\left\| \left(\sum_{j=1}^{\infty} 2^{j(s-\sigma)q} |\mathcal{F}^{-1}(\mu_j \Phi^{(\sigma)} \mathcal{F} f)(\cdot)|^q \right)^{1/q} \mid \mathcal{M}^\varphi_p(\mathbb{R}^d) \right\|$$

$$\lesssim \left(\sup_{k \in \mathbb{N}} \| H_k(2^{k+R+1} \cdot) \mid H_2^\nu(\mathbb{R}^d) \| \right) \left\| \left(\sum_{j=1}^{\infty} 2^{jsq} |\mathcal{F}^{-1}(\mu_j \mathcal{F} f)(\cdot)|^q \right)^{1/q} \mid \mathcal{M}^\varphi_p(\mathbb{R}^d) \right\|$$

$$\lesssim \| f \mid \mathcal{E}^s_{\mathcal{M}^\varphi_{p,q}}(\mathbb{R}^d) \|,$$

where we used Lemma 9.1.1 to estimate the first term in the last inequality. The term corresponding to $j = 0$ can be dealt with in a similar way, so that we arrive at

$$\left\| J_\sigma f \mid \mathcal{E}^{s-\sigma}_{\mathcal{M}^\varphi_{p,q}}(\mathbb{R}^d) \right\| \lesssim \| f \mid \mathcal{E}^s_{\mathcal{M}^\varphi_{p,q}}(\mathbb{R}^d) \|.$$

The proof is completed by observing that $J_\sigma J_{-\sigma} f = f$. $\square$

For the support of $J_\sigma f$ we have the following result. A proof may be found in [Saw10, Prop. 4.6].

Proposition 9.1.3 *If $f \in \mathcal{S}'(\mathbb{R}^d)$ is supported in $\mathbb{R}^{d-1} \times (-\infty, 0]$, then so is $J_\sigma f$.*

We have seen that our family of lift operators satisfies all the required properties stated at the beginning of this section. This now enables us to prove the following corollary, which will be used in the next section in order to construct a suitable extension operator.

Corollary 9.1.4 (Properties of Lift Operator on $\mathbb{R}^d{}_+$) *Let* $0 < p < \infty, 0 < q \le \infty$, *and* $\varphi \in \mathcal{G}_p$. *Additionally assume that* φ *satisfies* (2.5.2) *when* $q < \infty$ *and* $\mathcal{A} = \mathcal{E}$.

(i) *Let* $f \in \mathcal{A}^s_{\mathcal{M}^{\varphi}_{p,q}}(\mathbb{R}^d{}_+)$. *Then* $J_{\sigma} f := (J_{\sigma} g)\big|_{\mathbb{R}^d{}_+}$ *does not depend on the choice of the representative* $g \in \mathcal{A}^s_{\mathcal{M}^{\varphi}_{p,q}}(\mathbb{R}^d)$ *of* f.

(ii) J_{σ} *is an isomorphism from* $\mathcal{A}^s_{\mathcal{M}^{\varphi}_{p,q}}(\mathbb{R}^d{}_+)$ *to* $\mathcal{A}^{s-\sigma}_{\mathcal{M}^{\varphi}_{p,q}}(\mathbb{R}^d{}_+)$. *Furthermore,* $J_{-\sigma}$ *is the inverse of* J_{σ}.

Proof We will prove (i). Let $g_1, g_2 \in \mathcal{A}^s_{\mathcal{M}^{\varphi}_{p,q}}(\mathbb{R}^d)$ satisfy $f = g_1\big|_{\mathbb{R}^d{}_+} = g_2\big|_{\mathbb{R}^d{}_+}$. Then we have

$$(J_{\sigma} g_1)\big|_{\mathbb{R}^d{}_+} - (J_{\sigma} g_2)\big|_{\mathbb{R}^d{}_+} = (J_{\sigma}(g_1 - g_2))\big|_{\mathbb{R}^d{}_+} = 0,$$

by linearity of J_{σ}, the fact that $(g_1 - g_2)\big|_{\mathbb{R}^d{}_+} = 0$ and Proposition 9.1.3. Thus, we obtain

$$(J_{\sigma} g_1)\big|_{\mathbb{R}^d{}_+} = (J_{\sigma} g_2)\big|_{\mathbb{R}^d{}_+},$$

which means that $J_{\sigma} f$ does not depend on $g \in \mathcal{A}^s_{\mathcal{M}^{\varphi}_{p,q}}(\mathbb{R}^d)$ satisfying $g\big|_{\mathbb{R}^d{}_+} = f$. Assertion (ii) follows immediately from the properties of J_{σ} as an operator on $\mathcal{A}^s_{\mathcal{M}^{\varphi}_{p,q}}(\mathbb{R}^d)$. $\qquad\square$

9.2 Construction of an Extension Operator

Having constructed the lift operator J_{σ}, we are now able to establish the extension theorem in this section. For this, we deal with the following set in the sequel:

$$R(N) := \left\{ (p, q, s) : \frac{1}{N} \le p < \infty, \frac{1}{N} \le q \le \infty, |s| < N \right\}.$$

Theorem 9.2.1 (Extension Operator) *Let* $0 < p < \infty, 0 < q \le \infty$, *and* $\varphi \in \mathcal{G}_p$. *Additionally assume that* φ *satisfies* (2.5.2) *when* $q < \infty$ *and* $\mathcal{A} = \mathcal{E}$. *Then for* $N \in \mathbb{N}$, *there exists an extension operator* Ext_N,

$$\mathrm{Ext}_N : \bigcup_{(p,q,s) \in R(N)} \mathcal{A}^s_{\mathcal{M}^{\varphi}_{p,q}}(\mathbb{R}^d{}_+) \longrightarrow \bigcup_{(p,q,s) \in R(N)} \mathcal{A}^s_{\mathcal{M}^{\varphi}_{p,q}}(\mathbb{R}^d),$$

that satisfies the properties: if $(p, q, s) \in R(N)$ *then the restriction* $\text{Ext}_N|_{\mathcal{A}^s_{\mathcal{M}^\varphi_{p,q}}(\mathbb{R}^d_+)}$
is a continuous mapping from $\mathcal{A}^s_{\mathcal{M}^\varphi_{p,q}}(\mathbb{R}^d_+)$ *to* $\mathcal{A}^s_{\mathcal{M}^\varphi_{p,q}}(\mathbb{R}^d)$ *satisfying* $\text{Ext}_N f|_{\mathbb{R}^d_+} =$
f *for all* $f \in \mathcal{A}^s_{\mathcal{M}^\varphi_{p,q}}(\mathbb{R}^d_+)$.

Proof *Step 1.* We start with the general set up. Details may be found in [AF03, Ch. 5]. Let $M \in \mathbb{N}$ be large enough. We define $\lambda_1, \ldots, \lambda_M$ so that

$$\sum_{j=0}^{M} (-j)^l \lambda_j = \delta_{0,l}, \tag{9.2.1}$$

for all $l = 0, \ldots, M$. Here $\delta_{0,l}$ denotes the Kronecker-symbol, i.e., $\delta_{0,0} = 1$ and $\delta_{0,l} = 0$ for $l \geq 1$, and it is assumed that $0^l = \delta_{0,l}$. The determinant of this system of linear equations is a constant multiple of the Vandermonde determinant $\{j^i\}_{i,j=0,1,\ldots,M}$, which is never 0. Therefore, the unknowns $\lambda_1, \ldots, \lambda_M$ are determined uniquely by (9.2.1). Given a function $f : \mathbb{R}^{d-1} \times [0, \infty) \to \mathbb{C}$, we define

$$f^*(x) := \begin{cases} f(x), & \text{if } x_d \geq 0, \\ \sum_{j=0}^{M} \lambda_j f(x', -jx_d), & \text{if } x_d \leq 0. \end{cases}$$

By definition of the λ_j we see that $\sum_{j=0}^{M} \lambda_j f(x', 0) = f(x', 0)$. Furthermore, if f is in $C^{M-1}(\overline{\mathbb{R}^d_+})$, then $f^* : \mathbb{R}^d \to \mathbb{C}$ also belongs to $C^{M-1}(\mathbb{R}^d)$.

Step 2. We define Ext_N for $s > \frac{d}{p}$. By Proposition 2.5.17 and Remark 2.5.18, in this case we have

$$\mathcal{A}^s_{\mathcal{M}^\varphi_{p,q}}(\mathbb{R}^d) \hookrightarrow C(\overline{\mathbb{R}^d}). \tag{9.2.2}$$

We assume now that

$$\frac{1}{N} \leq p < \infty, \quad \frac{1}{N} \leq q \leq \infty, \quad \frac{d}{p} < s \leq N. \tag{9.2.3}$$

Let $M \gg (d + 1)N$ be large enough ($M > s$ for all s in (9.2.3)), where M is the integer from Step 1. Given $f \in \mathcal{A}^s_{\mathcal{M}^\varphi_{p,q}}(\mathbb{R}^d_+)$, we pick a representative $g \in \mathcal{A}^s_{\mathcal{M}^\varphi_{p,q}}(\mathbb{R}^d)$ such that

$$f = g|_{\mathbb{R}^d_+}, \quad \|g|\mathcal{A}^s_{\mathcal{M}^\varphi_{p,q}}(\mathbb{R}^d)\| \leq 2\|f|\mathcal{A}^s_{\mathcal{M}^\varphi_{p,q}}(\mathbb{R}^d_+)\|.$$

Taking the quarkonial decomposition of g,

$$g = \sum_{\beta \in \mathbb{N}_0} \sum_{j \in \mathbb{N}_0} \sum_{m \in \mathbb{Z}^n} \lambda_{j,m}^{\beta} (\beta q u)_{j,m},$$

the coefficients satisfy

$$\|\lambda | a_{\mathcal{M}_{p,q}^{\varphi}}^{s}\|_{\rho} \leq c \|g | \mathcal{A}_{\mathcal{M}_{p,q}^{\varphi}}^{s}(\mathbb{R}^d)\| \leq c' \|f | \mathcal{A}_{\mathcal{M}_{p,q}^{\varphi}}^{s}(\mathbb{R}^d{}_+)\|$$

with $\rho > R$, cf. Theorem 2.5.12. Since (9.2.2) holds, we see that

$$g^* := \sum_{\beta \in \mathbb{N}_0} \sum_{j \in \mathbb{N}_0} \sum_{m \in \mathbb{Z}^n} \lambda_{j,m}^{\beta} (\beta q u)_{j,m}^*$$

does not depend on the particular choice of the representative g. Define

$$\mathrm{Ext}_N f := \sum_{\beta \in \mathbb{N}_0} \sum_{j \in \mathbb{N}_0} \sum_{m \in \mathbb{Z}^n} \lambda_{j,m}^{\beta} (\beta q u)_{j,m}^*$$

and its β-partial sum

$$\mathrm{Ext}_N^{\beta} f := \sum_{j \in \mathbb{N}_0} \sum_{m \in \mathbb{Z}^n} \lambda_{j,m}^{\beta} (\beta q u)_{j,m}^*.$$

Although the sum defining $\mathrm{Ext}_N f$ is not a quarkonial decomposition, we are still able to regard $2^{-(R+\varepsilon)|\beta|} \mathrm{Ext}_N^{\beta} f$ as an atomic decomposition if $0 < \varepsilon < \rho - R$. Putting $\delta := -R - \varepsilon + \rho > 0$ we have

$$\left\| \mathrm{Ext}_N^{\beta} f | \mathcal{A}_{\mathcal{M}_{p,q}^{\varphi}}^{s}(\mathbb{R}^d) \right\| \leq 2^{(R+\varepsilon)|\beta|} \|\lambda^{\beta} | a_{\mathcal{M}_{p,q}^{\varphi}}^{s}\|$$

$$\leq 2^{(R+\varepsilon)|\beta|} 2^{-\rho|\beta|} \|\lambda | a_{\mathcal{M}_{p,q}^{\varphi}}^{s}\|_{\rho}$$

$$= 2^{-\delta|\beta|} \|\lambda | a_{\mathcal{M}_{p,q}^{\varphi}}^{s}\|_{\rho} \lesssim 2^{-\delta|\beta|} \|f | \mathcal{A}_{\mathcal{M}_{p,q}^{\varphi}}^{s}(\mathbb{R}^d{}_+)\|.$$

With this for $\varkappa := \min(1, p, q)$ we compute

$$\|\mathrm{Ext}_N f | \mathcal{A}_{\mathcal{M}_{p,q}^{\varphi}}^{s}(\mathbb{R}^d)\|^{\varkappa} = \left\| \sum_{\beta} \mathrm{Ext}_N^{\beta} f | \mathcal{A}_{\mathcal{M}_{p,q}^{\varphi}}^{s}(\mathbb{R}^d) \right\|^{\varkappa}$$

$$\leq \sum_{\beta} \|\mathrm{Ext}_N^{\beta} f | \mathcal{A}_{\mathcal{M}_{p,q}^{\varphi}}^{s}(\mathbb{R}^d)\|^{\varkappa}$$

$$\lesssim \sum_{\beta} 2^{-\delta|\beta|\varkappa} \|f | \mathcal{A}_{\mathcal{M}_{p,q}^{\varphi}}^{s}(\mathbb{R}^d{}_+)\|^{\varkappa} \lesssim \|f | \mathcal{A}_{\mathcal{M}_{p,q}^{\varphi}}^{s}(\mathbb{R}^d{}_+)\|^{\varkappa}.$$

Thus, we see that Ext_N is a continuous mapping with the desired properties.

Step 3.   We deal with the construction of Ext_N in general. For $(p, q, s) \in R(N)$, choose $\sigma \in \mathbb{R}$ and $L \in \mathbb{N}$ large enough so that

$$\frac{d}{p} \le dN < -N + \sigma < s + \sigma < N + \sigma < L.$$

Hence, $s + \sigma$ satisfies the assumptions of Step 2, so that $\mathrm{Ext}_L \big| \mathcal{A}^{s+\sigma}_{\mathcal{M}^{\varphi}_{p,q}}(\mathbb{R}^d{}_+)$ is a continuous mapping from $\mathcal{A}^{s+\sigma}_{\mathcal{M}^{\varphi}_{p,q}}(\mathbb{R}^d{}_+)$ to $\mathcal{A}^{s+\sigma}_{\mathcal{M}^{\varphi}_{p,q}}(\mathbb{R}^d)$ and $\mathrm{Ext}_L f \big|_{\mathbb{R}^d_+} = f$ for all $f \in \mathcal{A}^{s+\sigma}_{\mathcal{M}^{\varphi}_{p,q}}(\mathbb{R}^d{}_+)$. Since $J_{-\sigma}$ maps $\mathcal{A}^{s}_{\mathcal{M}^{\varphi}_{p,q}}(\mathbb{R}^d{}_+)$ to $\mathcal{A}^{s+\sigma}_{\mathcal{M}^{\varphi}_{p,q}}(\mathbb{R}^d{}_+)$ and $\mathcal{A}^{s}_{\mathcal{M}^{\varphi}_{p,q}}(\mathbb{R}^d)$ to $\mathcal{A}^{s+\sigma}_{\mathcal{M}^{\varphi}_{p,q}}(\mathbb{R}^d)$ continuously, the following composite mapping

$$\mathrm{Ext}_N := J_\sigma \circ \mathrm{Ext}_L \circ J_{-\sigma} : \mathcal{A}^{s}_{\mathcal{M}^{\varphi}_{p,q}}(\mathbb{R}^d{}_+) \to \mathcal{A}^{s}_{\mathcal{M}^{\varphi}_{p,q}}(\mathbb{R}^d)$$

makes sense.

We verify that $\mathrm{Ext}_N f \big|_{\mathbb{R}^d_+} = f$ for $f \in \mathcal{A}^{s}_{\mathcal{M}^{\varphi}_{p,q}}(\mathbb{R}^d{}_+)$. For this we pick a smooth test function $\psi \in \mathcal{D}(\mathbb{R}^d{}_+)$ and denote by $E\psi$ its extension to $\mathcal{S}(\mathbb{R}^d)$ obtained by setting $\psi(x) \equiv 0$ outside $\mathbb{R}^d{}_+$. We see that

$$\langle \mathrm{Ext}_N f \big|_{\mathbb{R}^d_+}, \psi \rangle = \langle \mathrm{Ext}_N f, E\psi \rangle = \langle \mathrm{Ext}_L J_{-\sigma} f, \mathcal{F}[\Phi^{(\sigma)} \mathcal{F}^{-1} E\psi] \rangle$$

$$= \langle J_{-\sigma} f, \mathcal{F}[\Phi^{(\sigma)} \mathcal{F}^{-1} E\psi] \big|_{\mathbb{R}^d_+} \rangle$$

from the property of Ext_L and Proposition 9.1.3. We further obtain for a representative $g \in \mathcal{A}^{s}_{\mathcal{M}^{\varphi}_{p,q}}(\mathbb{R}^d)$ of f,

$$\langle \mathrm{Ext}_N f \big|_{\mathbb{R}^d_+}, \psi \rangle = \langle J_{-\sigma} g, \mathcal{F}[\Phi^{(\sigma)} \mathcal{F}^{-1} E\psi] \rangle = \langle g, E\psi \rangle = \langle f, \psi \rangle.$$

Therefore, $\mathrm{Ext}_N f \big|_{\mathbb{R}^d_+} = f$ for all $f \in \mathcal{A}^{s}_{\mathcal{M}^{\varphi}_{p,q}}(\mathbb{R}^d{}_+)$ and the proof is finished.

$\square$

9.3   Trace Results

In this section we present the main results of this chapter which can be found in Theorem 9.3.2 (Trace theorem for spaces of type $\mathcal{A}^{s}_{\mathcal{M}^{\varphi}_{p,q}}$) and Theorem 9.3.9 (Trace theorem for spaces of type $B^{s,\tau}_{p,q}$). Our methods to solve the trace problem on domains in these functions spaces are quite standard. We briefly outline—with references for the scale of spaces $\mathcal{A}^{s}_{\mathcal{M}^{\varphi}_{p,q}}$—what tools we rely on in order to achieve our goal:

(A) We decompose the C^k domain with the resolution of unity from Remark 2.1.4.
     Then multiplier and diffeomorphism properties for the function spaces accord-
     ing to Theorem 2.5.19 (which in turn lead to the equivalent quasi-norm in
     Proposition 2.5.20) reduce the trace problem from Ω to $\mathbb{R}^d_+$.
(B) The extension operator according to Theorem 9.2.1 (constructed with the help
     of the suitable lift operator from Sect. 9.1 with 'good' mapping properties on
     $\mathbb{R}^d_+$, cf. Corollary 9.1.4), translates the problem from $\mathbb{R}^d_+$ to studying traces
     on the hyperplane $\mathbb{R}^{d-1}$.
(C) Already existing results for traces on hyperplanes (and extension operators), cf.
     Theorem 9.3.1 below, finally give the desired result.

For the Besov-type spaces $B^{s,\tau}_{p,q}$ we proceed in the same way finding suitable
substitutes for the items listed in (A)–(C).

As already explained above, one of the main cornerstones in order to prove trace
theorems on domains will be the following trace theorem for hyperplanes, which
was obtained in [NNS16, Thm. 5.1, 5.3].

Theorem 9.3.1 (Traces on Hyperplanes) *Let $d \geq 2$, $0 < p < \infty$, $0 < q \leq \infty$,
and $\varphi \in \mathcal{G}_p$, where the class $\mathcal{G}_p$ was defined in Remark 2.5.2. Additionally assume
that φ satisfies (2.5.2) when $q < \infty$ and $\mathcal{A} = \mathcal{E}$. Define s^* and φ^* by*

$$s^* := s - \frac{1}{p} \qquad and \qquad \varphi^*(t) := \varphi(t)\, t^{-1/p}, \quad t > 0.$$

Assume that

$$s > \frac{1}{p} + (d-1) \cdot \begin{cases} \left(\frac{1}{\min(1,p)} - 1\right), & \text{if } \mathcal{A} = \mathcal{N}, \\ \left(\frac{1}{\min(1,p,q)} - 1\right), & \text{if } \mathcal{A} = \mathcal{E}. \end{cases}$$

and that φ^* is increasing and satisfies

$$\sum_{j=0}^{\infty} \frac{1}{\varphi^*(2^j s)} \lesssim \frac{1}{\varphi^*(s)}, \quad 0 < s \leq 1.$$

Then $\mathrm{Tr}_{\mathbb{R}^{d-1}}$ is a bounded linear operator from $A^s_{\mathcal{M}^{\varphi}_{p,q}}(\mathbb{R}^d)$ onto $A^{s^*}_{\mathcal{M}^{\varphi^*}_{p,r}}(\mathbb{R}^{d-1})$,

$$\mathrm{Tr}_{\mathbb{R}^{d-1}} A^s_{\mathcal{M}^{\varphi}_{p,q}}(\mathbb{R}^d) = A^{s^*}_{\mathcal{M}^{\varphi^*}_{p,r}}(\mathbb{R}^{d-1}), \qquad where \quad r = \begin{cases} q, & \text{if } \mathcal{A} = \mathcal{N}, \\ p, & \text{if } \mathcal{A} = \mathcal{E}. \end{cases}$$

9.3.1 Trace Theorem for Spaces $\mathcal{A}^s_{\mathcal{M}^\varphi_{p,q}}(\Omega)$

With the preliminary work from the previous sections we are now able to state and proof our main theorem concerning traces of the generalized smoothness Morrey spaces on domains.

Theorem 9.3.2 (Trace Theorem on C^k Domains) *Let $d \geq 2$, $0 < p < \infty$, $0 < q \leq \infty$, and $\varphi \in \mathcal{G}_p$. Additionally assume that φ satisfies (2.5.2) when $q < \infty$ and $\mathcal{A} = \mathcal{E}$. Furthermore, let $\Omega \subset \mathbb{R}^d$ with boundary Γ be a C^k domain for k large enough. Define s^* and φ^* by*

$$s^* := s - \frac{1}{p} \quad \text{and} \quad \varphi^*(t) := \varphi(t)\,t^{-1/p}, \quad t > 0.$$

Assume that

$$s > \frac{1}{p} + (d-1) \cdot \begin{cases} \left(\frac{1}{\min(1,p)} - 1\right), & \text{if } \mathcal{A} = \mathcal{N}, \\ \left(\frac{1}{\min(1,p,q)} - 1\right), & \text{if } \mathcal{A} = \mathcal{E}, \end{cases} \tag{9.3.1}$$

and that φ^* is increasing and satisfies

$$\sum_{j=0}^{\infty} \frac{1}{\varphi^*(2^j s)} \lesssim \frac{1}{\varphi^*(s)}, \quad 0 < s \leq 1.$$

Then Tr_Γ is a linear and bounded operator from $\mathcal{A}^s_{\mathcal{M}^\varphi_{p,q}}(\Omega)$ onto $\mathcal{A}^{s^*}_{\mathcal{M}^{\varphi^*}_{p},r}(\Gamma)$,

$$\mathrm{Tr}_\Gamma \mathcal{A}^s_{\mathcal{M}^\varphi_{p,q}}(\Omega) = \mathcal{A}^{s^*}_{\mathcal{M}^{\varphi^*}_{p},r}(\Gamma), \quad \text{where} \quad r = \begin{cases} q, & \text{if } \mathcal{A} = \mathcal{N}, \\ p, & \text{if } \mathcal{A} = \mathcal{E}. \end{cases}$$

Proof Our understanding of the trace operator on Γ is as follows: If f is smooth, using the partition of unity from (2.1.5), (2.1.6) we can write

$$\mathrm{Tr}_\Gamma f = \sum_{j=1}^{N} \varphi_j\,(\mathrm{Tr}_\Gamma f).$$

Note that the term with φ_0 is unimportant because only the values of f near the boundary are of interest. Locally we see that for $x \in K_j \cap \Omega$ with $\psi^{(j)}(x) = y \in V_j \cap \mathbb{R}^d_+$, we have

$$\varphi_j \mathrm{Tr}_\Gamma f(x) = \varphi_j \mathrm{Tr}_\Gamma f \circ (\psi^{(j)})^{-1} \circ \psi^{(j)}(x)$$

$$= \varphi_j \mathrm{Tr}_\Gamma f \circ (\psi^{(j)})^{-1}(y) = \mathrm{Tr}_{\mathbb{R}^{d-1}} g_j(y) = g_j(y', 0),$$

where in the second step we extended $g_j := \varphi_j f \circ (\psi^{(j)})^{-1}$ by zero outside $V_j \cap \mathbb{R}^d_+$ for $\mathrm{Tr}_{\mathbb{R}^{d-1}}$ to make sense (concerning the notation we refer to Definition 2.1.1). Thus, the trace is well defined for smooth f. For general f we use the fact that $\mathrm{Tr}_{\mathbb{R}^{d-1}} g_j$ can be understood as explained at the beginning of Chap. 9 (cf. formula (9.0.1) and below), since $g_j \in \mathcal{A}^s_{\mathcal{M}^\varphi_{p,q}}(\mathbb{R}^d)$. Therefore, the trace makes sense in this case as well and by the definition of the spaces on the boundary Γ, we have that $\mathrm{Tr}_\Gamma f \in \mathcal{A}^{s^*}_{\mathcal{M}^{\varphi^*}_p,r}(\Gamma)$, if $\mathrm{Tr}_{\mathbb{R}^{d-1}} g_j \in \mathcal{A}^{s^*}_{\mathcal{M}^{\varphi^*}_p,r}(\mathbb{R}^{d-1})$ for all $j = 1, \ldots, N$.

Step 1.   We wish to prove in this step that

$$\mathrm{Tr}_\Gamma \mathcal{A}^s_{\mathcal{M}^\varphi_{p,q}}(\Omega) \subset \mathcal{A}^{s^*}_{\mathcal{M}^{\varphi^*}_p,r}(\Gamma). \tag{9.3.2}$$

According to Theorem 9.2.1 there exists a bounded extension operator

$$\mathrm{Ext} : \mathcal{A}^s_{\mathcal{M}^\varphi_{p,q}}(\mathbb{R}^d_+) \longrightarrow \mathcal{A}^s_{\mathcal{M}^\varphi_{p,q}}(\mathbb{R}^d) \quad \text{with} \quad \|\mathrm{Ext} f | \mathcal{A}^s_{\mathcal{M}^\varphi_{p,q}}(\mathbb{R}^d)\| \sim \|f | \mathcal{A}^s_{\mathcal{M}^\varphi_{p,q}}(\mathbb{R}^d_+)\|.$$

In particular, for the trace operator $\mathrm{Tr}_{\mathbb{R}^{d-1}}$ we see that

$$\mathrm{Tr}_{\mathbb{R}^{d-1}}(\mathrm{Ext}\, h)(x) = \mathrm{Tr}_{\mathbb{R}^{d-1}} h(x) = h(x', 0),$$

whenever the pointwise trace for h makes sense. Using Theorem 9.3.1 we have

$$\|\mathrm{Tr}_{\mathbb{R}^{d-1}} h \mid \mathcal{A}^{s^*}_{\mathcal{M}^{\varphi^*}_p,r}(\mathbb{R}^{d-1})\| = \|\mathrm{Tr}_{\mathbb{R}^{d-1}}(\mathrm{Ext}\, h) \mid \mathcal{A}^{s^*}_{\mathcal{M}^{\varphi^*}_p,r}(\mathbb{R}^{d-1})\|$$

$$\leq c\|\mathrm{Ext}\, h \mid \mathcal{A}^s_{\mathcal{M}^\varphi_{p,q}}(\mathbb{R}^d)\| \sim \|h \mid \mathcal{A}^s_{\mathcal{M}^\varphi_{p,q}}(\mathbb{R}^d_+)\|, \tag{9.3.3}$$

which shows that

$$\mathrm{Tr}_{\mathbb{R}^{d-1}} \mathcal{A}^s_{\mathcal{M}^\varphi_{p,q}}(\mathbb{R}^d_+) \subset \mathcal{A}^{s^*}_{\mathcal{M}^{\varphi^*}_p,r}(\mathbb{R}^{d-1}). \tag{9.3.4}$$

With this we calculate

$$\|\mathrm{Tr}_\Gamma f \mid \mathcal{A}^{s^*}_{\mathcal{M}^{\varphi^*}_p,r}(\Gamma)\| = \sum_{j=1}^N \|\mathrm{Tr}_{\mathbb{R}^{d-1}} g_j \mid \mathcal{A}^{s^*}_{\mathcal{M}^{\varphi^*}_p,r}(\mathbb{R}^{d-1})\|$$

$$= \sum_{j=1}^N \|\varphi_j f \circ (\psi^{(j)})^{-1}(\cdot, 0) \mid \mathcal{A}^{s^*}_{\mathcal{M}^{\varphi^*}_p,r}(\mathbb{R}^{d-1})\|$$

$$\leq \sum_{j=1}^N \|\varphi_j f \circ (\psi^{(j)})^{-1} \mid \mathcal{A}^s_{\mathcal{M}^\varphi_{p,q}}(\mathbb{R}^d_+)\|$$

$$\leq c\|f \mid \mathcal{A}^s_{\mathcal{M}^\varphi_{p,q}}(\Omega)\|, \tag{9.3.5}$$

where in the third step we used (9.3.3) and the last step is a consequence of Proposition 2.5.20. In fact the calculations in (9.3.5) show that our problem (9.3.2) reduces to (9.3.4).

Step 2.   In order to see that the trace operator Tr_Γ is onto $\mathcal{A}^{s^*}_{\mathcal{M}^{\varphi^*}_p,r}(\Gamma)$, we establish the existence of a bounded extension operator

$$\widetilde{\mathrm{Ex}} : \mathcal{A}^{s^*}_{\mathcal{M}^{\varphi^*}_p,r}(\Gamma) \longrightarrow \mathcal{A}^{s}_{\mathcal{M}^{\varphi}_{p,q}}(\Omega), \qquad \widetilde{\mathrm{Ex}}\, g\big|_\Gamma = g,$$

such that for $g \in \mathcal{A}^{s^*}_{\mathcal{M}^{\varphi^*}_p,r}(\Gamma)$ we have

$$\|\widetilde{\mathrm{Ex}}\, g \mid \mathcal{A}^{s}_{\mathcal{M}^{\varphi}_{p,q}}(\Omega)\| \le c\, \|g \mid \mathcal{A}^{s^*}_{\mathcal{M}^{\varphi^*}_p,r}(\Gamma)\|.$$

We choose functions $\eta_j \in D(\mathbb{R}^d)$, $j = 1, \dots, N$ with

$$\mathrm{supp}\,\eta_j \subset K_j, \qquad \eta_j = 1, \quad \text{if} \quad x \in \mathrm{supp}\,\varphi_j.$$

Put

$$\widetilde{\mathrm{Ex}}\, g(x) := \sum_{j=1}^{N} \eta_j(x) \cdot \mathrm{Ex}\left((\varphi_j g)(\psi^{(j)})^{-1}(\cdot, 0)\right)\left(\psi^{(j)}(x)\right), \qquad x \in \Omega,$$

where

$$\mathrm{Ex} : \mathcal{A}^{s^*}_{\mathcal{M}^{\varphi^*}_p,r}(\mathbb{R}^{d-1}) \longrightarrow \mathcal{A}^{s}_{\mathcal{M}^{\varphi}_{p,q}}(\mathbb{R}^d)$$

stands for the extension operator coming out from Theorem 9.3.1, cf. [NNS16, Thm. 5.3]. In particular, our construction can be extended from Ω to $\mathbb{R}^d$ by putting $\eta_j(x) \cdot \mathrm{Ex}\,(\dots)\left(\psi^{(j)}(x)\right) = 0$ outside K_j. This yields

$$\|\widetilde{\mathrm{Ex}}\, g \mid \mathcal{A}^{s}_{\mathcal{M}^{\varphi}_{p,q}}(\Omega)\| = \inf\left\{\|h|\mathcal{A}^{s}_{\mathcal{M}^{\varphi}_{p,q}}(\mathbb{R}^d)\| : h \in \mathcal{A}^{s}_{\mathcal{M}^{\varphi}_{p,q}}(\mathbb{R}^d),\ h\big|_\Omega = \widetilde{\mathrm{Ex}}\, g\right\}$$

$$\le \left\|\sum_{j=1}^{N} \eta_j(\cdot)\mathrm{Ex}\left((\varphi_j g)(\psi^{(j)})^{-1}(\cdot, 0)\right)(\psi^{(j)}(\cdot)) \mid \mathcal{A}^{s}_{\mathcal{M}^{\varphi}_{p,q}}(\mathbb{R}^d)\right\|$$

$$\sim \sum_{j=1}^{N} \left\|\eta_j(\cdot)\mathrm{Ex}\left((\varphi_j g)(\psi^{(j)})^{-1}(\cdot, 0)\right)(\psi^{(j)}(\cdot)) \mid \mathcal{A}^{s}_{\mathcal{M}^{\varphi}_{p,q}}(\mathbb{R}^d)\right\|$$

$$\sim \sum_{j=1}^{N} \left\|\mathrm{Ex}\left((\varphi_j g)(\psi^{(j)})^{-1}(\cdot, 0)\right) \mid \mathcal{A}^{s}_{\mathcal{M}^{\varphi}_{p,q}}(\mathbb{R}^d)\right\|$$

$$\leq c \sum_{j=1}^{N} \left\| (\varphi_j g)(\psi^{(j)})^{-1}(\cdot, 0) \mid \mathcal{A}^{s^*}_{\mathcal{M}_p^{\varphi^*}, r}(\mathbb{R}^{d-1}) \right\|$$

$$= c \left\| g \mid \mathcal{A}^{s^*}_{\mathcal{M}_p^{\varphi^*}, r}(\Gamma) \right\|,$$

where in the 4th step we used Theorem 2.5.19(i), (ii), since $\psi^{(j)}$ is a k-diffeomorphism from $\mathbb{R}^d$ onto itself, and $\eta_j \in D(\mathbb{R}^d)$ if we put $\eta_j(x) = 0$ whenever $x \in \mathbb{R}^d \setminus K_j$. This completes the proof. $\square$

Remark 9.3.3 Note that the above proof relies on the available diffeomorphism property, cf. Theorem 2.5.19(ii), and this assertion does in general not apply to variable exponent spaces. In the latter case, the values of the exponent depend on the point of the domain and therefore a diffeomorphism assertion like the one referred to above cannot be expected in the context of these variable exponent spaces, unless a strict condition is imposed. In this regard we refer for instance to [Gon17, Thm. 6.5.6].

Furthermore, it would also be interesting to clarify the trace in the limiting case, which corresponds in the classical case to $s = 1/p$, see e.g. [Tri97]. This study will be postponed to future work.

9.3.2 Trace Theorem for Spaces $B^{s,\tau}_{p,q}(\Omega)$

We now turn our attention to the Besov-type spaces $B^{s,\tau}_{p,q}(\Omega)$, which by (2.5.4) are related but not included in the scale of generalized Besov-Morrey spaces $\mathcal{N}^s_{\mathcal{M}_{p,q}^{\varphi}}(\Omega)$ considered so far. Our aim is to obtain trace results on C^k domains similar to Theorem 9.3.2 for these spaces. Our understanding of the trace operator is in the same spirit as explained at the beginning of Chap. 9, we also refer to [YSY10, p. 164] in this context. A close inspection of the proof of Theorem 9.3.2 reveals that we need to collect suitable substitutes for (A), (B), and (C) listed on page 285 in order to establish corresponding trace results for the scale $B^{s,\tau}_{p,q}$.

Concerning (A), in terms of diffeomorphisms and multipliers the results are stated in Theorem 2.5.25. As for (B), using the quarkonial decomposition from Theorem 2.5.23 we can construct the following extension operator.

Theorem 9.3.4 (Extension Operator) *Let* $0 < p < \infty$, $0 < q \leq \infty$, *and* $0 \leq \tau \leq \frac{1}{p}$. *Then for* $N \in \mathbb{N}$, *there exists an extension operator* Ext_N,

$$\mathrm{Ext}_N : \bigcup_{(p,q,s) \in R(N)} B^{s,\tau}_{p,q}(\mathbb{R}^d{}_+) \longrightarrow \bigcup_{(p,q,s) \in R(N)} B^{s,\tau}_{p,q}(\mathbb{R}^d),$$

that satisfies the properties: if $(p, q, s) \in R(N)$ then the restriction $\text{Ext}_N |_{B_{p,q}^{s,\tau}(\mathbb{R}^d_+)}$ is a continuous mapping from $B_{p,q}^{s,\tau}(\mathbb{R}^d_+)$ to $B_{p,q}^{s,\tau}(\mathbb{R}^d)$ satisfying $\text{Ext}_N f |_{\mathbb{R}^d_+} = f$ for all $f \in B_{p,q}^{s,\tau}(\mathbb{R}^d_+)$.

Proof The proof follows along the same lines as the proof of Theorem 9.2.1. We use the same construction to obtain an extended function f^* as described in Step 1 of the proof. In Step 2 instead of (9.2.2) we now make use of [YHMSY15, Prop. 4.1], i.e.,

$$B_{p,q}^{s,\tau}(\mathbb{R}^d) \hookrightarrow C(\overline{\mathbb{R}^d}), \quad \text{if, and only if,} \quad s > d\left(\frac{1}{p} - \tau\right).$$

In particular, this enables us to define Ext_N for $s > \frac{d}{p}$ with the help of the quarkonial decomposition from Theorem 2.5.23. Finally, in Step 3 of Theorem 9.2.1 we note that the lift operator from Corollary 9.1.4 can also be generalized to the Besov-type spaces $B_{p,q}^{s,\tau}$. □

Remark 9.3.5 Our results on the extension operator generalize [YSY10, Thm. 6.11] to the case when $p \leq 1$.

An assertion for traces on hyperplanes as required in (C) is also available from [YSY10, Thm. 6.8].

Theorem 9.3.6 (Traces on Hyperplanes) *Let $d \geq 2$, $0 < p, q \leq \infty$, $0 \leq \tau \leq \frac{1}{p}$, and*

$$s > \frac{1}{p} + (d - 1)\left(\frac{1}{\min(1, p)} - 1\right). \tag{9.3.6}$$

Then $\text{Tr}_{\mathbb{R}^{d-1}}$ is a linear and bounded operator from $B_{p,q}^{s,\tau}(\mathbb{R}^d)$ onto $B_{p,q}^{s-\frac{1}{p}, \frac{d\tau}{d-1}}(\mathbb{R}^{d-1})$,

$$\text{Tr}_{\mathbb{R}^{d-1}} B_{p,q}^{s,\tau}(\mathbb{R}^d) = B_{p,q}^{s-\frac{1}{p}, \frac{d\tau}{d-1}}(\mathbb{R}^{d-1}).$$

Remark 9.3.7 The proof in [YSY10, Thm. 6.8] also establishes the existence of a linear and bounded extension operator $\widetilde{\text{Ex}} : B_{p,q}^{s-\frac{1}{p}, \frac{d\tau}{d-1}}(\mathbb{R}^{d-1}) \to B_{p,q}^{s,\tau}(\mathbb{R}^d)$.

Remark 9.3.8 (Restriction on Smoothness k of Domain) Note that for the diffeomorphism and pointwise multiplier results in $B_{p,q}^{s,\tau}(\mathbb{R}^d)$ from Theorem 2.5.25(i), (ii) we require

$$k \geq \max([s + d\tau + 1], 0). \tag{9.3.7}$$

This follows from a closer look at the proof of [YSY10, Thm. 6.1], whenever $s >$ σ_p. This is always the case using the restriction (9.3.6) on s from Theorem 9.3.6, since then we have

$$\sigma_p = d\left(\frac{1}{\min(1,\,p)} - 1\right)$$

$$= \frac{1}{p} + (d-1)\left(\frac{1}{\min(1,\,p)} - 1\right) \underbrace{-\frac{1}{p} + \frac{1}{\min(1,\,p)} - 1}_{<0}$$

$$< \frac{1}{p} + (d-1)\left(\frac{1}{\min(1,\,p)} - 1\right) < s.$$

The restriction (9.3.7) on k comes from the atomic decomposition of the Besov-type spaces as established in [YSY10, Thm. 3.3]. Moreover, in [YSY10] we have the same restriction for the $F_{p,q}^{s,\tau}$ scale. However, the fact that τ comes into play in (9.3.7) is a little confusing: for the atomic decomposition of the spaces $\mathcal{E}_{u,p,q}^s$ we only need $k \geq \max\{[s+1],0\}$, cf. Theorem 2.5.9, which is independent of τ. By the coincidence $\mathcal{E}_{u,p,q}^s = F_{p,q}^{s,\tau}$ with $u = \frac{p}{1-p\tau}$ and $0 \leq \tau < \frac{1}{p}$, the dependence on τ in (9.3.7) can be removed for the $F_{p,q}^{s,\tau}$ scale. This was also noted in [NNS16, Thm. 4.4, 4.5] (mentioned in the proof given there). This raises the question whether in (9.3.7) the dependence on τ can be removed for $B_{p,q}^{s,\tau}$ as well.

Putting together our substitutes for (A)–(C), we obtain the following result concerning traces on C^k domains for Besov-type spaces.

Theorem 9.3.9 (Trace Theorem on C^k Domains) *Let $d \geq 2$, $0 \leq p < \infty$, $0 < q \leq \infty$, $0 \leq \tau \leq \frac{1}{p}$, and*

$$s > \frac{1}{p} + (d-1)\left(\frac{1}{\min(1,\,p)} - 1\right). \tag{9.3.8}$$

Furthermore, let $\Omega \subset \mathbb{R}^d$ be a bounded C^k domain with boundary Γ, where

$$k \geq [s + d\tau + 1].$$

Then Tr_Γ *is a linear and bounded operator from* $B_{p,q}^{s,\tau}(\Omega)$ *onto* $B_{p,q}^{s-\frac{1}{p},\frac{d\tau}{d-1}}(\Gamma)$,

$$\mathrm{Tr}_\Gamma \, B_{p,q}^{s,\tau}(\Omega) = B_{p,q}^{s-\frac{1}{p},\frac{d\tau}{d-1}}(\Gamma).$$

9.4   Application: A Priori Estimates for Solutions of Elliptic Boundary Value Problems

In [ElB05] the author obtained *a priori* estimates for solutions of elliptic boundary value problems in the spaces $\mathcal{L}^{p,\lambda,s}(\mathbb{R}^d)$, $s \in \mathbb{R}$, $\lambda \geq 0$, $1 \leq p < \infty$, which are linked with our Besov-type spaces via

$$\mathcal{L}^{p,\lambda,s}(\mathbb{R}^d) = B_{p,p}^{s,\frac{\lambda}{dp}}(\mathbb{R}^d).$$

His estimates are based on trace results for the respective spaces on hyperplanes. In the present section, we improve [ElB05, Thm. 1.8] by applying our trace results for the Besov-type spaces from Theorem 9.3.9 and establish corresponding *a priori* estimates on C^k domains.

We consider the following elliptic Dirichlet problem:

$$\begin{cases} Lu = f & \text{on } \Omega, \\ \text{Tr}_\Gamma u = g & \text{on } \Gamma = \partial\Omega, \end{cases}$$

where L is a differential operator of second order with smooth coefficients in $\overline{\Omega}$, i.e.,

$$L = \sum_{|\alpha|\leq 2} a_\alpha(x) D_x^\alpha,$$

which is properly elliptic. By this we mean that the following conditions are satisfied:

(H1) For any $x \in \overline{\Omega}$,

$$\sum_{|\alpha|=2} a_\alpha(x)\xi^\alpha \neq 0, \qquad \xi \in \mathbb{R}^d \setminus \{0\}.$$

(H2) For any $x \in \Gamma$, $\xi_x \in \mathbb{R}^d \setminus \{0\}$ tangent to Γ at x, the polynomial in the complex variable z,

$$P(z) = \sum_{|\alpha|=2} a_\alpha(x)(\xi_x + z\nu_x)^\alpha,$$

has exactly one root with positive imaginary part (and therefore exactly one root lying in the lower half plane). Here ν_x denotes the inward unit normal vector to the boundary Γ at x.

In this setting [ElB05, Thm. 1.8] can be generalized (and reformulated) in terms of our Besov-type spaces as follows.

Theorem 9.4.1 *Let $s > 0$, $1 \leq p < \infty$, and $0 \leq \tau \leq \frac{1}{p}$. Furthermore, let $\Omega \subset \mathbb{R}^d$ be a bounded C^k domain with boundary Γ, where*

$$k \geq [s + d\tau + 1].$$

Then, under the hypotheses (H1) and (H2), there is a constant $C > 0$ such that

$$\|u|B_{p,p}^{s+2,\tau}(\Omega)\| \leq C \left(\|Lu|B_{p,p}^{s,\tau}(\Omega)\| + \|\mathrm{Tr}_\Gamma u|B_{p,p}^{s+2-\frac{1}{p},\frac{d\tau}{d-1}}(\Gamma)\| + \|u|B_{p,p}^{s+1,\tau}(\Omega)\| \right)$$

holds for any $u \in B_{p,p}^{s+2,\tau}(\Omega)$.

Remark 9.4.2 It would be interesting to study whether the *a priori* estimates from [ElB05] can be generalized to the spaces $B_{p,q}^{s,\tau}(\Omega)$ when $p \neq q$ or even to the spaces $\mathcal{A}_{\mathcal{M}_{p,q}^{\varphi}}^{s}(\Omega)$ with the help of our trace results obtained in Theorem 9.3.2. This is out of the scope of this thesis but will be investigated in a forthcoming paper.

Chapter 10
Traces on Riemannian Manifolds

In the final chapter of this manuscript we turn our attention towards traces on submanifolds N in fractional Sobolev, Besov- and Triebel-Lizorkin spaces defined on noncompact Riemannian manifolds M. Here M is of dimension d and N of dimension k with $k < d$, respectively. Usually, these spaces are defined via geodesic normal coordinates. However, for our investigations we require more flexibility in the choice of coordinates and therefore rely on the results obtained in Chap. 2, where we studied which conditions are sufficient such that different local coordinates give equivalent norms and therefore yield the same spaces. In this context we introduced the notion of admissible trivializations, cf. Definition 2.6.10, and formulated our results in Theorem 2.6.12. With this preliminary work we can now proceed as follows:

In Sect. 10.1 we introduce Fermi coordinates and show that for a certain cover there is a subordinated partition of unity such that the resulting trivialization is admissible. These coordinates have the advantage that they are well adapted to the submanifold N (compared to geodesic coordinates).

Afterwards, in Sect. 10.2, we present the trace theorems on N for all function spaces mentioned above. We focus particularly on fractional Sobolev spaces for they are easier to handle than B- and F-spaces, respectively, and give a detailed proof in this case. It turns out that all of our results are in good agreement with the trace results in the Euclidean setting (i.e., on hyperplanes $\mathbb{R}^k$). In particular, the trace space of a fractional Sobolev space on M turns out to be a Besov space on N. The analogy of the results is not surprising since when using Fermi coordinates, via localization and pull-back the trace problem on N immediately transfers to traces on hyperplanes $\mathbb{R}^k$.

After having established the traces for the Sobolev spaces we show how to obtain corresponding results for B- and F-spaces. Our findings improve [Skr90, Thm. 1], since we do not make any unnecessary assumptions on the manifolds M and N. We only require that the pair (M, N) is of bounded geometry due to the Fermi

© The Author(s), under exclusive license to Springer Nature Switzerland AG 2021
C. Schneider, *Beyond Sobolev and Besov*, Lecture Notes in Mathematics 2291,
https://doi.org/10.1007/978-3-030-75139-5_10

coordinates we use. Thus, the admissible trivializations are well suited in order to tackle these kinds of problems.

Finally, Sect. 10.3 contains another application of the admissible trivializations to spaces with symmetries. Here we consider a countable discrete group G acting in an adequate way on M and show that Sobolev spaces on the orbit M/G can be characterized via weighted Sobolev spaces of G-invariant functions on M. This is in the spirit of [Tri83, Sect. 9.2.1], where Sobolev spaces on the torus were studied.

Before we start let us briefly explain our understanding of the trace operator when dealing with function spaces on manifolds: Let (M^d, g) be a Riemannian manifold of dimension d an let N^k be an embedded submanifold of dimension k where $k < d$. Then for a function $f \in \mathcal{D}(M)$ the trace operator can be defined by pointwise restriction,

$$\mathrm{Tr}_N f := f\big|_N.$$

Now let $X(M)$ and $Y(N)$ be some function or distribution spaces on M and N, respectively. If Tr_N extends to a continuous map from $X(M)$ into $Y(N)$, we say that the trace exists in $Y(N)$. Moreover, if this extension is onto, we write $\mathrm{Tr}_N X(M) = Y(N)$.

10.1   Fermi Coordinates on Submanifolds

In this section we introduce Fermi coordinates, which are special coordinates adapted to a submanifold N of M, where the pair (M, N) is of bounded geometry according to Definition 10.1.1 below. Moreover, we show in Theorem 10.1.9 that for a certain cover with Fermi coordinates there is a subordinated partition of unity such that the resulting trivialization denoted by $\mathcal{T}^{\mathrm{FC}}$ is admissible. This enables us to use this trivialization in Sect. 10.2 for proving the trace theorems.

From now on let $N^k \subset M^d$ be an *embedded submanifold*, meaning, there is a k-dimensional manifold N' and an injective immersion $f : N' \to M$ with $f(N') = N$. It is our aim to prove a trace theorem for M and N. We restrict ourselves to submanifolds of bounded geometry in the following sense.

Definition 10.1.1 (Bounded Geometry of (M, N)) Let (M^d, g) be a Riemannian manifold with a k-dimensional embedded submanifold $(N^k, g|_N)$. We say that (M, N) *is of bounded geometry* if the following is fulfilled:

(i) (M, g) is of bounded geometry.
(ii) The injectivity radius r_N of $(N, g|_N)$ is positive.

Fig. 10.1 Collar around N

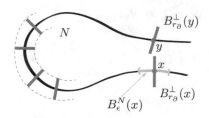

(iii) There is a collar around N (a tubular neighbourhood of fixed radius, see Fig. 10.1), i.e., there is $r_\partial > 0$ such that for all $x, y \in N$ with $x \neq y$ the normal balls $B_{r_\partial}^\perp(x)$ and $B_{r_\partial}^\perp(y)$ are disjoint where

$$B_{r_\partial}^\perp(x) := \{z \in M \mid \operatorname{dist}_M(x, z) \leq r_\partial, \ \exists \epsilon_0 \forall \epsilon < \epsilon_0 :$$

$$\operatorname{dist}_M(x, z) = \operatorname{dist}_M(B_\epsilon^N(x), z)\}$$

with

$$B_\epsilon^N(x) = \{u \in N \mid \operatorname{dist}_N(u, x) \leq \epsilon\}$$

and dist_M and dist_N denote the distance functions in M and N, respectively.

(iv) The mean curvature l of N given by

$$l(X, Y) := \nabla_X^M Y - \nabla_X^N Y \quad \text{for all } X, Y \in TN,$$

and all its covariant derivatives are bounded. Here, ∇^M is the Levi-Civita connection of (M, g) and ∇^N the one of $(N, g|_N)$.

Remark 10.1.2

(i) If the normal bundle of N in M is trivial, condition (iii) in Definition 10.1.1 simply means that $\{z \in M \mid \operatorname{dist}_M(z, N) \leq r_\partial\}$ is diffeomorphic to $B_{r_\partial}^{d-k} \times N$. Then

$$F : B_{r_\partial}^{d-k} \times N \to M; \ (t, z) \mapsto \exp_z^M\left(t^i v_i\right)$$

is a diffeomorphism onto its image, where $(t^1, \dots, t^{d-k})$ are the coordinates for t with respect to a standard orthonormal basis on $\mathbb{R}^{d-k}$ and $(v_1, \dots, v_{d-k})$ is an orthonormal frame for the normal bundle of N in M.

If the normal bundle is not trivial (e.g. consider a noncontractible circle N in the infinite Möbius strip M, see Fig. 10.2), F still exists locally, which means that for all $x \in N$ and ϵ smaller than the injectivity radius of N, the map $F : B_{r_\partial}^{d-k} \times B_\epsilon^N(x) \to M; \ (t, z) \mapsto \exp_z^M\left(t^i v_i\right)$ is a diffeomorphism onto its image. All included quantities are as in the case of a trivial vector bundle, but v_i is now just a local orthonormal frame of the normal bundle. By abuse

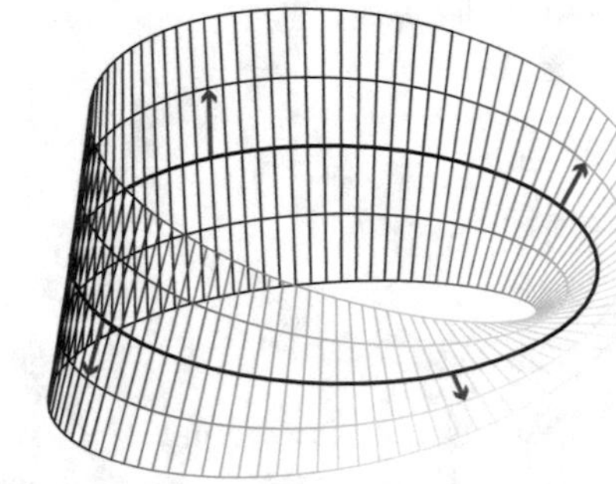

Fig. 10.2 Möbius strip

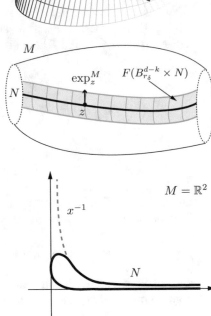

Fig. 10.3 N admits a collar

M

$\exp_z^M$ $F(B_{r\delta}^{d-k} \times N)$

N

z

Fig. 10.4 N does not admit a collar

x^{-1}

$M = \mathbb{R}^2$

N

of notation, we suppress here and in the following the dependence of F on ϵ and x.

(ii) Figure 10.3 shows a submanifold N of a manifold M that admits a collar. On the other hand in Fig. 10.4 one sees that for $M = \mathbb{R}^2$ the submanifold N describing the curve which for large enough x contains the graph of $x \mapsto x^{-1}$ together with the x-axis does not have a collar. This situation is therefore excluded by Definition 10.1.1. However, to a certain extend, manifolds as in Fig. 10.4 can still be treated, cf. Example 10.2.6 and Remark 10.2.7.

(iii) Although our notation (M, N) hides the underlying metric g, this is obviously part of the definition and fixed when talking about M.

(iv) If N is the boundary of the manifold M, the counterpart of Definition 10.1.1 can be found in [Sch01, Def. 2.2], where also Fermi coordinates are introduced and certain properties discussed. In this section we adapt some of the methods from [Sch01] to our situation. Note that the normal bundle of the boundary of

Fig. 10.5 Illustration of cover $(U_\gamma)_{\gamma \in I}$

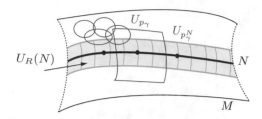

a manifold is always trivial, which explains why in [Sch01, Def. 2.2] condition (iii) of Definition 10.1.1 reads as (i) above.

We now introduce Fermi coordinates adapted to a submanifold N.

Definition 10.1.3 (Fermi Coordinates)   We use the notations from Definition 10.1.1. Let (M^d, N^k) be of bounded geometry. Let $R = \min\left\{\frac{1}{2}r_N, \frac{1}{4}r_M, \frac{1}{2}r_\partial\right\}$, where r_N is the injectivity radius of N and r_M the one of M. Let there be countable index sets $I_N \subset I$ and sets of points $\{p_\alpha^N\}_{\alpha \in I_N}$ and $\{p_\beta\}_{\beta \in I \setminus I_N}$ in N and $M \setminus U_R(N)$, respectively, where $U_R(N) := \bigcup_{x \in N} B_{\frac{1}{R}}(x)$. Those sets are chosen such that

(i) The collection of the metric balls $(B_R^N(p_\alpha^N))_{\alpha \in I_N}$ gives a uniformly locally finite cover of N. Here the balls are meant to be metric with respect to the induced metric $g|_N$.

(ii) The collection of metric balls $(B_R(p_\beta))_{\beta \in I \setminus I_N}$ covers $M \setminus U_R(N)$ and is uniformly locally finite on all of M.

We consider the covering $(U_\gamma)_{\gamma \in I}$ with $U_\gamma = B_R(p_\gamma)$ for $\gamma \in I \setminus I_N$ and $U_\gamma = U_{p_\gamma^N} := F(B_{2R}^{d-k} \times B_{2R}^N(p_\gamma^N))$ with $\gamma \in I_N$ (Fig. 10.5). Coordinates on U_γ are chosen to be geodesic normal coordinates around p_γ for $\gamma \in I \setminus I_N$. Otherwise, if $\gamma \in I_N$, coordinates are given by Fermi coordinates

$$\kappa_\gamma : V_{p_\gamma^N} := B_{2R}^{d-k} \times B_{2R}^k \to U_{p_\gamma^N}, \quad (t, x) \mapsto \exp_{\exp_{p_\gamma^N}^N(\lambda_\gamma^N(x))}^M \left(t^i v_i\right) \quad (10.1.1)$$

where $(t^1, \ldots, t^{d-k})$ are the coordinates for t with respect to a standard orthonormal basis on $\mathbb{R}^{d-k}$, $(v_1, \ldots, v_{d-k})$ is an orthonormal frame for the normal bundle of $B_{2R}^N(p_\gamma^N)$ in M, $\exp^N$ is the exponential map on N with respect to the induced metric $g|_N$, and $\lambda_\gamma^N : \mathbb{R}^k \to T_{p_\gamma^N} N$ is the choice of an orthonormal frame on $T_{p_\gamma^N} N$ (Fig. 10.6).

Before giving a remark on the existence of the points $\{p_\gamma\}_{\gamma \in I}$ claimed in the Definition above, we prove two lemmas.

Lemma 10.1.4   *Let (M^d, N^k) be of bounded geometry, and let $C > 0$ be such that the Riemannian curvature tensor fulfills $|R^M| \le C$ and for the mean curvature of N it holds $|l| \le C$. Fix $z \in N$ and let R be as in Definition 10.1.3. Let $U = F(B_{2R}^{d-k} \times B_{2R}^N(z))$ and define a chart κ for U as in (10.1.1). Then there is a constant*

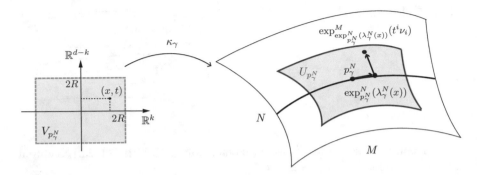

Fig. 10.6 Fermi coordinates adapted to submanifold N

$C' > 0$ *only depending on C, d and k, such that $|g_{ij}| \le C'$ and $|g^{ij}| \le C'$ where* g_{ij} *denotes the metric g with respect to κ.*

Proof For N being the boundary of M this was shown in [Sch01, Lem. 2.6]. We follow the idea given there and use the extension of the Rauch comparison theorem to submanifolds of arbitrary codimension given by Warner in [War66, Thm. 4.4]. For the comparison, let M_C and M_{-C} be two complete d-dimensional Riemannian manifolds of constant sectional curvature C and $-C$, respectively. In each of them we choose a k-dimensional submanifold N_C and N_{-C}, points $p_{\pm C} \in N_{\pm C}$ and a chart of $M_{\pm C}$ around $p_{\pm C}$ given by Fermi coordinates such that all eigenvalues of the second fundamental form with respect to these coordinates at $p_{\pm C}$ are given by $\pm C$ (this is always possible, cf. [Spi79, Ch. 7]). Let $(\nu_i)_{1 \le i \le d-k}$ be an orthonormal frame of the normal bundle of $U \cap N$ and $(e_i)_{1 \le i \le k}$ be an orthonormal frame of $T|_{U \cap N} N$ obtained via geodesic flow on N. Let the frame $(\nu_1, \ldots, \nu_{d-k}, e_1, \ldots, e_k)$ be transported to all of U via parallel transport along geodesics normal to N (the transported vectors are also denoted by ν_i and e_i, respectively).

Then, we are in the situation to apply [War66, Thm. 4.4]: Let now $p \in U$ and $v \in T_p U$ with $v \perp \nu_i$ for all $1 \le i \le d - k$. Then, the comparison theorem yields constants $C_1, C_2 > 0$, depending only on C, n, and k, such that $C_1 |v|_E^2 \le g_p(v, v) \le C_2 |v|_E^2$, where $| \cdot |_E$ denotes the Euclidean metric with respect to the basis (e_i). Moreover, we have $g_p(\nu_i, \nu_j) = \delta_{ij}$ and $g_p(\nu_i, e_l) = 0$ for all $1 \le i, j \le d - k$ and $1 \le l \le k$, since this is true for $p \in U \cap N$, and this property is preserved by parallel transport. Altogether this implies the claim.   □

The previous lemma enables us to show that $(N, g|_N)$ is also of bounded geometry.

Lemma 10.1.5 (Bounded Geometry of Submanifold N) *If (M, N) is of bounded geometry, then $(N, g|_N)$ is of bounded geometry.*

Proof Since Definition 10.1.1 already includes the positivity of the injectivity radius of N, it is enough to show that $(\nabla^N)^k R^N$, where R^N is the Riemannian curvature of $(N, g|_N)$, is bounded for all $k \in \mathbb{N}_0$: Let $z \in N$. We consider geodesic normal coordinates $\kappa^{\text{geo}} : B_{2R}^k \to U^{\text{geo}} = B_{2R}^N(z)$ on N around z and Fermi coordinates $\kappa : B_{2R}^{d-k} \times B_{2R}^k \to U = F(B_{2R}^{d-k} \times B_{2R}^N(z))$ on M around z, cf. Definition 10.1.3. Let g_{ij} be the metric with respect to the coordinates given by κ, and let g^{ij} be its inverse. Since M is of bounded geometry, Lemma 10.1.4 yields a constant C independent on z such that we have $|g_{ij}| \leq C$ and $|g^{ij}| \leq C$. Together with the uniform boundedness of R^M, l and their covariant derivatives, we obtain that their representations R^M_{ijkl}, l_{rs} and their derivatives in the coordinates given by κ are uniformly bounded for all $1 \leq i, j, k, l \leq d$, $d - k + 1 \leq r, s \leq d - k$. Then the claim follows by Gauss' equation [Spi79, p. 47],

$$g(R^N(U, V)W, Z) = g(R^M(U, V)W, Z) + g(l(U, Z), l(V, W))$$
$$- g(l(U, W), l(V, Z))$$

for all $U, V, W, Z \in TN$ and the formulas for covariant derivatives of tensors along N. We refer to [Sch01, Lem. 2.22], where everything is stated for hypersurfaces but the formulas remain true for arbitrary codimension subject to obvious modifications.

□

Remark 10.1.6

(i) By construction the covering $(U_\gamma)_{\gamma \in I}$ is uniformly locally finite and $(U'_\gamma :=$ $U_\gamma \cap N, \kappa'_\gamma = \kappa_\gamma|_{\kappa_\gamma^{-1}(U'_\gamma)})_{\gamma \in I_N}$ gives a geodesic atlas on N. Moreover, none of the balls $B_R(p_\beta)$ with $\beta \in I \setminus I_N$ intersects N.

(ii) Existence of points p_α^N and p_β as claimed in Definition 10.1.3 (note that the proofs of Lemmas 10.1.4 and 10.1.5 only use the definition of Fermi coordinates on a single chart and the existence of the points p_α^N and p_β): We choose a maximal set $\{p_\alpha^N\}_{\alpha \in I_N}$ of points in N such that the metric balls $B_{\frac{R}{2}}^N(p_\alpha^N)$ are pairwise disjoint. Then, the balls $B_R^N(p_\alpha^N)$ cover N. Since by Lemma 10.1.5 the submanifold $(N, g|_N)$ is of bounded geometry, the volume of metric balls in N with fixed radius is uniformly bounded from above and from below away from zero. Let a ball $B_{2R}^N(p_\alpha^N)$ be intersected by L balls $B_{2R}^N(p_{\alpha'}^N)$. Then the union of the L balls $B_{2R}^N(p_{\alpha'}^N)$ forms a subset of $B_{4R}^N(p_\alpha^N)$. Comparison of the volumes gives an upper bound on L. Hence, the balls in $(B_{2R}^N(p_\alpha^N))_{\alpha \in I_N}$ cover N uniformly locally finite. Moreover, choose a maximal set of points $\{p_\beta\}_{\beta \in I \setminus I_N} \subset M \setminus U_R(N)$ such that the metric balls $B_{\frac{R}{2}}(p_\beta)$ are pairwise disjoint in M. Then the balls $B_R(p_\beta)$ cover $M \setminus U_R(N)$. Trivially the balls $B_R(p_\beta)$ for $\beta \in I \setminus I_N$ cover $\cup_\beta B_R(p_\beta)$, and by volume comparison as above this cover is uniformly locally finite.

Next we show that there is a trivialization $\mathcal{T}^{\text{FC}} = (U_\gamma, \kappa_\gamma, h_\gamma)_{\gamma \in I}$ with the Fermi coordinates from Definition 10.1.3 and a suitable partition of unity of M subordinate to the cover $(U_\gamma)_{\gamma \in I}$, which is admissible according to Definition 2.6.10. For this we need to verify that the conditions (B1) and (B2) from Definition 2.6.10 are satisfied for $\mathcal{T}^{\text{FC}}$. This will be done in the next two lemmas.

Lemma 10.1.7 *The atlas* $(U_\gamma, \kappa_\gamma)_{\gamma \in I}$ *introduced in Definition 10.1.3 fulfills condition (B1).*

Proof For all $\gamma \in I \setminus I_N$ the chart κ_γ is given by geodesic normal coordinates and, thus, condition (B1) follows from Remark 2.6.5(ii).

Let now $\gamma \in I_N$. Then the claim follows from [Sch01, Lem. 3.9]. We sketch the proof: Consider a chart $(B_{4R}(p_\alpha^N), \kappa^{\mathrm{geo}})$ in M and a chart $(B_{2R}^N(p_\alpha^N), \kappa^{N,\mathrm{geo}})$ in $(N, g|_N)$ both given by geodesic normal coordinates around p_α^N for $\alpha \in I_N$. Note that $4R < r_M$ by Definition 10.1.3. Let Φ_2 be the geodesic flow in $(N, g|_N)$ with respect to the coordinates given by $\kappa^{N,\mathrm{geo}}$, cf. Example 2.6.2. Let Φ_1 be the corresponding geodesic flow in (M, g) given by κ^{geo}. Then, $\Phi_2(1, 0, x) = (\kappa^{N,\mathrm{geo}})^{-1} \circ \exp_{p_\alpha^N}^N(\lambda_\alpha^N(x))$ and by (10.1.1) we have $\kappa_\alpha(t, x) = \kappa^{\mathrm{geo}} \circ \Phi_1(1, \Phi_2(1, 0, x), (\kappa^{\mathrm{geo}})^*(t^i v_i))$ with $t = (t^1, \ldots, t^{d-k}) \in \mathbb{R}^{d-k}$. Since (M, g) is of bounded geometry, the coefficient matrix g_{ij} of g with respect to κ^{geo}, its inverse and all its derivatives are uniformly bounded by (2.6.3). Moreover, by Lemma 10.1.5 we know that $(N, g|_N)$ is also of bounded geometry and, thus, we get an analogous statement for the coefficient matrix of $g|_N$ with respect to $\kappa^{N,\mathrm{geo}}$. Hence, applying Lemma 2.6.3 to the differential equation of the geodesic flows, see Example 2.6.2 and formula (2.6.1), we obtain that $(\kappa^{\mathrm{geo}})^{-1} \circ \kappa_\alpha$ and all its derivatives are bounded independent on α. Conversely, $(\kappa_\alpha)^{-1} \circ \kappa^{\mathrm{geo}}$: $(\kappa^{\mathrm{geo}})^{-1} \circ \kappa_\alpha(B_{2R}^{d-k} \times B_{2R}^k) \subset B_{4R}^d \to B_{2R}^{d-k} \times B_{2R}^k$ is bounded independent on α. Hence, using the chain rule on $((\kappa_\alpha)^{-1} \circ \kappa^{\mathrm{geo}}) \circ ((\kappa^{\mathrm{geo}})^{-1} \circ \kappa_\alpha) = \mathrm{Id}$ one sees that also the derivatives of $(\kappa_\alpha)^{-1} \circ \kappa^{\mathrm{geo}}$ are uniformly bounded, which gives the claim.   □

Lemma 10.1.8 *There is a partition of unity* $(h_\gamma)_{\gamma \in I}$ *subordinated to the Fermi coordinates introduced in Definition 10.1.3 fulfilling condition (B2).*

Proof By Lemma 10.1.5, $(N, g|_N)$ is of bounded geometry. Then, by Example 2.6.6, there is a partition of unity h_α' subordinated to a geodesic atlas $(U_\alpha' := U_\alpha \cap N = B_{2R}^N(p_\alpha^N), \kappa_\alpha' = \kappa_\alpha|_{\kappa_\alpha^{-1}(U_\alpha')})_{\alpha \in I_N}$ of N such that for each $\mathfrak{a} \in \mathbb{N}_0^k$ the derivatives $D^\mathfrak{a}(h_\alpha' \circ \kappa_\alpha')$ are uniformly bounded independent of α. Since by construction the balls $B_R^N(p_\alpha^N)$ already cover N the functions h_α' can be chosen such that supp $h_\alpha' \subset B_R^N(p_\alpha^N)$.

Choose a function $\psi : \mathbb{R}^{d-k} \to [0, 1]$ that is compactly supported on $B_{\frac{3}{2}R}^{d-k} \subset \mathbb{R}^{d-k}$ and $\psi|_{B_R^{d-k}} = 1$. Set $h_\alpha = (\psi \times (h_\alpha' \circ \kappa_\alpha')) \circ \kappa_\alpha^{-1}$ on U_α and zero outside. Then, supp $h_\alpha \subset U_\alpha$ and all $D^\mathfrak{a}(h_\alpha \circ \kappa_\alpha)$ are uniformly bounded by a constant depending on $|\mathfrak{a}|$ but not on $\alpha \in I_N$.

Let $S \subset M$ be a maximal set of points containing the set $\{p_\beta\}_{\beta \in I \setminus I_N}$ of Definition 10.1.3 such that the metric balls in $\{B_{\frac{R}{2}}(p)\}_{p \in S}$ are pairwise disjoint. Then $(B_R(p))_{p \in S}$ forms a uniformly locally finite cover of M. We equip this cover with a geodesic trivialization $(B_R(p), \kappa_p^{\text{geo}}, h_p^{\text{geo}})_{p \in S}$, see Example 2.6.6. For $\beta \in I \setminus I_N$ we have by construction $\kappa_\beta = \kappa_{P_\beta}^{\text{geo}}$ and set

$$
h_\beta = \begin{cases} \left(1 - \displaystyle\sum_{\alpha \in I_N} h_\alpha\right) \dfrac{h_{P_\beta}^{\text{geo}}}{\sum_{\beta' \in I \setminus I_N} h_{P_{\beta'}}^{\text{geo}}}, & \text{where } \sum_{\beta' \in I \setminus I_N} h_{P_{\beta'}}^{\text{geo}} \neq 0 \\[4mm] 0, & \text{otherwise.} \end{cases}
$$

Next, we will argue that all h_β are smooth: It suffices to prove the smoothness in points $x \in M$ on the boundary of $\{\sum_{\beta' \in I \setminus I_N} h_{P_{\beta'}}^{\text{geo}} \neq 0\}$. For all other x smoothness follows by smoothness of the functions h_α and h_p^{geo}. Let now $x \in M$ as specified above. Then $\sum_{\beta' \in I \setminus I_N} h_{P_{\beta'}}^{\text{geo}}(x) = 0$ and, thus, $x \in U_R(N)$ (cf. Remark 10.1.6(ii)). Together with $\psi|_{B_R^{d-k}} = 1$ this implies that for ϵ small enough there is a neighbourhood $B_\epsilon(x) \subset U_R(N)$ such that $\sum_{\alpha \in I_N} h_\alpha(y) = 1$ for all $y \in B_\epsilon(x)$. Therefore, $h_\beta|_{B_\epsilon(x)} = 0$ and h_β is smooth in x for all $\beta \in I \setminus I_N$.

Moreover, by construction $\sum_{\beta \in I \setminus I_N} h_\beta + \sum_{\alpha \in I_N} h_\alpha = 1$. Hence, $(h_\gamma)_{\gamma \in I}$ gives a partition of unity subordinated to the Fermi coordinates. The uniform boundedness of all $D^\alpha(h_\beta \circ \kappa_\beta)$ follows from the uniform boundedness of all $D^\alpha(h_\alpha \circ \kappa_\alpha)$, $D^\alpha(h_{P_{\beta'}}^{\text{geo}} \circ \kappa_{\beta'})$, $D^\alpha(\kappa_\alpha^{-1} \circ \kappa_\beta)$ and $D^\alpha((\kappa_{\beta'})^{-1} \circ \kappa_\beta)$ together with Remark 2.6.11(ii). $\qquad\square$

Collecting the last two lemmas gives the following.

Theorem 10.1.9 (Admissibility of Trivialization $\mathcal{T}^{\text{FC}}$ with Fermi Coordinates) *Let (M, N) be of bounded geometry. Let $\mathcal{T}^{\text{FC}} = (U_\gamma, \kappa_\gamma, h_\gamma)_{\gamma \in I}$ be a trivialization of M given by an atlas $(U_\gamma, \kappa_\gamma)_{\gamma \in I}$ with Fermi coordinates as in Definition 10.1.3 together with the subordinated partition of unity $(h_\gamma)_{\gamma \in I}$ from Lemma 10.1.8. Then, $\mathcal{T}^{\text{FC}}$ is an admissible trivialization.*

10.2 Trace Results

This section contains the main results of this chapter. In order to keep the presentation as simple as possible we start in Sect. 10.2.1 with traces of fractional Sobolev spaces on submanifolds N of (noncompact) Riemannian manifolds M. The only assumption on the manifolds we require is that (M, N) is of bounded geometry. In particular, we show that the trace operator Tr_N in fact is onto and has a linear and bounded right inverse Ex_M (mapping from the trace space into the original one). For the proof, we use the trivialization $\mathcal{T}^{\text{FC}}$ build upon Fermi coordinates instead of geodesic coordinates (originally used for defining the spaces on manifolds). This is possible, since Fermi coordinates are admissible and gives us more flexibility in what follows. Then via pull-back and localization the problem reduces to well-known trace results on hyperplanes $\mathbb{R}^k$ of $\mathbb{R}^d$. When just asking for Tr_N to be linear and bounded, one can reduce the assumptions on (M, N) further even, cf. Remark 10.2.7.

Afterwards, in Sect. 10.2.1, we show how the trace theorem for Sobolev spaces can be transferred to the more complicated B- and F-spaces defined on M. Our results generalize [Skr90, Thm. 1], since in our case no unnecessary assumptions on M and N are made. In particular, in Example 10.2.4 we consider a submanifold N which is covered by our assumptions but not totally geodesic as required in [Skr90].

10.2.1   Trace Theorem for Spaces $H_p^s(M)$

For fractional Sobolev spaces on manifolds we obtain the following trace theorem.

Theorem 10.2.1 *Let* (M^d, g) *be a Riemannian manifold together with an embedded k-dimensional submanifold* N *and let* (M, N) *be of bounded geometry. If* $1 < p < \infty$ *and* $s > \frac{d-k}{p}$, *then* Tr_N *is a linear and bounded operator from* $H_p^s(M)$ *onto* $B_{p,p}^{s-\frac{d-k}{p}}(N)$, *i.e.,*

$$\mathrm{Tr}_N H_p^s(M) = B_{p,p}^{s-\frac{d-k}{p}}(N). \tag{10.2.1}$$

Remark 10.2.2 For $(M, N) = (\mathbb{R}^d, \mathbb{R}^k)$ this is a classical result, cf. [Tri83, p. 138, Rem. 1] and the references given there. Here we think of $\mathbb{R}^k \cong \{0\}^{d-k} \times \mathbb{R}^k \subset \mathbb{R}^d$. Furthermore, in [Tri83, p. 138, Rem. 1] it is also shown that $\mathrm{Tr}_{\mathbb{R}^k}$ has a linear bounded right inverse, i.e., an extension operator $\mathrm{Ex}_{\mathbb{R}^d}$. Note that $\mathrm{Tr}_{\mathbb{R}^k}$ respects products with test functions, i.e., for $f \in H_p^s(\mathbb{R}^d)$ and $\eta \in \mathcal{D}(\mathbb{R}^d)$ we have $\mathrm{Tr}_{\mathbb{R}^k}(\eta f) = \eta|_{\mathbb{R}^k} \mathrm{Tr}_{\mathbb{R}^k} f$. Moreover, if κ is a diffeomorphism on $\mathbb{R}^d$ such that $\kappa(\mathbb{R}^k) = \mathbb{R}^k$, then $\mathrm{Tr}_{\mathbb{R}^k}(f \circ \kappa) = \mathrm{Tr}_{\mathbb{R}^k} f \circ \kappa|_{\mathbb{R}^k}$.

Proof of Theorem 10.2.1 Via localization and pull-back we will reduce (10.2.1) to the classical problem of traces on hyperplanes $\mathbb{R}^k$ in $\mathbb{R}^d$. The proof is similar to [Skr90, Thm. 1], but the Fermi coordinates enable us to drop some of the restricting assumptions made there.

By Theorem 10.1.9 we have an admissible trivialization $\mathcal{T} = (U_\alpha, \kappa_\alpha, h_\alpha)_{\alpha \in I}$ of M by Fermi coordinates and the subordinated partition of unity from Lemma 10.1.8. Moreover, from the construction of the Fermi coordinates it is clear that $\kappa_\alpha^{-1}(N \cap U_{p_\alpha^N}) = \{0\}^{d-k} \times B_{2R}^k$ and, thus, their restriction to N gives a geodesic trivialization $\mathcal{T}^{N,\mathrm{geo}} = (U_\alpha' := U_\alpha \cap N, \kappa_\alpha' := \kappa_\alpha|_{\kappa_\alpha^{-1}(U_\alpha')}, h_\alpha' := h_\alpha|_{U_\alpha'})_{\alpha \in I_N}$ of N.

Step 1.   Let $f \in H_p^s(M)$. We define the trace operator via

$$(\mathrm{Tr}_N f)(x) := \sum_{\alpha \in I_N} \mathrm{Tr}_{\mathbb{R}^k}[(h_\alpha f) \circ \kappa_\alpha] \circ (\kappa_\alpha')^{-1}(x), \qquad x \in N.$$

Note that Tr_N is well-defined since $(h_\alpha f) \circ \kappa_\alpha \in H_p^s(\mathbb{R}^d)$ and $\mathrm{supp}\,\mathrm{Tr}_{\mathbb{R}^k}((h_\alpha f) \circ \kappa_\alpha) \subset V_\alpha' = V_\alpha \cap \mathbb{R}^k$. Moreover, for fixed $x \in N$ the summation is meant to run only over those α for which $x \in U'_\alpha$. Hence, the summation only runs over finitely many α due to the uniform locally finite cover. Obviously, Tr_N is linear and $\mathrm{Tr}_N|_{\mathcal{D}(M)}$ is given by the pointwise restriction. In order to show that Tr_N : $H_p^s(M) \to B_{p,p}^{s-\frac{d-k}{p}}(N)$ is bounded, we set $A(\alpha) := \{\beta \in I_N \mid U_\alpha \cap U_\beta \neq \varnothing\}$. Since the cover is uniformly locally finite, the number of elements in $A(\alpha)$ is bounded independent of α. Using Lemma 2.2.2 and Remark 10.2.2 we obtain

$$
\left\| \mathrm{Tr}_N f \, | B_{p,p}^{s-\frac{d-k}{p}}(N) \right\|^p = \sum_{\beta \in I_N} \left\| (h'_\beta \mathrm{Tr}_N f) \circ \kappa'_\beta \, | B_{p,p}^{s-\frac{d-k}{p}}(\mathbb{R}^k) \right\|^p
$$

$$
\lesssim \sum_{\beta \in I_N;\, \alpha \in A(\beta)} \left\| (h'_\beta \circ \kappa'_\beta) \left(\mathrm{Tr}_{\mathbb{R}^k}[(h_\alpha f) \circ \kappa_\alpha] \circ \left[(\kappa'_\alpha)^{-1} \circ \kappa'_\beta \right] \right) \right.
$$

$$
\left. \times | B_{p,p}^{s-\frac{d-k}{p}}(\mathbb{R}^k) \right\|^p
$$

$$
= \sum_{\beta \in I_N;\, \alpha \in A(\beta)} \left\| \mathrm{Tr}_{\mathbb{R}^k}[(h_\alpha h_\beta f) \circ \kappa_\alpha] \circ \left[(\kappa'_\alpha)^{-1} \circ \kappa'_\beta \right] | B_{p,p}^{s-\frac{d-k}{p}}(\mathbb{R}^k) \right\|^p
$$

$$
\lesssim \sum_{\beta \in I_N;\, \alpha \in A(\beta)} \left\| \mathrm{Tr}_{\mathbb{R}^k}[(h_\alpha h_\beta f) \circ \kappa_\alpha] | B_{p,p}^{s-\frac{d-k}{p}}(\mathbb{R}^k) \right\|^p
$$

$$
\lesssim \sum_{\alpha \in I_N;\, \beta \in A(\alpha)} \left\| (h_\alpha h_\beta f) \circ \kappa_\alpha | H_p^s(\mathbb{R}^d) \right\|^p \lesssim \sum_{\alpha \in I_N} \left\| (h_\alpha f) \circ \kappa_\alpha | H_p^s(\mathbb{R}^d) \right\|^p,
$$

$$(10.2.2)$$

therefore,

$$
\left\| \mathrm{Tr}_N f \, | B_{p,p}^{s-\frac{d-k}{p}}(N) \right\|^p \lesssim \left\| f | H_p^s(M) \right\|^p,
$$

where the involved constants do not depend on f.

Step 2. Now we show that Tr_N is onto by constructing a right inverse, i.e., an extension operator Ex_M. Let $\psi_1 \in \mathcal{D}(\mathbb{R}^k)$ and $\psi_2 \in \mathcal{D}(\mathbb{R}^{d-k})$ such that supp $\psi_1 \in B_{2R}^k$ and supp $\psi_2 \in B_{2R}^{d-k}$, $\psi_1 \equiv 1$ on B_R^k and $\psi_2 \equiv 1$ on $B_{\frac{3}{2}R}^{d-k}$.
Moreover, we put $\psi := \psi_1 \times \psi_2 \in \mathcal{D}(\mathbb{R}^d)$.
Let $f' \in B_{p,p}^{s-\frac{d-k}{p}}(N)$. Then we define the extension operator by

$$
(\mathrm{Ex}_M f')(x) := \begin{cases} \sum_{\alpha \in I_N} \left[\psi \mathrm{Ex}_{\mathbb{R}^d}((h'_\alpha f') \circ \kappa'_\alpha) \right] \circ \kappa_\alpha^{-1}(x), & x \in U_{2R}(N) \\ 0, & \text{otherwise.} \end{cases}
$$

Note that the use of ψ is to ensure that $\psi \mathrm{Ex}_{\mathbb{R}^d}((h'_\alpha f') \circ \kappa'_\alpha)$ is compactly supported in $V_\alpha = B^k_{2R} \times B^{d-k}_{2R}$ for all $\alpha \in I_N$. Hence, one sees immediately that Ex_M is well-defined and calculates $\mathrm{Tr}_N(\mathrm{Ex}_M f') = f'$. Thus, Tr_N is onto. Moreover, in order to show that $\mathrm{Ex}_M : B^{s-\frac{d-k}{p}}_{p,p}(N) \to H^s_p(M)$ is bounded, we use Lemma 2.2.2 and Remark 10.2.2 again, which give

$$\left\| \mathrm{Ex}_M f' | H^s_p(M) \right\|^p = \sum_{\alpha \in I} \left\| (h_\alpha \mathrm{Ex}_M f') \circ \kappa_\alpha | H^s_p(\mathbb{R}^d) \right\|^p$$

$$\lesssim \sum_{\alpha \in I;\, \beta \in A(\alpha)} \left\| \left(h_\alpha \left(\left[\psi \mathrm{Ex}_{\mathbb{R}^d}((h'_\beta f') \circ \kappa'_\beta) \right] \circ \kappa_\beta^{-1} \right) \right) \circ \kappa_\alpha | H^s_p(\mathbb{R}^d) \right\|^p$$

$$\lesssim \sum_{\beta \in I_N;\, \alpha \in I;\, U_\alpha \cap U_\beta \neq \varnothing} \left\| (h_\alpha \circ \kappa_\beta) \psi \mathrm{Ex}_{\mathbb{R}^d}((h'_\beta f') \circ \kappa'_\beta) | H^s_p(\mathbb{R}^d) \right\|^p$$

$$\lesssim \sum_{\beta \in I_N} \left\| \mathrm{Ex}_{\mathbb{R}^d}((h'_\beta f') \circ \kappa'_\beta) | H^s_p(\mathbb{R}^d) \right\|^p$$

$$\lesssim \sum_{\beta \in I_N} \left\| (h'_\beta f') \circ \kappa'_\beta | B^{s-\frac{d-k}{p}}_{p,p}(\mathbb{R}^k) \right\|^p = \left\| f' | B^{s-\frac{d-k}{p}}_{p,p}(N) \right\|^p .$$

Note that the estimate in the second but last line uses $(h_\alpha \circ \kappa_\beta)\psi \in \mathcal{D}(V_\beta)$. This finishes the proof. □

Since the classical Sobolev spaces $W^k_p(M)$ are included as special cases in the scale of fractional Sobolev spaces whenever M is of bounded geometry, cf. (2.6.5), we obtain the following corollary.

Corollary 10.2.3 *Let (M^d, g) be a Riemannian manifold together with an embedded k-dimensional submanifold N and let (M, N) be of bounded geometry. If $m \in \mathbb{N}$, $1 < p < \infty$, and $m > \frac{d-k}{p}$, then Tr_N is a linear and bounded operator from $W^m_p(M)$ onto $B^{m-\frac{d-k}{p}}_{p,p}(N)$, i.e.,*

$$\mathrm{Tr}_N W^m_p(M) = B^{m-\frac{d-k}{p}}_{p,p}(N).$$

Example 10.2.4 Our results generalize [Skr90] where traces were restricted to submanifolds N which had to be totally geodesic. By using Fermi coordinates we can drop this extremely restrictive assumption and cover more (sub-)manifolds.

For example, consider the case where M is a surface of revolution of a curve γ and N a circle obtained by the revolution of a fixed point $p \in M$ (Fig. 10.7). This resulting circle is a geodesic if and only if the rotated curve has an extremal point at p. But there is always a collar around N, hence, this situation is also covered by our assumptions.

Fig. 10.7 Submanifold N
which is not totally geodesic

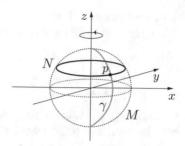

Remark 10.2.5 We proved even more than stated in Theorem 10.2.1 above: In Step 2 of the theorem it was actually shown that there exists a linear and bounded extension operator Ex_M from the trace space into the original space such that

$$\mathrm{Tr}_N \circ \mathrm{Ex}_M = \mathrm{Id},$$

where Id stands for the identity in $B_{p,p}^{s-\frac{d-k}{p}}(N)$.

The first part of Theorem 10.2.1, which deals with the boundedness of the trace operator, can be extended to an even broader class of submanifolds. We give an example to illustrate the idea.

Example 10.2.6 Let (M^d, N_1^k) and (M^d, N_2^k) be manifolds of bounded geometry with $N_1 \cap N_2 = \varnothing$. Set $N := N_1 \cup N_2$. Clearly, $\mathrm{Tr}_N = \mathrm{Tr}_{N_1} + \mathrm{Tr}_{N_2}$ (where $\mathrm{Tr}_{N_1} f$ and $\mathrm{Tr}_{N_2} f$ are viewed as functions on N that equal zero on N_2 and N_1, respectively), and Tr_N is a linear bounded operator from $H_p^s(M)$ to $B_{p,p}^{s-\frac{d-k}{p}}(N)$. One may think of $N = \mathrm{Graph}(x \mapsto x^{-1}) \cup x - \mathrm{axis} \subset \mathbb{R}^2$, where $(\mathbb{R}^2, N)$ does not posses a uniform collar, cf. Fig. 10.4 and Definition 10.1.1(iii).

The boundedness of Tr_N is no longer expectable for an arbitrary infinite union of N_i, e.g. consider $N_i = \mathbb{R} \times \{i^{-1}\} \subset \mathbb{R}^2$, $i \in \mathbb{N}$ and put $f = \psi_1 \times \psi_2$ with $\psi_1, \psi_2 \in \mathcal{D}(\mathbb{R})$. Then $N = \sqcup_i N_i \hookrightarrow \mathbb{R}^2$ is an embedding, when N is equipped with the standard topology on each copy of $\mathbb{R}$. But one cannot expect the trace operator to be bounded, since not every function $f \in C_0^\infty(\mathbb{R}^2)$ restricts to a compactly supported function on N (N as a subset of $\mathbb{R}^2$ is not intersection compact).

This problem can be circumvented when requiring that the embedded submanifold N has to be a closed subset of M. However, even in this situation on can find submanifolds N for which the trace operator is not bounded in the sense of (10.2.2), e.g. consider $N = \sqcup_{i \in \mathbb{N}} \sqcup_{j=0}^{i-1} \mathbb{R} \times \{i + \frac{j}{i}\} \hookrightarrow \mathbb{R}^2$.

Remark 10.2.7 The above considerations give rise to the following generalization of Step 1 of the Theorem 10.2.1. Assume that N is a k-dimensional embedded submanifold of (M^d, g) fulfilling (i), (ii) and (iv) of Definition 10.1.1—but not (iii). Lemmas 10.1.4 and 10.1.5 remain valid, since their proofs do not use (iii). We replace (iii) with the following weaker version:

(iii)′ Let $(U'_\alpha = B_{2R}(p^N_\alpha))_{\alpha \in I_N}$ be a uniformly locally finite cover of N. Set $U_\alpha = F(B^{d-k}_{2R} \times B^N_{2R}(p^N_\alpha))$ as before. Then $(U_\alpha)_{\alpha \in I_N}$ is a uniformly locally finite cover of $\cup_{\alpha \in I_N} U_\alpha$.

Condition (iii)′ excludes the negative examples from above. Furthermore, (iii)′ together with the completeness of N implies that N is a closed subset of M.

With this modification, one can still consider Fermi coordinates as in Definition 10.1.3 but in general $U_{p^N_\alpha} \cap N \neq B^N_{2R}(p^N_\alpha) =: U'_{p^N_\alpha}$. Also the partition of unity can be constructed as in Lemma 10.1.8 when making the following step in between: Following the proof of Lemma 10.1.8 we define the map $\tilde{h}_\alpha = (\psi \times (h'_\alpha \circ \kappa'_\alpha)) \circ \kappa_\alpha^{-1}$ for $\alpha \in I_N$. Since in general $\sum_{\alpha \in I_N} \tilde{h}_\alpha(x)$ can be bigger than one, those maps cannot be part of the desired partition of unity. Hence, we put $h_\alpha = \tilde{h}_\alpha (\sum_{\alpha' \in I_N} \tilde{h}_{\alpha'})^{-1}$ where $\sum_{\alpha' \in I_N} \tilde{h}_{\alpha'} \neq 0$ and $h_\alpha = 0$ otherwise. Smoothness and uniform boundedness of the derivatives of h_α follow as in Lemma 10.1.8. Then one proceeds as before, defining h_β for $I \setminus I_N$.

Now the proof of Step 1 of Theorem 10.2.1 carries over when replacing h_α by $\tilde{h}_\alpha$ in the definition of Tr_N and in the estimate (10.2.2). This leads to

$$\left\| \mathrm{Tr}_N f \,|\, B^{s-\frac{d-k}{p}}_{p,p}(N) \right\|^p \lesssim \sum_{\alpha \in I_N} \left\| (\tilde{h}_\alpha f) \circ \kappa_\alpha \,|\, H^s_p(\mathbb{R}^d) \right\|^p .$$

Finally,

$$\left\| \mathrm{Tr}_N f \,|\, B^{s-\frac{d-k}{p}}_{p,p}(N) \right\|^p \lesssim \sum_{\alpha \in I_N} \left\| (\tilde{h}_\alpha f) \circ \kappa_\alpha \,|\, H^s_p(\mathbb{R}^d) \right\|^p$$

$$= \sum_{\alpha \in I_N} \left\| \left(h_\alpha \Big(\sum_{\beta \in I_N} \tilde{h}_\beta \Big) f \right) \circ \kappa_\alpha \,|\, H^s_p(\mathbb{R}^d) \right\|^p$$

$$\lesssim \sum_{\alpha \in I_N;\ \beta \in A(\alpha)} \left\| (h_\alpha \tilde{h}_\beta f) \circ \kappa_\alpha \,|\, H^s_p(\mathbb{R}^d) \right\|^p$$

$$\lesssim \sum_{\alpha \in I_N} \left\| (h_\alpha f) \circ \kappa_\alpha \,|\, H^s_p(\mathbb{R}^d) \right\|^p \leq \left\| f \,|\, H^s_p(M) \right\|^p ,$$

demonstrates the boundedness of the trace operator Tr_N under this generalized assumptions on the submanifold N.

10.2.2 Trace Theorem for Spaces $F_{p,q}^s(M)$, $B_{p,q}^s(M)$

In order to keep our considerations as easy as possible, we have been concentrating on fractional Sobolev spaces on Riemannian manifolds M so far. This last subsection is devoted to Besov and Triebel-Lizorkin spaces and their traces. We briefly want to sketch how the trace results from Theorem 10.2.1 can be generalized to this setting. The results are stated below and improve [Skr90, Thm. 1, Cor. 1].

Theorem 10.2.8 *Let (M^d, g) be a Riemannian manifold together with an embedded k-dimensional submanifold N and let (M^d, N^k) be of bounded geometry. Furthermore, let $0 < p, q \le \infty$ ($0 < p, q < \infty$ or $p = q = \infty$ for F-spaces) and let*

$$s - \frac{d-k}{p} > k\left(\frac{1}{p} - 1\right)_+ . \tag{10.2.3}$$

Then Tr_N is a linear and bounded operator from $F_{p,q}^s(M)$ onto $B_{p,p}^{s-\frac{d-k}{p}}(N)$ and $B_{p,q}^s(M)$ onto $B_{p,q}^{s-\frac{d-k}{p}}(N)$, respectively, i.e.,

$$\mathrm{Tr}_N F_{p,q}^s(M) = B_{p,p}^{s-\frac{d-k}{p}}(N) \quad and \quad \mathrm{Tr}_N B_{p,q}^s(M) = B_{p,q}^{s-\frac{d-k}{p}}(N). \tag{10.2.4}$$

Proof The proof of (10.2.4) runs along the same lines as the proof of Theorem 10.2.1. Choosing Fermi coordinates, via pull back and localization the problem can be reduced to corresponding trace results in $\mathbb{R}^d$ on hyperplanes $\mathbb{R}^k$, cf. [Tri92, Thm. 4.4.2], where the proof for $k = d - 1$ may be found. The result for general hyperplanes as well as condition (10.2.3) follows by iteration. The assertion for B-spaces then results from Definition 2.6.14(ii). □

10.3 Application: Spaces with Symmetries

In this section we give an application of admissible trivializations to spaces with symmetries. We consider manifolds M, where a countable discrete group G acts in a convenient way and show that the Sobolev spaces of functions on the resulting orbit space M/G and the weighted Sobolev spaces of G-invariant functions on M coincide. This is in spirit of the following theorem which can be found in [Tri83, Sect. 9.2.1], where the author characterizes Sobolev spaces on the tori $\mathbb{T}^d := \mathbb{R}^d \backslash \mathbb{Z}^d$ via weighted Sobolev spaces on $\mathbb{R}^d$ containing $\mathbb{Z}^d$ periodic distributions only.

Theorem 10.3.1 *Let $1 < p < \infty$ and consider the weight $\rho(x) = (1 + |x|)^{-\varkappa}$ on Euclidean space $\mathbb{R}^d$ where $\varkappa p > d$. Let $\mathbb{T}^d := \mathbb{R}^d / \mathbb{Z}^d$ denote the torus and*

$\pi : \mathbb{R}^d \to \mathbb{T}^d$ the natural projection. Put $H^s_{p,\pi}(\mathbb{R}^d, \rho) := \{f \in \mathcal{D}'(\mathbb{R}^d) : | \rho f \in H^s_p(\mathbb{R}^d)$ and f is π-periodic$\}$, then

$$H^s_p(\mathbb{T}^d) = H^s_{p,\pi}(\mathbb{R}^d, \rho).$$

This is just a special case of the theorem given in [Tri83, Sect. 9.2.1], where the more general Besov and Triebel-Lizorkin spaces are treated. The proof uses Fourier series.

With the help of admissible trivializations, we now want to present a small generalization of this result for manifolds with G-actions. We start by introducing our setup. In order to avoid any confusion with the metric g, elements of the group G are denoted by h.

Definition 10.3.2 (G-Manifold) Let (M, g) be a Riemannian manifold, and let G be a countable discrete group that acts freely and properly discontinuously on M. If, additionally, g is invariant under the G-action (which means that $h : p \in M \mapsto h \cdot p \in M$ is an isometry for all $h \in G$), we call (M, g) a G-manifold.

By [Lee11, Cor. 12.27] the orbit space $\widetilde{M} := M/G$ of a G-manifold is again a manifold. From now on we restrict ourselves to the case where $\widetilde{M}$ is closed. Let $\pi : M \to \widetilde{M}$ be the corresponding projection. If (M, g) is a G-manifold, then there is a Riemannian metric $\tilde{g}$ on $\widetilde{M}$ such that $\pi^*\tilde{g} = g$. Let now $\widetilde{\mathcal{T}} = (U_\alpha, \kappa_\alpha, h_\alpha)_{\alpha \in I}$ be an admissible trivialization of $\widetilde{M}$. In particular, this means we assume that $(\widetilde{M}, \tilde{g})$ is of bounded geometry and, hence, so is (M, g). Then there are $U_{\alpha,h} \subset M$ with $\pi^{-1}(U_\alpha) = \sqcup_{h \in G} U_{\alpha,h}$ and $U_{\alpha,h} = h \cdot U_{\alpha,e}$ for all $\alpha \in I$. Here e is the identity element of G. Let $\pi_{\alpha,h} := \pi|_{U_{\alpha,h}} : U_{\alpha,h} \to U_\alpha$ denote the corresponding diffeomorphism. Setting $\kappa_{\alpha,h} := \pi^{-1}_{\alpha,h} \circ \kappa_\alpha : V_\alpha \to U_{\alpha,h}$ and

$$h_{\alpha,h} := \begin{cases} h_\alpha \circ \pi_{\alpha,h} & \text{on } U_{\alpha,h}, \\ 0 & \text{else,} \end{cases}$$

we have $h_{\alpha,h} \circ \kappa_{\alpha,h} = h_\alpha \circ \kappa_\alpha$ for all $\alpha \in I$, $h \in G$. This way we obtain an admissible trivialization $\mathcal{T} = (U_{\alpha,h}, \kappa_{\alpha,h}, h_{\alpha,h})_{\alpha \in I, h \in G}$ of M, which we call G-adapted trivialization.

Definition 10.3.3 (G-Adapted Weight) Let (M, g) be a G-manifold with a G-adapted trivialization $\mathcal{T}$ as above. A function $\rho : M \to (0, \infty)$ on M is called G-adapted weight, if there exists a constant $C_k > 0$ for all $k \in \mathbb{N}_0$ such that for $\mathfrak{a} \in \mathbb{N}^d_0$ with $|\mathfrak{a}| \leq k$ and all $\alpha \in I$,

$$\sum_{h \in G} |D^\mathfrak{a}(\rho \circ \kappa_{\alpha,h})| \leq C_k.$$

Remark 10.3.4 The notion of a G-adapted weight is independent on the chosen admissible trivialization on M/G. This follows immediately from the compatibility of two admissible trivializations, cf. Remark 2.6.11(ii).

Example 10.3.5 We give an example of a weight adapted to the G-action. Take a geodesic trivialization on $\widetilde{M}$ as in Example 2.6.1 and let $\mathcal{T}$ be an admissible trivialization of M constructed from $\widetilde{\mathcal{T}}$ on $\widetilde{M}$ as above. There is an injection $\iota : G \to \mathbb{N}$, since G is countable, and we set

$$\rho(p) = \sum_{(\alpha,h)\in I\times G;\; p\in U_{\alpha,h}} \iota(h)^{-2} h_{\alpha,h}(p).$$

Since the covering is locally finite, the summation is always finite. Moreover, Definition 2.6.10 and the uniform finiteness of the cover yield for fixed $\alpha \in I$ and all $\mathfrak{a} \in \mathbb{N}_0^d$ with $|\mathfrak{a}| \leq k, k \in \mathbb{N}_0$,

$$\sum_{h\in G} |D^{\mathfrak{a}}(\rho \circ \kappa_{\alpha,h})| \leq \sum_{h\in G} \sum_{\substack{(\alpha',h')\in I\times G;\\ U_{\alpha,h}\cap U_{\alpha',h'}\neq\varnothing}} \iota(h')^{-2} \left| D^{\mathfrak{a}}(h_{\alpha',h'} \circ \kappa_{\alpha,h}) \right|$$

$$\leq C_k' \sum_{h\in G} \sum_{|\mathfrak{a}'|\leq|\mathfrak{a}|} \sum_{\substack{(\alpha',h')\in I\times G;\\ U_{\alpha,h}\cap U_{\alpha',h'}\neq\varnothing}} \iota(h')^{-2} \left| D^{\mathfrak{a}'}(h_{\alpha'} \circ \kappa_{\alpha'}) \right|$$

$$\leq C_k'' \sum_{h\in G} \sum_{\substack{(\alpha',h')\in I\times G;\\ U_{\alpha,h}\cap U_{\alpha',h'}\neq\varnothing}} \iota(h')^{-2} = C_k'' \sum_{h\in G} \sum_{\substack{(\alpha',h')\in I\times G,\\ U_{\alpha,e}\cap U_{\alpha',h^{-1}h'}\neq\varnothing}} \iota(h')^{-2}$$

$$= C_k'' \sum_{h\in G} \sum_{\substack{(\alpha',h')\in I\times G,\\ U_{\alpha,e}\cap U_{\alpha',h'}\neq\varnothing}} \iota(hh')^{-2}$$

$$\leq C_k'' L \sum_{h\in G} \iota(h)^{-2} \leq C_k'' L \sum_{i\in\mathbb{N}} i^{-2} < \infty,$$

where L is the multiplicity of the cover and the constants C_k', C_k'' do not depend on $h \in G$ and $\alpha \in I$. In particular, together with Remark 10.3.4, this example demonstrates that each G-manifold admits a G-adapted weight.

We fix some more notation. Let $s \in \mathbb{R}$ and $1 < p < \infty$. Then the space $H_p^s(M, \rho)$ consists of all distributions $f \in \mathcal{D}'(M)$ such that

$$\|f|H_p^s(M,\rho)\| := \|\rho f|H_p^s(M)\| < \infty.$$

Moreover, we call a distribution $f \in \mathcal{D}'(M)$ G-invariant, if $f(\phi) = f(h^*\phi)$ holds for all $\phi \in \mathcal{D}(M)$ and $h \in G$. The space of all G-invariant distributions in $H_p^s(M, \rho)$ is denoted by $H_p^s(M, \rho)^G$.

Theorem 10.3.6 *Let (M, g) be a G-manifold of bounded geometry where $\widetilde{M} = M/G$ is closed, and let ρ be a G-adapted weight on M. Furthermore, let $s \in \mathbb{R}$ and $1 < p < \infty$. Then*

$$H_p^s(\widetilde{M}) = H_p^s(M, \rho)^G.$$

Proof It suffices to show that the norms of the corresponding spaces are equivalent. We work with a geodesic trivialization $\widetilde{\mathcal{T}}^{\text{geo}}$ of $(M/G, \tilde{g})$ and a G-adapted trivialization $\mathcal{T}$ of M constructed from $\widetilde{\mathcal{T}}^{\text{geo}}$ as described above. Note that the closedness of M/G implies that $\rho|_{\cup_\alpha U_{\alpha,e}} \geq c > 0$ for some constant $c > 0$ (since then $\cup_\alpha U_{\alpha,e}$ is compact). Let $f' \in H_p^s(M/G)$ and set $f = f' \circ \pi$. Then

$$\|f|H_p^s(M, \rho)\|^P = \sum_{\alpha \in I, h \in G} \|(h_{\alpha,h}\rho f) \circ \kappa_{\alpha,h} | H_p^s(\mathbb{R}^d)\|^P$$

$$= \sum_{\alpha \in I, h \in G} \|(\rho \circ \kappa_{\alpha,h}) \left((h_\alpha f') \circ \kappa_\alpha\right) | H_p^s(\mathbb{R}^d)\|^P$$

$$\lesssim \sum_{\alpha \in I} \|(h_\alpha f') \circ \kappa_\alpha | H_p^s(\mathbb{R}^d)\|^P = \|f' | H_p^s(M/G)\|^P.$$

Let now $f \in H_p^s(M, \rho)^G$. Since f is G-invariant, there is a unique f' with $f = f' \circ \pi$. Therefore,

$$\|f' | H_p^s(M/G)\|^P = \sum_{\alpha \in I} \|(h_\alpha f') \circ \kappa_\alpha | H_p^s(\mathbb{R}^d)\|^P$$

$$= \sum_{\alpha \in I} \left\| \left(\frac{1}{\rho} \circ \kappa_{\alpha,e}\right) \left((h_{\alpha,e}\rho f) \circ \kappa_{\alpha,e}\right) | H_p^s(\mathbb{R}^d) \right\|^P$$

$$\lesssim \sum_{\alpha \in I} \| \left((h_{\alpha,e}\rho f) \circ \kappa_{\alpha,e}\right) | H_p^s(\mathbb{R}^d)\|^P$$

$$\leq \sum_{\alpha \in I, h \in G} \| \left((h_{\alpha,h}\rho f) \circ \kappa_{\alpha,h}\right) | H_p^s(\mathbb{R}^d)\|^P = \|f|H_p^s(M, \rho)\|^P.$$

Here we used the uniform boundedness of $\frac{1}{\rho} \circ \kappa_{\alpha,e}$ and its derivatives, which follows from the corresponding statement for $\rho \circ \kappa_{\alpha,e}$ and the lower bound $\rho \circ \kappa_{\alpha,e} \geq c > 0$. $\square$

Remark 10.3.7 The restriction to closed manifolds $\widetilde{M}$ (i.e., compact manifolds without boundary) in Theorem 10.3.6 should not be necessary. In case that $\widetilde{M}$ is noncompact, one needs to modify the definition of G-adapted weights in a suitable way to assure the weight is bounded away from zero with respect to the 'noncompact directions' of $\widetilde{M}$.

We conclude our considerations with an example of a G-manifold other than the torus, which is covered by Theorem 10.3.6.

Example 10.3.8 Let $(\widetilde{M}, \widetilde{g})$ be a closed manifold. Let G be a subgroup of the fundamental group $\pi_1(\widetilde{M})$ of $\widetilde{M}$. Note that G is countable since $\pi_1(\widetilde{M})$ is. Let (M, g) be the G-cover of $(\widetilde{M}, \widetilde{g})$ where $g = \pi^* \widetilde{g}$. Then (M, g) is a G-manifold with $\widetilde{M} = M/G$.

Bibliography

[Ada15] D.R. Adams, *Morrey Spaces*. Lecture Notes in Applied and Numerical Harmonic Analysis (Birkhäuser, Springer, Cham, 2015)

[AX04] D.R. Adams, J. Xiao, Nonlinear potential analysis on Morrey spaces and their capacities. Ind. Univ. Math. J. **53**, 1629–1663 (2004)

[AX11] D.R. Adams, J. Xiao, Morrey potentials and harmonic maps. Commun. Math. Phys. **308**, 439–456 (2011)

[AX12a] D.R. Adams, J. Xiao, Morrey spaces in harmonic analysis. Ark. Mat. **50**, 201–230 (2012)

[AX12b] D.R. Adams, J. Xiao, Regularity of Morrey commutators. Trans. Am. Math. Soc. **364**, 4801–4818 (2012)

[AF03] R.A. Adams, J.J.F. Fournier, *Sobolev Spaces*. Pure and Applied Mathematics, vol. 140, 2nd edn. (Elsevier, Academic Press, New York, 2003)

[ADN59] S. Agmon, A. Douglis, L. Nierenberg, Estimates near the boundary for solutions of elliptic partial differential equations satisfying general boundary conditions I. Commun. Pure Appl. Math. **12**, 623–727 (1959)

[AGI08] H. Aimar, I. Gómez, B. Iaffei, Parabolic mean values and maximal estimates for gradients of temperatures. J. Funct. Anal. **255**, 1939–1956 (2008)

[AGI10] H. Aimar, I. Gómez, B. Iaffei, On Besov regularity of temperatures. J. Fourier Anal. Appl. **16**, 1007–1020 (2010)

[AG12] H. Aimar, I. Gómez, Parabolic Besov regularity for the heat equation. Constr. Approx. **36**, 145–159 (2012)

[AH08] N.T. Anh, N.M. Hung, Regularity of solutions of initial-boundary value problems for parabolic equations in domains with conical points. J. Differ. Equ. **245**(7), 1801–1818 (2008)

[ALL16] N.T. Anh, D.V. Loi, V.T. Luong, L_p-regularity for the Cauchy-Dirichlet problem for parabolic equations in convex polyhedral domains. Acta Math. Vietnam. **41**(4), 731–742 (2016)

[AZ90] J. Appell, P.P. Zabrejko, *Nonlinear Superposition Operators*. Cambridge Tracts in Mathematics, vol. 95 (Cambridge Univ. Press, Cambridge, 1990)

[Aub76] T. Aubin, Espaces de Sobolev sur les variétés Riemanniennes. Bull. Sci. Math. **100**, 149–173 (1976)

[Aub82] T. Aubin, *Nonlinear Analysis on Manifolds*. Monge-Ampère Equations (Springer, New York, 1982)

© The Author(s), under exclusive license to Springer Nature Switzerland AG 2021
C. Schneider, *Beyond Sobolev and Besov*, Lecture Notes in Mathematics 2291,
https://doi.org/10.1007/978-3-030-75139-5

[BG97] I. Babuska, B. Guo, Regularity of the solutions for elliptic problems on nonsmooth domains in $\mathbb{R}^3$, Part I: countably normed spaces on polyhedral domains. Proc. Roy. Soc. Edinburgh Sect. A **127**, 77–126 (1997)

[BMNZ10] C. Bacuta, A. Mazzucato, V. Nistor, L. Zikatanov, Interface and mixed boundary value problems on n-dimensional polyhedral domains. Doc. Math. **15**, 687–745 (2010)

[BNZ05] C. Bacuta, V. Nistor, L. Zikatanov, Impoving the rate of convergence of 'high order finite elements' on polygons and domains with cusps. Numer. Math. **100**(2), 165–184 (2005)

[Bau86] H.F. Bauer, Tables of the roots of the associated Legendre function with respect to the degree. Math. Comp. **46**(174), 601–602, S29–S41 (1986)

[BL76] J. Bergh, J. Löfström, *Interpolation Spaces*. Grundlehren der Mathematischen Wissenschaften, vol. 223 (Springer, Berlin, 1976)

[Bou83] G. Bourdaud, *Sur les opérateurs pseudo-différentiels à coefficients peu réguliers.* Habilitation thesis, Université de Paris-Sud, Paris (1983)

[Bou91] G. Bourdaud, Le calcul fonctionnel dans le espace de Sobolev. Invent. Math. **104**, 435–446 (1991)

[BL08] G. Bourdaud, M. Lanza de Cristoforis, Regularity of the symbolic calculus in Besov algebras. Studia Math. **184**, 271–298 (2008)

[BS11] G. Bourdaud, W. Sickel, Composition operators on function spaces with fractional order of smoothness. RIMS Kokyuroko Bessatsu **B26**, 93–132 (2011)

[BM01] H. Brezis, P. Mironescu, Gagliardo-Nirenberg, compositions and products in fractional Sobolev spaces. J. Evol. Equ. **1**, 387–404 (2001)

[Bur98] V.I. Burenkov, *Sobolev Spaces on Domains.* Teubner Texte zur Mathematik (Teubner, Stuttgart, 1998)

[CLT07] A.M. Caetano, S. Lopes, H. Triebel, A homogeneity property for Besov spaces. J. Funct. Spaces Appl. **5**(2), 123–132 (2007)

[CL09] A.M. Caetano, S. Lopes, Homogeneity, non-smooth atoms and Besov spaces of generalised smoothness on quasi-metric spaces. Dissertationes Math. (Rozprawy Mat.) **460**, pp. 44 (2009)

[Cap14] A. Capatina, *Variational Inequalities and Frictional Contact Problems.* Advances in Mechanics and Mathematics, vol. 31 (Springer, New York, 2014)

[CH98] T. Cazenave, A. Haraux, *An Introduction to Semilinear Evolution Equations.* Oxford Lecture Series in Mathematics and Its Applications (Clarendon Press, Oxford, 1998)

[Cio13] P. Cioica, *Besov regularity of stochastic partial differential equations on bounded Lipschitz domains.* Ph.D. Thesis, Philipps-University Marburg (2013)

[CKLL13] P. Cioica, K.-H. Kim, K. Lee, F. Lindner, On the $L_q(L_p)$-regularity and Besov smoothness of stochastic parabolic equations on bounded Lipschitz domains. Electron. J. Probab. **18**(82), 1–41 (2013)

[Coh03] A. Cohen, *Numerical Analysis of Wavelet Methods.* Studies in Mathematics and Its Applications, vol. 32, 1st edn. (Elsevier, Amsterdam, 2003)

[CDD01] A. Cohen, W. Dahmen, R.A. DeVore, Adaptive wavelet methods for elliptic operator equation: Convergence rates. Math. Comp. **70**, 27–75 (2001)

[CDD03] A. Cohen, W. Dahmen, R.A. DeVore, Adaptive wavelet schemes for nonlinear variational problems. SIAM. J. Numer. Anal. **41**(5), 1785–1823 (2003)

[Coh13] D.L. Cohn, *Measure Theory.* Birkhäuser Advances Texts: Basel Textbooks, 2nd edn. (Springer, New York, 2013)

[CDN10] M. Costabel, M. Dauge, S. Nicaise, Mellin analysis of weighted Sobolev spaces with nonhomogeneous norms on cones. *Around the Research of Vladimir Maz'ya I, Int. Math. Ser. (N. Y.)*, vol. 11 (Springer, New York, 2010), pp. 105–136

[CK95] M. Cwikel, N. Kalton, Interpolation of compact operators by the methods of Calderón and Gustavsson-Peetre. Proc. Edinburgh Math. Soc. **38**(2), 261–276 (1995)

[Dah98] S. Dahlke, Besov regularity for elliptic boundary value problems with variable coefficients. Manuscripta Math. **95**, 59–77 (1998)

[Dah99a] S. Dahlke, Besov regularity for interface problems. Z. Angew. Math. Mech. **79**(6), 383–388 (1999)

[Dah99b] S. Dahlke, Besov regularity for elliptic boundary value problems on polygonal domains. Appl. Math. Lett. **12**(6), 31–38 (1999)

[Dah02] S. Dahlke, Besov regularity of edge singularities for the Poisson equation in polyhedral domains. Num. Linear Algebra Appl. **9**(6-7), 457–466 (2002)

[DDD97] S. Dahlke, W. Dahmen, R.A. DeVore, Nonlinear approximation and adaptive techniques for solving elliptic operator equations. *Multiscale Wavelet Methods for Partial Differential Equations (W. Dahmen, A.J. Kurdila, and P. Oswald, eds.), Wavelet Analysis and Applications*, vol. 6 (Academic Press, San Diego, 1997), pp. 237–283

[DD97] S. Dahlke, R.A. DeVore, Besov regularity for elliptic boundary value problems. Comm. Partial Differ. Equ. **22**(1-2), 1–16 (1997)

[DDHSW16] S. Dahlke, L. Diening, C. Hartmann, B. Scharf, M. Weimar, Besov regularity of solutions to the p–Poisson equation. Nonlinear Anal. **130**, 298–329 (2016)

[DHSS18a] S. Dahlke, M. Hansen, C. Schneider, W. Sickel, On Besov regularity of solutions to nonlinear elliptic partial differential equations. *Preprint-Reihe Philipps-University Marburg*, Bericht Mathematik Nr. **2018-04** (2018)

[DHSS18b] S. Dahlke, M. Hansen, C. Schneider, W. Sickel, Properties of Kondratiev spaces. *Preprint-Reihe Philipps-University Marburg*, Bericht Mathematik Nr. **2018-06** (2018)

[DNS06] S. Dahlke, E. Novak, W. Sickel, Optimal approximation of elliptic problems by linear and nonlinear mappings. II. J. Complexity **22**, 549–603 (2006)

[DS18] S. Dahlke, C. Schneider, Describing the singular behaviour of parabolic equations on cones in fractional Sobolev spaces. Int. J. Geomath. **9**(2), 293–315 (2018)

[DS19] S. Dahlke, C. Schneider, Besov regularity of parabolic and hyperbolic PDEs. Anal. Appl. **17**(2), 235–291 (2019)

[DS09] S. Dahlke, W. Sickel, Besov regularity for the Poisson equation in smooth and polyhedral cones. Sobolev Spaces Math. II Int. Math. Ser. (N.Y.) **9**, 123–145 (2009)

[DS13] S. Dahlke, W. Sickel, On Besov regularity of solutions to nonlinear elliptic partial differential equations. Rev. Mat. Complut. **26**(1), 115–145 (2013)

[Dac04] B. Dacorogna, *Introduction to the Calculus of Variations* (Imperial College Press, London, 2004). Translated from the 1992 French original

[Dau92] I. Daubechies, *Ten Lectures on Wavelets*. CBMS-NSF Regional Conference Series in Applied Mathematics, vol. 61 (SIAM, Philladelphia, 1992)

[Dau98] I. Daubechies, Orthonormal bases of compactly supported wavelets. Commun. Pure Appl. Math. **41**(7), 909–996 (1998)

[Dau88] M. Dauge, *Elliptic Boundary Value Problems in Corner Domains*. Lecture Notes in Mathematics, vol. 1341 (Springer, Berlin, 1988)

[Dau08] M. Dauge, Regularity and singularities in polyhedral domains: the case of Laplace and Maxwell equations. Talk in Karlsruhe (2008): https://perso.univ-rennes1.fr/monique.dauge/publis/Talk_Karlsruhe08.pdf

[DeV98] R.A. DeVore, Nonlinear approximation. Acta Numer. **7**, 51–150 (1998)

[DJP92] R.A. DeVore, B. Jawerth, V. Popov, Compression of wavelet decompositions. Am. J. Math. **114**, 737–785 (1992)

[DL93] R.A. DeVore, G.G. Lorentz, *Constructive Approximation*. Grundlehren der Mathematischen Wissenschaften, vol. 303 (Springer, Berlin, 1993)

[DL90] R.A. DeVore, B. Lucier, High order regularity for conservation laws. Ind. Math. J. **39**(2), 413–430 (1990)

[DP88] R.A. DeVore, V.A. Popov, Interpolation of Besov spaces. Trans. Am. Math. Soc. **305**(1), 397–414 (1988)

[DS93] R.A. DeVore, R.C. Sharpley, Besov spaces on domains in $\mathbf{R}^d$. Trans. Am. Math. Soc. **335**(2), 843–864 (1993)

[Dis03] S. Dispa, Intrinsic characterizations of Besov spaces on Lipschitz domains. Math. Nachr. **260**, 21–33 (2003)

[Dob10] M. Dobrowolski, *Angewandte Funktionalanalysis*. Springer-Lehrbuch Masterclass (Springer, Berlin, 2010)

[DM07] P. Drabek, J. Milota, *Methods of Nonlinear Analysis. Applications to Differential Equations*. Birkhäuser Advanced Texts Basler Lehrbücher (Springer, Basel, 2007)

[ET96] D.E. Edmunds, H. Triebel, *Function Spaces, Entropy Numbers, Differential Operators* (Cambridge Univ. Press, Cambridge, 1996)

[Eic91] J. Eichhorn, The boundedness of connection coefficients and their derivatives. Math. Nachr. **152**, 144–158 (1991)

[Eic07] J. Eichhorn, *Global Analysis on Open Manifolds*. (Nova Science Publishers, New York, 2007), p. x+644

[ElB05] A. El Baraka, Optimal BMO and $\mathcal{L}^{p,\lambda}$ estimates for solutions of elliptic boundary value problems. Arab. J. Sci. Eng. Sect. A Sci. **30**(1), 85–116 (2005)

[Eva10] L.C. Evans, Partial differential equations. *Graduate Studies in Mathematics*, vol. 19, 2nd edn. (American Mathematical Society, Providence, RI, 2010)

[Fal90] K. Falconer, *Fractal Feometry*. Mathematical Foundations and Applications (Wiley, Chichester, 1990)

[FR95] J. Franke, T. Runst, Regular elliptic boundary value problems in Besov-Triebel-Lizorkin spaces. Math. Nachr. **174**, 113–149 (1995)

[FJ90] M. Frazier, B. Jawerth, A discrete transform and decompositions of distribution spaces. J. Funct. Anal. **93**(1), 34–170 (1990)

[Fuc80] S. Fučík, *Solvability of Nonlinear Equations and Boundary Value Problems* (D. Reidel Publishing Company, Dordrecht, 1980)

[Gag57] E. Gagliardo, Caratterizzazioni delle tracce sulla frontiera relative ad alcune classi di funzioni in n variabili (Italian). Rend. Sem. Mat. Univ. Padova **27**, 284–305 (1957)

[GM09] F.D. Gaspoz, P. Morin, Convergence rates for adaptive finite elements. IMA J. Numer. Anal. **29**(4), 917–936 (2009)

[GM14] F. Gaspoz, P. Morin, Approximation classes for adaptive higher order finite element approximation. Math. Comp. **83**(289), 2127–2160 (2014)

[GS13] N. Große, C. Schneider, Sobolev spaces on Riemannian manifolds with bounded geometry: general coordinates and traces. Math. Nachr. **286**(16), 1586–1613 (2013)

[GT01] D. Gilbarg, N.S. Trudinger, *Elliptic Partial Differential Equations of Second Order.* Classics in Mathematics, Reprint of the 1998 edition (Springer, Berlin, 2001)

[Gon17] H. Gonçalves, *2-microlocal spaces with variable integrability*. Ph.D. Thesis, University of Chemnitz (2017)

[Gri85] P. Grisvard, *Elliptic Problems in Nonsmooth Domains* (Pitman, Boston, 1985)

[Gri92] P. Grisvard, *Singularities in Boundary Value Problems*. Recherches en mathématiques appliquées, vol. 22, Masson, Paris; (Springer, Berlin, 1992)

[Gri11] P. Grisvard, *Elliptic Problems in Nonsmooth Domains*. Classics in Applied Mathematics, vol. 69. Reprint of the 1985 original (SIAM, Philadelphia, 2011)

[Gul84] A.B. Gulisashvili, On multipliers in Besov spaces (Russian). Sapiski Nautch. Sem. LOMI **135**, 36–50 (1984)

[Gul85] A.B. Gulisashvili, Multipliers in Besov spaces and traces of functions on subsets of the Euclidean n-space (Russian). DAN SSSR **281**, 777–781 (1985)

[Hac92] W. Hackbusch, *Ellliptic Differential Equations: Theory and Numerical Treatment* (Springer, Berlin-Heidelberg, 1992)

[Han14] M. Hansen, New embedding results for Kondratiev spaces and application to adaptive approximation of elliptic PDEs. *Preprint: SAM-report* **2014-30**. Seminar for Applied Mathematics, ETH Zürich (2014)

[Han15] M. Hansen, Nonlinear approximation rates and Besov regularity for elliptic PDEs on polyhedral domains. Found. Comput. Math. **15**, 561–589 (2015)

[Han17] M. Hansen, Refined localization spaces. Preprint (2017)

[HS18] M. Hansen, B. Scharf, Relations between Kondratiev spaces and refined localization Triebel-Lizorkin spaces. Preprint (2018)

[HS09] D.D. Haroske, C. Schneider, Besov spaces with positive smoothness on $\mathbb{R}^n$, embeddings and growth envelopes. J. Approx. Theory **161**(2), 723–747 (2009)

[HN07] L.I. Hedberg, Y. Netrusov, An axiomatic approach to function spaces, spectral synthesis, and Luzin approximation. Mem. Am. Math. Soc. **188**(882), 97p. (2007)

[JW84] A. Jonsson, H. Wallin, Function spaces on subsets of $\mathbf{R}^n$. Math. Rep. **2**(1), xiv+221pp. (1984)

[JK95] D. Jerison, C.E. Kenig, The inhomogeneous Dirichlet problem in Lipschitz domains. J. Funct. Anal. **130**, 161–219 (1995)

[Kim04] K.-H. Kim, On stochastic partial differential equations with variable coefficients in C^1 domains. Stochastic Process. Appl. **112**(2), 261–283 (2004)

[Kim09] K.-H. Kim, A W_n^p-theory of parabolic equations with unbounded leading coefficients on non-smooth domains. J. Math. Anal. Appl. **350**, 294–305 (2009)

[Kim12] K.-H. Kim, A weighted Sobolev space theory of parabolic stochastic PDEs on non-smooth domains. J. Theor. Probab. **27**(1), 107–136 (2012)

[KK04] K.-H. Kim, N.V. Krylov, On the Sobolev space theory of parabolic and elliptic equations in C^1 domains. SIAM J. Math. Anal. **36**(2), 618–642 (2004)

[KS02] H. Koch, W. Sickel, Pointwise multipliers of Besov spaces of smoothness zero and spaces of continuous functions. Rev. Mat. Iberoamericana **18**(3), 587–626 (2002)

[Kon67] V.A. Kondratiev, Boundary value problems for elliptic equations in domains with conical and angular points. Trudy Moskov. Mat. Obshch. **16**, 209–292 (1967). English translation in Trans. Moscow Math. Soc. **16**, 227–313

[Kon77] V.A. Kondratiev, Singularities of the solution of the Dirichlet problem for a second order elliptic equation in the neighborhood of an edge. Differ. Uravn. **13**(11), 2026–2032 (1977)

[KO83] V.A. Kondratiev, A.O. Oleinik, Boundary value problems for partial differential equations in non-smooth domains. Russ. Math. Surv. **8**, 1–86 (1983)

[Koz91] V.A. Kozlov, On the spectrum of the pencil generated by the Dirichlet problem for an elliptic equation in an angle. Siberian Math. J. **32**(2), 238–251 (1991)

[KM87] V.A. Kozlov, V.G. Maz'ya, Singularities of solutions of the first boundary value problem for the heat equation in domains with conical points II. Izv. Vyssh. Uchebn. Zaved. Mat. **3**, 37–44 (1987)

[KMR97] V.A. Kozlov, V.G. Maz'ya, J. Rossmann, *Elliptic Boundary Value Problems in Domains with Point Singularities* (American Mathematical Society, Providence, RI, 1997)

[KMR01] V.A. Kozlov, V.G. Mazya, J. Rossman, *Spectral Problems Associated with Corner Singularities of Solutions to Elliptic Equations*. Mathematical Surveys and Monographs, vol. 85 (American Mathematical Society, Providence, RI, 2001)

[KN14] V.A. Kozlov, A. Nazarov, The Dirichlet problem for non-divergence parabolic equations with discontinuous in time coefficients in a wedge. Math. Nachr. **287**(10), 1142–1165 (2014)

[KY94] H. Kozono, M. Yamazaki, Semilinear heat equations and the Navier-Stokes equation with distributions in new function spaces as initial data. Commun. Partial Differ. Equ. **19**, 959–1014 (1994)

[Kre15] M. Kreuter, *Sobolev spaces of vector-valued functions*. Master Thesis, University of Ulm (2015)

[KO84] A. Kufner, B. Opic, How to define reasonably weighted Sobolev spaces. Comment. Math. Univ. Carolin. **25**(3), 537–554 (1984)

[KO86] A. Kufner, B. Opic, Some remarks on the definition of weighted Sobolev spaces. *Partial Differential Equations (Proceedings of an International Conference)* (Nauka, Novosibirsk, 1986), pp. 120–126

[KS87] A. Kufner, A.-M. Sändig, *Some Applications of Weighted Sobolev Spaces*. Teubner-Texte Math., vol. 100 (Teubner, Leipzig, 1987)

[Lan01] J. Lang, *Adaptive Multilevel Solution of Nonlinear Parabolic PDE Systems.* Lecture Notes in Computational Science and Engineering; Theory, Algorithm, and Applications, vol. 16 (Springer, Berlin, 2001)

[Lee11] J.M. Lee, *Introduction to Topological Manifolds.* Graduate Texts in Mathematics, vol. 202, 2nd edn. (Springer, New York, 2011)

[LeR07] P.G. Lemarié-Rieusset, The Navier-Stokes equations in the critical Morrey-Campanato space. Rev. Mat. Iberoamericana **23**, 897–930 (2007)

[LeR12] P.G. Lemarié-Rieusset, The role of Morrey spaces in the study of Navier-Stokes and Euler equations. Eurasian Math. J. **3**, 62–93 (2012)

[LeR13] P.G. Lemarié-Rieusset, Multipliers and Morrey spaces. Potential Anal. **38**, 741–752 (2013)

[LeR18] P.G. Lemarié-Rieusset, Sobolev multipliers, maximal functions and parabolic equations with a quadratic nonlinearity. J. Funct. Anal. **274**(3), 659–694 (2018)

[LSUYY12] Y. Liang, Y. Sawano, T. Ullrich, D. Yang, W. Yuan, New characterizations of Besov-Triebel-Lizorkin-Hausdorff spaces including coorbits and wavelets. J. Fourier Anal. Appl. **18**, 1067–1111 (2012)

[LM72] J.L. Lions, E. Magenes, *Non-homogeneous Boundary Value Problems and Applications.* Grundlehren der mathematischen Wissenschaften, vol. 1 (Springer, Berlin, 1972)

[Lot00] S.V. Lototsky, Sobolev spaces with weights in domains and boundary value problems for degenerate elliptic equations. Methods Appl. Anal. **7**(1), 195–204 (2000)

[LL15] V.T. Luong, D.V. Loi, The first initial-boundary value problem for parabolic equations in a cone with edges. Vestn. St.-Petersbg. Univ. Ser. 1. Mat. Mekh. Asron. 2 **60**(3), 394–404 (2015)

[LT15] V.T. Luong, N.T. Tung, The Dirichlet-Cauchy problem for nonlinear hyperbolic equations in a domain with edges. Nonlinear Anal. **125**, 457–467 (2015)

[May05] S. Mayboroda, *The Poisson problem on Lipschitz domains.* Ph.D. thesis, University of Missouri-Columbia, USA (2005)

[Maz85] V.G. Maz'ya, *Sobolev Spaces.* Springer Series in Soviet Mathematics (Springer, Berlin, 1985)

[MR10] V.G. Maz'ya, J. Rossmann, *Elliptic Equations in Polyhedral Domains.* Mathematical Surveys and Monographs, vol. 162 (American Mathematical Society, Providence, RI, 2010)

[MS85] V.G. Maz'ya, T.O. Shaposhnikova, *Theory of Multipliers in Spaces of Differentiable Functions.* Monographs and Studies in Mathematics, vol. 23 (Pitman (Advanced Publishing Program), Boston, MA, 1985)

[MS02] V.G. Maz'ya, T.O. Shaposnikova, An elementary proof of the Brezis and Mironescu theorem on the composition operator in fractional Sobolev spaces. J. Evol. Equ. **2**, 113–125 (2002)

[MS09] V.G. Maz'ya, T.O. Shaposhnikova, *Theory of Sobolev Multipliers.* Grundlehren der Mathematischen Wissenschaften, vol. 337 (Springer, Berlin, 2009)

[Maz03a] A. Mazzucato, Besov-Morrey spaces: Function spaces theory and applications to non-linear PDE. Trans. Am. Math. Soc. **355**, 1297–1369 (2003)

[Maz03b] A. Mazzucato, Decomposition of Besov-Morrey spaces, Contemp. Math. **320**, 279–294 (2003)

[MN10] A. Mazzucato, V. Nistor, Well posedness and regularity for the elasticity equation with mixed boundary conditions on polyhedral domains and domains with cracks. Arch. Ration. Mech. Anal. **195**, 25–73 (2010)

[Mey92] Y. Meyer, *Wavelets and Operators.* Cambridge Studies in Advances Mathematics, vol. 37 (Cambridge Univ. Press, Cambridge, 1992)

[MPP07] S.D. Moura, I. Piotrowska, M. Piotrowski, Non-smooth atomic decompositions of anisotropic function spaces and some applications. Studia Math. **180**(2), 169–190 (2007)

[MNS19] S.D. Moura, J. Neves, C. Schneider, Traces and extensions of generalized smoothness Morrey spaces on domains. Nonlinear Anal. **181**, 311–339 (2019)

[Mor38] C.B. Morrey, On the solutions of quasi-linear elliptic partial differential equations. Trans. Am. Math. Soc. **43**, 126–166 (1938)

[Nak94] E. Nakai, Hardy-Littlewood maximal operator, singular integral operators and the Riesz potentials on generalized Morrey spaces. Math. Nachr. **166**, 95–103 (1994)

[Nak00] E. Nakai, A characterization of pointwise multipliers on the Morrey spaces. Sci. Math. **3**(3), 445–454 (2000)

[NNS16] S. Nakamura, T. Noi, Y. Sawano, Generalized Morrey spaces and trace operator. Sci. China Math. **59**(2), 281–336 (2016)

[Osw94] P. Oswald, *Multilevel Finite Element Approximation.* Teubner Skripten zur Numerik (Teubner, Stuttgart, 1994)

[Pee57] J. Peetre, A generalization of Courant's nodal domain theorem. Math. Scand. **5**, 15–20 (1957)

[Pee76] J. Peetre, *New Thoughts on Besov Spaces.* Duke University Mathematics Series (Duke University, Durham, N.C., 1976)

[RS96] T. Runst, W. Sickel, *Sobolev Spaces of Fractional Order, Nemytskij Operators, and Nonlinear Partial Differential Equations.* De Gruyter Series in Nonlinear Analysis and Applications (1996)

[Ryc00] V.S. Rychkov, Linear extension operators for restrictions of function spaces to irregular open sets. Studia Math. **140**(2), 141–162 (2000)

[Saw10] Y. Sawano, Besov-Morrey and Triebel-Lizorkin-Morrey spaces on domains. Math. Nachr. **283**(10), 1456–1487 (2010)

[SW13] Y. Sawano, H. Wakade, On the Gagliardo-Nirenberg type inequality in the critical Sobolev-Morrey space. J. Fourier Anal. Appl. **19**, 20–47 (2013)

[Sch01] T. Schick, Manifolds with boundary and of bounded geometry. Math. Nachr. **223**, 103–120 (2001)

[Sch09a] C. Schneider, Spaces of Sobolev type with positive smoothness on $\mathbb{R}^d$, embeddings and growth envelopes. J. Funct. Spaces Appl. **7**(3), 251–288 (2009)

[Sch09b] C. Schneider, On Dilation operators in Besov spaces. Rev. Mat. Complut. **22**(1), 111–128 (2009)

[Sch10] C. Schneider, Trace operators in Besov and Triebel-Lizorkin spaces. Z. Anal. Anwendungen **29**(3), 275–302 (2010)

[Sch11a] C. Schneider, Trace operators on fractals, entropy and approximation numbers. Georgian Math. J. **18**(3), 549–575 (2011)

[Sch11b] C. Schneider, Traces in Besov and Triebel-Lizorkin spaces on domains. Math. Nachr. **284**(5–6), 572–586 (2011)

[SV12] C. Schneider, J. Vybíral, Homogeneity in Besov and Triebel-Lizorkin spaces. J. Funct. Spaces Appl., Art. ID 281085, 17 pp. (2012)

[SV13] C. Schneider, J. Vybíral, Non-smooth atomic decompositions, traces on Lipschitz domains, and pointwise multipliers in function spaces. J. Funct. Anal. **264**(5), 1197–1237 (2013)

[Shu92] M.A. Shubin, *Spectral Theory of Elliptic Operators on Noncompact Manifolds.* Méthodes Semi-Classiques, vol. 1 (Nantes, 1991), Astérisque, No. 207, pp. 35–108 (1992)

[Sic12] W. Sickel, Smoothness spaces related to Morrey spaces – a survey. I. Eurasian Math. J. **3**, 110–149 (2012)

[Sic13] W. Sickel, Smoothness spaces related to Morrey spaces – a survey. II. Eurasian Math. J. **4**, 82–124 (2013)

[SS99] W. Sickel, I. Smirnow, *Localization properties of Besov spaces and its associated multiplier spaces.* Jenaer Schriften Math/Inf **21/99**, Jena (1999)

[SYY10] W. Sickel, D.C. Yang, W. Yuan, *Morrey and Campanato Meet Besov, Lizorkin and Triebel* (Springer, Berlin, 2010)

[Sim90] J. Simon, Sobolev, Besov and Nikol' skiĭ fractional spaces: imbeddings and comparisons for vector valued spaces on an interval. Ann. Mat. Pura Appl. **157**(4), 117–148 (1990)

[Skr90] L. Skrzypczak, Traces of function spaces of $F_{p,q}^s - B_{p,q}^s$ type on submanifolds. Math. Nachr. **46**, 137–147 (1990)

[Skr98] L. Skrzypczak, Atomic decompositions on manifolds with bounded geometry. Forum Math. **10**, 19–38 (1998)

[Spi79] M. Spivak, *A Comprehensive Introduction to Differential Geometry*, vol. IV, 2nd edn. (Publish or Perish Inc., Wilmington, Del., 1979), p. viii+561

[Ste70] E.M. Stein, *Singular Integrals and Differentiability Properties of Functions* (Princeton Univ. Press, Princeton, NJ, 1970)

[Ste93] E.M. Stein, *Harmonic Analysis: Real-Variable Methods, Orthogonality, and Oscillatory Integrals*. Princeton Mathematical Series, vol. 43 (Princeton University Press, Princeton, NJ, 1993)

[Ste09] R. Stevenson, Adaptive wavelet methods for solving operator equations: an overview, in *Multiscale, Nonlinear and Adaptive Approximation*, dedicated to Wolfgang Dahmen on the occasion of his 60th birthday, ed. by R. DeVore, et al. (Springer, Berlin, 2009), pp. 543–597

[SS09] R. Stevenson, C. Schwab, Space-time adaptive wavelet methods for parabolic evolution problems. Math. Comput. **78**, 1293–1318 (2009)

[Str83] R.S. Strichartz, Analysis of the Laplacian on the complete Riemannian manifold. J. Funct. Anal. **52**, 48–79 (1983)

[Str08] M. Struwe, *Variational Methods: Applications to Nonlinear Partial Differential Equations and Hamiltonian Systems* (Springer, Berlin, Heidelberg, 2008)

[Tay11] M.E. Taylor, *Partial Differential Equations I. Basic Theory*. Applied Mathematical Sciences, vol. 115, 2nd edn. (Springer, New York, 2011)

[Tho06] V. Thomée, *Galerkin Finite Element Methods for Parabolic Problems*. Springer Series in Computational Mathematics, vol. 25, 2nd edn. (Springer, Berlin, 2006)

[Tri78] H. Triebel, *Interpolation Theory, Function Spaces, Differential Operators*. North-Holland Mathematical Library, vol. 18 (North-Holland Publisher, Amsterdam-New York, 1978)

[Tri83] H. Triebel, *Theory of Function Spaces*. Monographs in Mathematics, vol. 78 (Birkhäuser Verlag, Basel, 1983)

[Tri92] H. Triebel, *Theory of Function Spaces II*. Monographs in Mathematics, vol. 84 (Birkhäuser, Basel, 1992)

[Tri97] H. Triebel, *Fractals and Spectra*. Monographs in Mathematics, vol. 91 (Birkhäuser, Basel, 1997)

[Tri02] H. Triebel, Function spaces in Lipschitz domains and on Lipschitz manifolds. Characteristic functions as pointwise multipliers. Rev. Mat. Complut. **15**(2), 475–524 (2002)

[Tri03] H. Triebel, Non-smooth atoms and pointwise multipliers in function spaces. Ann. Mat. Pura Appl. **182**(4), 457–486 (2003)

[Tri06] H. Triebel, *Theory of Function Spaces III*. Monographs in Mathematics, vol. 100 (Birkhäuser, Basel, 2006)

[Tri08a] H. Triebel, *Function Spaces and Wavelets on Domains*. EMS Tracts on Mathematics, vol. 7 (EMS Publishing House, Zürich, 2008)

[Tri08b] H. Triebel, The dichotomy between traces on d-sets Γ in $\mathbb{R}^n$ and the density of $D(\mathbb{R}^n \setminus \Gamma)$ in function spaces. Acta Math. Sin. (Engl. Ser.) **24**(4), 539–554 (2008)

[TW96] H. Triebel, H. Winkelvoß, Intrinsic atomic characterizations of function spaces on domains. Math. Z. **221**(4), 647–673 (1996)

[War66] F.W. Warner, Extensions of the Rauch comparison theorem to submanifolds. Trans. Am. Math. Soc. **122**, 341–356 (1966)

[Wlo82] J. Wloka, *Partielle Differentialgleichungen* (Teubner, Stuttgart, 1982)

[Woo07] I. Wood, Maximal L^p-regularity for the Laplacian on Lipschitz domains. Math. Z. **255**(4), 855–875 (2007)

[YZY15] D. Yang, C. Zhuo, W. Yuan, Triebel-Lizorkin type spaces with variable exponent. Banach J. Math. Anal. **9**(4), 146–202 (2015)

[YHMSY15] W. Yuan, D.D. Haroske, S.D. Moura, L. Skrzypczak, D. Yang, Limiting embeddings in smoothness Morrey spaces, continuity envelopes and applications. J. Approx. Theory **192**, 306–335 (2015)

[YY13] W. Yuan, D. Yang, Relations among Besov-type spaces, Triebel-Lizorkin-type spaces and generalized Carleson measure spaces. Appl. Anal. **92**, 549–561 (2013)

[YSY10] W. Yuan, W. Sickel, D. Yang, *Morrey and Campanato Meet Besov, Lizorkin and Triebel*. Lecture Notes in Mathematics 2005 (Springer, Berlin, 2010)

Index

LECTURE NOTES IN MATHEMATICS 🐴 Springer

Editors in Chief: J.-M. Morel, B. Teissier;

Editorial Policy

1. Lecture Notes aim to report new developments in all areas of mathematics and their applications – quickly, informally and at a high level. Mathematical texts analysing new developments in modelling and numerical simulation are welcome.

   Manuscripts should be reasonably self-contained and rounded off. Thus they may, and often will, present not only results of the author but also related work by other people. They may be based on specialised lecture courses. Furthermore, the manuscripts should provide sufficient motivation, examples and applications. This clearly distinguishes Lecture Notes from journal articles or technical reports which normally are very concise. Articles intended for a journal but too long to be accepted by most journals, usually do not have this "lecture notes" character. For similar reasons it is unusual for doctoral theses to be accepted for the Lecture Notes series, though habilitation theses may be appropriate.

2. Besides monographs, multi-author manuscripts resulting from SUMMER SCHOOLS or similar INTENSIVE COURSES are welcome, provided their objective was held to present an active mathematical topic to an audience at the beginning or intermediate graduate level (a list of participants should be provided).

   The resulting manuscript should not be just a collection of course notes, but should require advance planning and coordination among the main lecturers. The subject matter should dictate the structure of the book. This structure should be motivated and explained in a scientific introduction, and the notation, references, index and formulation of results should be, if possible, unified by the editors. Each contribution should have an abstract and an introduction referring to the other contributions. In other words, more preparatory work must go into a multi-authored volume than simply assembling a disparate collection of papers, communicated at the event.

3. Manuscripts should be submitted either online at www.editorialmanager.com/lnm to Springer's mathematics editorial in Heidelberg, or electronically to one of the series editors. Authors should be aware that incomplete or insufficiently close-to-final manuscripts almost always result in longer refereeing times and nevertheless unclear referees' recommendations, making further refereeing of a final draft necessary. The strict minimum amount of material that will be considered should include a detailed outline describing the planned contents of each chapter, a bibliography and several sample chapters. Parallel submission of a manuscript to another publisher while under consideration for LNM is not acceptable and can lead to rejection.

4. In general, **monographs** will be sent out to at least 2 external referees for evaluation.

   A final decision to publish can be made only on the basis of the complete manuscript, however a refereeing process leading to a preliminary decision can be based on a pre-final or incomplete manuscript.

   Volume Editors of **multi-author works** are expected to arrange for the refereeing, to the usual scientific standards, of the individual contributions. If the resulting reports can be forwarded to the LNM Editorial Board, this is very helpful. If no reports are forwarded or if other questions remain unclear in respect of homogeneity etc, the series editors may wish to consult external referees for an overall evaluation of the volume.

5. Manuscripts should in general be submitted in English. Final manuscripts should contain at least 100 pages of mathematical text and should always include

   – a table of contents;
   – an informative introduction, with adequate motivation and perhaps some historical remarks: it should be accessible to a reader not intimately familiar with the topic treated;
   – a subject index: as a rule this is genuinely helpful for the reader.
   – For evaluation purposes, manuscripts should be submitted as pdf files.

6. Careful preparation of the manuscripts will help keep production time short besides ensuring satisfactory appearance of the finished book in print and online. After acceptance of the manuscript authors will be asked to prepare the final LaTeX source files (see LaTeX templates online: https://www.springer.com/gb/authors-editors/book-authors-editors/manuscriptpreparation/5636) plus the corresponding pdf- or zipped ps-file. The LaTeX source files are essential for producing the full-text online version of the book, see http://link.springer.com/bookseries/304 for the existing online volumes of LNM). The technical production of a Lecture Notes volume takes approximately 12 weeks. Additional instructions, if necessary, are available on request from lnm@springer.com.

7. Authors receive a total of 30 free copies of their volume and free access to their book on SpringerLink, but no royalties. They are entitled to a discount of 33.3 % on the price of Springer books purchased for their personal use, if ordering directly from Springer.

8. Commitment to publish is made by a *Publishing Agreement*; contributing authors of multiauthor books are requested to sign a *Consent to Publish form*. Springer-Verlag registers the copyright for each volume. Authors are free to reuse material contained in their LNM volumes in later publications: a brief written (or e-mail) request for formal permission is sufficient.

Addresses:
Professor Jean-Michel Morel, CMLA, École Normale Supérieure de Cachan, France
E-mail: moreljeanmichel@gmail.com

Professor Bernard Teissier, Equipe Géométrie et Dynamique,
Institut de Mathématiques de Jussieu – Paris Rive Gauche, Paris, France
E-mail: bernard.teissier@imj-prg.fr

Springer: Ute McCrory, Mathematics, Heidelberg, Germany,
E-mail: lnm@springer.com

Printed in the United States
by Baker & Taylor Publisher Services